Thin Film Transistor Technology 8

Editor:

Y. Kuo
Texas A&M University
College Station, Texas, USA

Sponsoring Division:

 Electronics and Photonics

Published by
The Electrochemical Society
65 South Main Street, Building D
Pennington, NJ 08534-2839, USA

tel   609 737 1902
fax 609 737 2743
www.electrochem.org

ecstransactions ™

Vol. 3 No. 8

Copyright 2006 by The Electrochemical Society, Inc.
All rights reserved.

This book has been registered with Copyright Clearance Center, Inc.
For further information, please contact the Copyright Clearance Center,
Salem, Massachusetts.

Published by:

The Electrochemical Society, Inc.
65 South Main Street
Pennington, New Jersey 08534-2839, USA

Telephone 609.737.1902
Fax 609.737.2743
e-mail: ecs@electrochem.org
Web: www.electrochem.org

Printed in the United States of America

PREFACE

This issue of *ECS Transactions* contains 46 papers presented orally and posted at the Eighth Symposium on Thin Film Transistor Technologies (TFT 8), held in Cancun, Mexico, October 29-November 3, 2006. This symposium was sponsored by the Electronics and Photonics Division of the Electrochemical Society.

This symposium was organized into nine sections: TFT Systems, Devices and Reliability, Thin Film Deposition and Characterization, Poly-Si TFTs from Laser Crystallization Processes, Poly-Si TFTs from Non-Laser Crystallization Processes, TFTs on Flexible Substrates, Non-Silicon TFTs, Non LCD Applications, and Posters. Totally, 52 papers contributed by authors from 11 countries or regions, i.e., Canada, France, Greece, Japan, Korea, Netherlands, Norway, Spain, Taiwan, UK and USA, are included in this proceedings volume. Authors are from industry/institutes (ALTEDEC, BOE Hydis Technology Co., Canon, ETRI AIST Tsukuba, Fairchild Semiconductor, Hitachi, HP, Innovavent Gmbh, LG Philips, Mitsui Engineering & Shipbuilding, National Institute of Advanced Industrial Science and Technology, National Nano Device Laboratories-Taiwan, NEC, NIST, Palo Alto Research Center, Philips, Sekisui Chemicals Co., Shimadzu, Sharp USA, Sharp Corp., and Toshiba) and universities (California Institute of Technology, CINVESTAV Mexico, Delft University of Technology, Harvard University, Korea Electronics Technology Institute, Kyushu University, Lehigh University, McMaster University, Nara Institute of Science and Technology, National Chiao-Tung University, National Taiwan University, NCSR Demokritos, Norwegian University of Science and Technology, Osaka University, Pennsylvania State University, Politecnica de Catalunya, Princeton University, Rennes I University, Rensselaer Polytechnic Institute, Ryukoku University, Seoul National University, Texas A&M University, Tohoku University, Tokyo Institute of Technology, University of Athens, University of Illinois, University of Surrey, University of Twente, Universitat Rovira i Virgili, University of Victoria, University of Wisconsin-Madison, and University of Waterloo).

In order to present all papers in a coherent manner, this issue is edited into eight sections. All papers are published as original received without alteration of their technical contents except for the format.

The editor wishes to express his sincere appreciation to the following people for their involvement in organizing and conducting the symposium:
- authors and presenters of papers for their participation.
- co-organizers for their contributions in planning and coordinating the program.
- section chairs and co-chairs for their conducting the meeting.
- H. Nominda for her assistance in editing the volume.
- staffs of the Electrochemical Society for their administrative assistance in organizing the symposium and in publishing this volume.

> Yue Kuo
> Dow Professor
> Thin Film Nano & Microelectronics Research Laboratory
> Texas A&M University
> College Station, TX
> October 29, 2006

Thin Film Transistor Technologies 8 Symposium Co-Organizers

D. Ast (Cornell University)
O. Bonnaud (Université de Rennes I)
S. Fonash (Pennsylvania State University)
M. Hatano (Hitachi Ltd.)
J. Jang (Kyung Hee University)
O.-K. Kwon (Hanyang University)
Y. Kuo (Texas A&M University)
E. Lueder (University of Stuttgart)
M. Matsumura (ALTEDEC)
P. Migliorato (Cambridge University)
M. Shur (Rensselaer Polytechnic Institute)
S. Uchikoga (Toshiba Corp.)
Y. Uraoka (Nara Institute of Science and Technology)

Section Chairs and Co-Chairs

O. Bonnaud (Université de Rennes I)
J. Daniel (Palo Alto Research Center)
S. Deane (Philips Research)
D. Gundlach (NIST)
M. Hatano (Hitachi)
H. Hayama (NEC Corp.)
H. Hosono (Tokyo Institute of Technology)
Y. Kuo (Texas A&M University)
T. Motooka (Kyushu University)
A. Nathan (University of Waterloo)
J. Rogers (University of Illinois)
T. Serikawa (Osaka University)
M. Shur (Rensselaer Polytechnic Institute)
S. Uchikoga (Toshiba Corp.)
Y. Uraoka (Nara Institute of Science and Technology)
V. Voutsas (Sharp USA)
Y. Yamamoto (Sharp Corp.)

ECS Transactions, Volume 3, Issue 8
Thin Film Transistor Technologies 8

Table of Contents

Preface iii

Chapter 1
TFT Systems

Panel-Sized LCD Drivers Using SOG Technology * 3
  H. Hayama

Technology Trend of AMLCDs for Mobile Application * 11
  Y. Yamamoto and A. T. Voutsas

Optical Feedback for AMOLED Pixel Circuits * 23
  S. Deane, D. Fish, N. Young, A. Steer, D. George, A. Giraldo, H. Lifka, W. Oepts and N. Bramante

High Reliability and Performance Poly-Si TFTs for System in Displays * 35
  M. Hatano, M. Matsumura, Y. Toyota, M. Tai, H. Hamamura, T. Miyazawa and M. Ohkura

Chapter 2
Devices and Reliability

Design of Ultra-High Performance Si TFTs * 45
  G. Kawachi

Electrical Hysteresis Behavior of Low Temperature Polycrystalline Silicon Thin Film Transistors 57
  D. Nam, H. Lee, S. Jung, T. Ahn, C. Kim, C. Kim and I. Chung

Analysis of the Hump Characteristics in Poly-Si Thin Film Transistor 63
  S. Kim, J. Oh, J. Yang, M. Yang and I. Chung

Analysis of Characteristics in Poly-Si Thin Film Transistor Crystallized by a New Alignment SLS Process 69
  H. Kwang Sik, J. Yang, M. Yang, Y. Kim, T. Ahn and I. Chung

Front and Back Channel Properties of Asymmetrical Double-Gate Polysilicon TFTs 75
  F. V. Farmakis, D. N. Kouvatsos, A. T. Voutsas, D. C. Moschou, G. P. Kontogiannopoulos and G. J. Papaioannou

Threshold-Voltage Instability of Single-Crystal Si Thin-Film Transistors Fabricated on Plastic Substrate   81
  H. Yuan, Z. Ma, M. G. Lagally and G. E. Celler

The Role of Grain Boundaries on the Performance of Poly-Si TFTs   87
  G. J. Papaioannou, D. N. Kouvatsos and A. T. Voutsas

Hydrogen Passivation and Channel Capping for Threshold Voltage Shift and Off-Current Reduction in Nanocrystalline Silicon TFTs   93
  M. Esmaeili Rad, A. Sazonov and A. Nathan

Chapter 3
Thin Films Deposition and Characterization

Development of ALD/PECVD Reactor for High Quality LTPS-TFTs Insulator *   101
  K. Murata, N. Miyatake, Y. Mori, H. Tachibana, Y. Uraoka and T. Fuyuki

Room-Temperature Sputter-Deposited Gate SiO2 Films for High Quality Poly-Si TFTs   107
  T. Serikawa, T. Miyamoto, H. Ueno, Y. Sugawara, Y. Uraoka and T. Fuyuki

Novel Characterization Technique for Oxidation Processes   113
  Y. V. Sokolov

Deposition of Highly Crystallized Poly-Si Thin Films on Polymer Substrates Using Pulsed-Plasma CVD under Near-Atmospheric Pressure   119
  M. Matsumoto, M. Suemitsu, T. Yara, N. Setsuo, U. Tuyoshi, T. Yasutake and I. Syun

Chapter 4
Poly-Si TFTs from Laser Crystallization Processes

Nanosecond Monitoring of Lateral Crystallization Dynamics Induced by ELA *   127
  Y. Takami, T. Warabisako and M. Matsumura

The Geometry Effect of A Counter-Doped Lateral Body Terminal on Poly-Si TFTs   137
  S. Han, I. Ji, J. Park, S. Choi, J. Yoo, B. Choi, K. Lee and M. Han

High-Performance Low Temperature Poly-Silicon Thin Film Transistors Fabricated by Excimer Laser Irradiation with Bottom-Gate Scheme   143
  C. Tsai, H. Chen, Y. Lee, H. Chen and H. Cheng

Excimer Laser Crystallization of Amorphous Silicon Film with Artificially Designed Spatial Intensity Profile Beam   149
  E. Kim, K. Kim, M. Ryu, H. Kwon, C. Kim, G. Son and J. Lee

Poly-Si TFT Technology: Advances in Material, Process and Device Technology * 157
  A. T. Voutsas

Preferred <100> Surface and In-Plane Orientations in Self-Assembled Poly-Si by 167
Multiple Excimer Laser Irradiation
  M. He, R. Ishihara, W. Metselaar and K. Beenakker

Crystallization of Double-Layered Silicon Thin Films by Solid Green Laser Annealing 173
and Its Application to Low Temperature poly-Si Thin Film Transistors
  Y. Sugawara, Y. Uraoka, H. Yano, T. Hatayama, T. Fuyuki and A. Mimura

Poly-Si Thin Fim Transistor with Multiple Nanowire Channels Prepared by Excimer 179
Laser Annealing
  S. Lee, C. Meng, M. Tsai and P. Yang

Green Laser Crystallization of a-Si Films Using Preformed a-Si Lines 185
  I. Brunets, J. Holleman, A. Y. Kovalgin, T. Aarnink, A. Boogaard, P. Oesterlin and
  J. Schmitz

Chapter 5
Poly-Si TFTs from Non-Laser Crystallization Processes

Low-Temperature Crystallization of Amorphous Si Films Using Ferritin Protein with 195
Ni Nanoparticles *
  Y. Uraoka, H. Kirimura, T. Fuyuki, M. Okuda and I. Yamashita

A Simple Method for Gettering of Nickel within the NILC Polycrystalline Silicon 203
Film Using a Gettering Substrate
  Y. S. Wu, C. Hou, C. Lin and C. Hu

Molecular-Dynamics Simulations of Recrystallization Processes in Silicon: 207
Nucleation and Crystal Growth in the Solid-Phase and Melt *
  T. Motooka, S. Munetoh, R. Kishikawa, T. Kuranaga, T. Ogata and T. Mitani

Ge Nuclei for Fabrication of Poly-Si Thin Films on Glass Substrates * 215
  K. Yasutake, H. Watanabe, H. Ohmi and H. Kakiuchi

Chapter 6
TFTs on Flexible Substrates

The Road Towards Large-Area Electronics Without Vacuum Tools * 229
  J. Daniel, A. Arias, B. Krusor, R. Lujan and R. Street

230 DPI High Resolution AMPLED Display on Flexible Metal Foils with Integrated 237
Row Drivers
  M. Troccoli, T. Chuang, A. Hamshidi, P. Kuo, J. Spirko, M. Hatalis, A. T. Voutsas,
  T. Afentakis and J. Hartzell

vii

Overlay Alignment in a-Si:H TFTs Fabricated on Foil Substrates 249
  H. Gleskova, I. Cheng, A. Z. Kattamis, S. Wagner and Z. Suo

Mist Deposition for TFT Technology 255
  K. Shanmugasundaram, S. Price, K. Chang, D. Lee and J. Ruzyllo

Chapter 7
Non-Silicon TFTs

Hole Mobility in Structurally-Different Pentacene Field-Sffect Transistors 263
  H. Kwok

Design, Fabrication and Characterization of Parylene-Packaged Thin-Film Transistors 273
  H. Lo and Y. Tai

The Improvement of Electrical Characteristic of Solution Processed Triisopropylsilyl 279
Pentacene Organic Thin-Film Transistors Employing Hexamethyldisilazane Treatment
  J. Han, J. Lee, M. Han and J. Han

Suppression of OLED Current Error Caused by the Hysteresis of a-Si:H TFT For 287
AMOLED display
  J. Lee, S. Park, J. Jeon, J. Goh, J. Choi, K. Chung and M. Han

Integrated Circuits Based on Amorphous Indium-Gallium-Zinc-Oxide-Channel Thin- 293
Film Transistors *
  M. Ofuji, K. Abe, N. Kaji, R. Hayashi, M. Sano, H. Kumomi, K. Nomura, T.
  Kamiya and H. Hosono

The Abnormal Degradation Behavior of ZnO TFT Under Gate Bias Stress 301
  C. Hwang, S. Ko Park, S. Chung, J. Lee, Y. Yang, L. Do and H. Chu

All-Solution-Processed Organic Thin Film Transistors Fabricated by Non- 307
Piezoelectric Inkjet Printing
  I. Takasu, K. Sugi, Y. Nomura, H. Nakao, K. Mori, I. Amemiya and S. Uchikoga

Thin-Film Transistors in Disordered Semiconductors for High Performance 313
Macroelectronic Circuits.
  F. Balon and J. M. Shannon

Chapter 8
Non LCD Applications

Artificial Retina using Thin-Film Photodiode and Thin-Film Transistor * 325
  M. Kimura, T. Shima and T. Yamashita

Nonvolatile Amorphous Silicon Thin Film Transistor Memories with the a-Si:H   333
Embedded Gate Dielectric Structure
Y. Kuo and H. Nominanda

Sensitivity of Suspended-Gate Polysilicon TFTs to Charge Variation and Application   341
to DNA Recognition
T. Mohammed-Brahim, F. Bendriaa, F. Le Bihan, A. Salaun, O. Bonnaud and M. Harnois

Process Technology for High-Resolution AM-PLED Displays on Flexible Metal Foil   349
Substrates
T. Chuang, M. Troccoli, P. Kuo, A. Jamshidi Roudbari, M. Hatalis, J. Spirko, K. Klier, I. Biaggio, A. T. Voutsas, T. Afentakis and J. Hartzell

Uniform OLED-Pixels Using Microcrystalline Silicon TFTs for Active-Matrix   361
Addressing
T. Mohammed-Brahim, A. Gaillard, R. Rogel, S. Crand, C. Prat and P. Leroy

Author Index   369

* invited paper

Facts about ECS

The Electrochemical Society (ECS) is an international, nonprofit, scientific, educational organization founded for the advancement of the theory and practice of electrochemistry, electrothermics, electronics, and allied subjects. The Society was founded in Philadelphia in 1902 and incorporated in 1930. There are currently over 7,000 scientists and engineers from more than 70 countries who hold individual membership; the Society is also supported by more than 100 corporations through Corporate Memberships.

The technical activities of the Society are carried on by Divisions. Sections of the Society have been organized in a number of cities and regions. Major international meetings of the Society are held in the spring and fall of each year. At these meetings, the Divisions and Groups hold general sessions and sponsor symposia on specialized subjects.

The Society has an active publications program that includes the following.

Journal of The Electrochemical Society — JES is the peer-reviewed leader in the field of electrochemical and solid-state science and technology. Articles are posted online as soon as they become available for publication. This archival journal is also available in a paper edition, published monthly following electronic publication.

Electrochemical and Solid-State Letters — ESL is the first and only rapid-publication electronic journal covering the same technical areas as JES. Articles are posted online as soon as they become available for publication. This peer-reviewed, archival journal is also available in a paper edition, published monthly following electronic publication. It is a joint publication of ECS and the IEEE Electron Devices Society.

Interface — *Interface* is ECS's quarterly news magazine. It provides a forum for the lively exchange of ideas and news among members of ECS and the international scientific community at large. Published online (with free access to all) and in paper, issues highlight special features on the state of electrochemical and solid-state science and technology. The paper edition is automatically sent to all ECS members.

Meeting Abstracts (formerly Extended Abstracts) — Abstracts of the technical papers presented at the spring and fall meetings of the Society are published on CD-ROM.

ECS Transactions — This online database provides access to full-text articles presented at ECS and ECS-sponsored meetings. Content is available through individual articles, or as collections of articles representing entire symposia.

Monograph Volumes — The Society sponsors the publication of hardbound monograph volumes, which provide authoritative accounts of specific topics in electrochemistry, solid-state science, and related disciplines.

For more information on these and other Society activities, visit the ECS website:

www.electrochem.org

CHAPTER 1

TFT SYSTEMS

2

Panel-sized LCD Drivers using SOG Technology

Hiroshi Hayama*

NEC Corporation, Kawasaki, Kanagawa 211-8668, Japan
(*Presently NEC Electronics Corporation)

"Panel-sized drivers" for medium- to large-sized LCDs has been proposed. They are TFT-LSIs fabricated on glass substrates whose length is nearly equal to the width or height of an LCD, and they have long narrow shapes and identical outputs for scan/column lines of the LCD. Scan- and column-driver prototypes have successfully been developed using System-on-glass technology for XGA amorphous silicon TFT-LCD panels.

Introduction

System-on-glass (SOG) TFT-LCDs are used for small LCDs because they have small module-volumes. As the LCD size increases, however, the proportion of the monolithically integrated driver-circuit area decreases rapidly. Because expected costs for present SOG LCDs do not seem to match market prices, amorphous silicon (a-Si) TFT-LCDs assembled with bulk silicon driver-LSIs are often used for medium to large LCDs. However, there are still strong demands for medium to large LCDs with small module-volumes in such portable terminal applications as laptop PCs.

We propose "Panel-Sized Drivers (PSDs) (1)." PSDs are LSIs whose length is nearly equal to the width or height of an LCD, and they have long narrow shapes and identical outputs for scan/column lines of the LCD. Conventional a-Si TFT-LCD modules need printed circuit boards (PCBs) for signal distribution and power supply, and tape carrier/film packages (TCPs) in many cases, as shown in Figure 1. On the other hand, PSD modules need only drivers assembled directly on the LCD panel. PSDs are fabricated with large rectangular glass substrates using SOG (LTPS TFT CMOS) technology at a reasonable cost. In addition, assembly reliability is better than that of conventional modules because PSDs have the same coefficient of thermal expansion as LCDs.

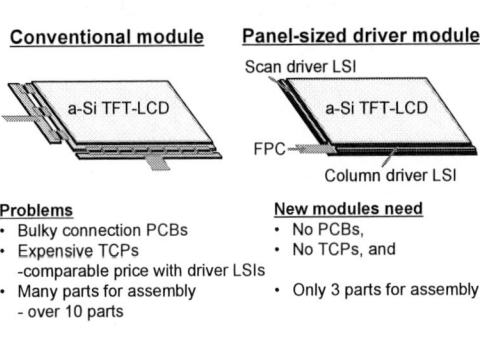

Figure 1. Panel-sized driver concept

Even if we fabricate PSDs with bulk silicon technology, very poor assembly reliability may be expected with them, since bulk silicon have a very large coefficient of thermal expansion compared with LCD glasses. In addition, very long silicon chips may become expensive, since an available number of such long chips laid on round silicon wafers is small.

SOG scan drivers have, so far, been used as integrated drivers for polycrystalline silicon (p-Si) TFT displays, and have never been used for a-Si TFT displays, since scan drivers for a-Si TFT displays need monolithically integrated low-voltage (ex. 5V) and high-voltage (ex. 40V) CMOS TFTs, and need very uniform bumping technology (1, 2). In addition, Panel-sized column drivers need precise tiny digital analog converters (DACs) and very low resistivity interconnections (3).

Panel-sized scan driver

We have successfully fabricated a prototype panel-sized scan driver (PSSD) for a 12-inch diagonal XGA TFT-LCD (1, 2). It is 189 mm long and 1.41 mm wide with 768 outputs. Chip micrographs and a specification table are shown in Figure 2 and Table I respectively. For the PSSDs, we developed such technologies as high-voltage CMOS TFTs, uniform bumps, and a new level shift circuit in order suppress SOI effects.

Chip view

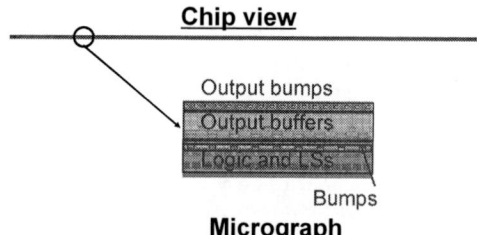

Micrograph

Figure 2. Chip micrograph for the PSSD

TABLE I. Specifications for the PSSD

Logic voltage	5V
VGon/VGoff	30V/-10V
Output pins	768
Output pin pitch	240μm (comparable with 12.1" XGA TFT-LCDs)
Driver size	189mm*1.41mm (without cutting margin)

High-Voltage TFT-CMOS and uniform bumps

In most cases, integrated scan circuits drive SOG LCDs at around 15V. However, PSSDs have to drive a-Si TFT-LCDs at higher voltages, up to 40V. For this scan purpose, we developed low-voltage (5V) and high-voltage (15V, 40V) CMOS TFTs with different device structures. The 15V CMOS TFTs were mainly used in level shift circuits, and the

40V CMOS TFTs were used in output circuits. We also developed a quite uniform bump forming technology using electroless-plating. Figure 3 shows an SEM image and Figure 4 shows height variation for the electroless-plated bumps.

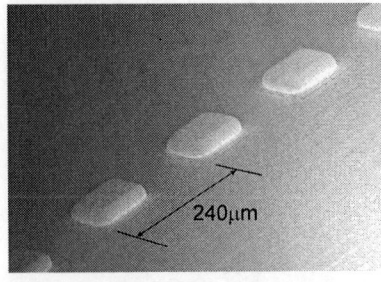
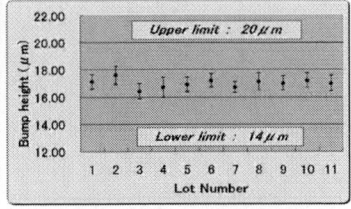

Figure 3. SEM image for PSSD bumps   Figure 4. Bump height variation

A New Level Shift Circuit

Since SOG PSSDs are SOI devices, and since a scan driver must be operated at a high output voltage such as 30-40V, SOI effects appear prominently (4). We have therefore fabricated PSSDs taking into account the floating body effect (FBE). PSSDs that use a conventional level shifter (LS) suffered from unexpected operation delays. Figures 5 and 6 show circuit diagram and pulse-duty versus propagation-delay measurement results for a PSSD with conventional LSs. Propagation delays for rising edges (tPLHs) sharply increased as the duty decreased and the tPLHs and propagation delays for falling edges (tPHLs) strongly depended on the duty. These results show that switching hysteresis affected the LS delay. Since the pulse duty of about 1/768 is used in scan drivers for XGA panels, this unexpected tPLH increase is actually very serious. No duty dependence is expected without taking into account the FBE. A pulse stretching phenomenon arose at this level shifter because of the FBE (5).

We developed a new LS that cannot be easily influenced by the FBE. It is shown in Figure 7, and it utilizes a current mirror circuit. The current mirror circuit boosts a conversion operation, and this circuit is tiny and consumes little power in the steady state.

Delay dependences on pulse duty were simulated. We used the BSIM Partially Depleted SOI model (6). All supply voltages and device parameters were scaled down to 1/10 based on the constant electric field law, because the present BSIM SOI model limits the body voltage from -5V to 1.5V (6). Figure 8 shows the simulation results for the conventional and new LSs. The delay of the new LS was much shorter than that of the conventional one at all duty values. The dependence of the delay on the duty was also small. This good characteristics were confirmed with the actual prototype PSSDs.

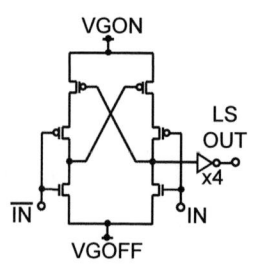

Figure 5. Conventional level shifter

Figure 6. Operation delay

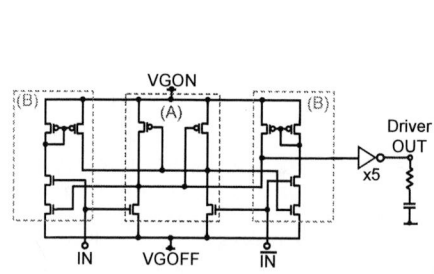

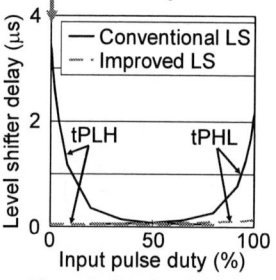

Figure 7. A new level shift circuit

Figure 8. Simulated delay

Panel-sized column driver

We have also developed prototype 6bit-color panel-sized column drivers (PSCDs) for 15-inch diagonal XGA TFT-LCDs (3). Two types of PSCDs have been fabricated. One is 3072-output full-sized PSCD, and the other is 1536-output half-sized PSCD. Considering the yield, half-sized PSCDs have been fabricated consequently. The full-sized PSCD is 298mm long and 7.92mm wide, and the half-sized PSCD is 151mm long and 7.92mm wide. Chip micrographs and specifications are shown in Figure 9 and Table II respectively. Full-sized and half-sized drivers are composed of approximately 2 and 1 million TFTs respectively. Copper-plated conductors are used to reduce degradation and latency for the internal signals, to reduce voltage-source bus-resistances, and are also used as bumps. For the voltage follower amplifier, we used with an offset cancellation scheme in order to reduce output deviation.

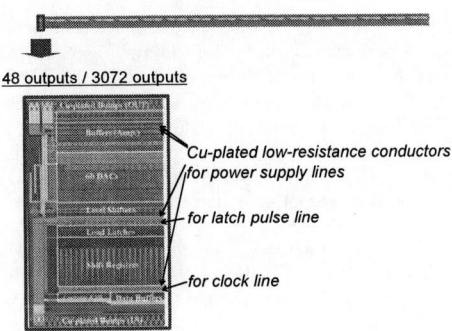

48 outputs / 3072 outputs

Cu-plated low-resistance conductors
for power supply lines

for latch pulse line

for clock line

Figure 9. Chip micrograph for the PSCD

TABLE II. Specifications for the PSCD

Application	15-inch XGA a-Si TFT-LCDs
Inputs	6b RGB data x 4 (0V/3.3V CMOS)
Outputs	64 gray levels (deviation: < +/-20mV)
Output pins	3072 (Full-sized), 1536 (Half-sized)
Output pin pitch	96μm (Zigzag alignment)
Power supply voltage	Analog 10V Logic 5V
Output voltage	Negative polarity: 4.1V~0.7V Positive polarity: 6.0V~9.4V
Driver size	298mm x 7.92mm (Full-sized) 151mm x 7.92mm (Half-sized)
Process, TFT design rule	LTPS TFT CMOS, Analog 4μm Logic 2μm
Number of TFTs	approx. 2 million (Full-sized) approx. 1 million (Half-sized)
Operating frequency	16.25MHz (Full-sized) 8.125MHz (Half-sized)
Functions	Bi-directional data transmission, Output voltage polarity inverting, Line inverting, Dot inverting
Low-resistance conductor	plated copper

Offset Cancellation for Amplifiers

Assuming 6-bit (64 gray-scales/color) driving for 5V liquid crystal cells, column driver's DACs have to have better output accuracy than about +/-30mV. Present SOG TFTs, unfortunately, have poor uniformity for their electrical characteristics. For example, offset voltages of SOG amplifiers, which are used in DACs, are larger about ten times than bulk Si LSI amplifiers in the worst case. They are due to a large difference between threshold voltages for a pair TFTs in differential input-stages.

We adopted offset cancellation technique for the amplifiers of SOG DACs. One horizontal (field) time allowed for XGA d~~~'~~~ is about 20μs. We divide the horizontal times into two periods. We allot the first 10μs as "offset detection periods," and allot the second 10μs as "output periods." During offset detection periods, offset voltages are detected and are stored in a capacitor. During output periods, offset cancelled voltages are output. Figure 10 shows actual output wave forms for a prototype PSCD. Output voltages are shown for 64 horizontal periods, in which 20μs horizontal periods gray-scale value is increased one by one sequentially. Voltage variations during the offset detection period were in a range of +/- 600mV with crude offset voltages. During the output periods, the driver achieved an output deviation of +/-20 mV. Figure 11 shows output voltage uniformity of DACs. Measured DAC outputs showed +/-20mV accuracy for typical 16th, 32nd, and 48th gray-scales. Since precise voltage is finally applied to liquid crystal cells for each horizontal periods, column line driving can be achieved in +/-20mV accuracy even if offset voltages are +/-600mV during the detection periods.

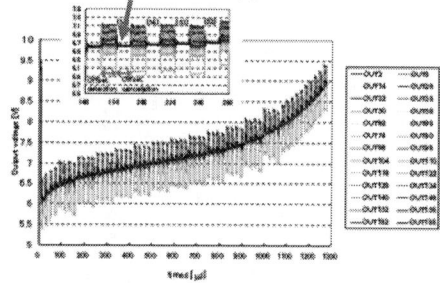

Figure 10. Output waveforms

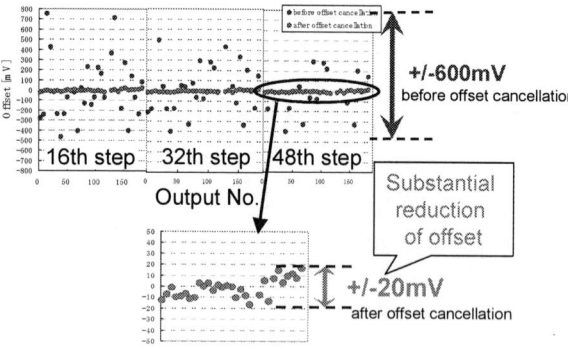

Figure 11. Output deviations

Plated Copper Bus

As mentioned in the previous section, column drivers must output voltage with +/- 30mV accuracy for 6-bit driving for 5V liquid crystal cells. Since the PSCDs is several hundred mm long, resistivity of voltage-source bus for DACs must be very low. If there is a voltage drop over 60mV between the nearest DAC and the farthest DAC, we cannot achieve enough accuracy even if we adopt sophisticated DACs. We used 17μm thick plated copper for voltage source & ground interconnections for DACs and for clock signal interconnections. 17μm thick plated copper is also used as bumps.

Summary

We proposed panel-sized drivers using system-on glass technology, and a prototype panel-sized scan driver with 768 outputs for XGA amorphous silicon TFT-LCDs was successfully fabricated. Fully functional scan drivers were achieved taking into account the floating body effect for TFT/SOI devices. We also fabricated panel-sized column drivers with 6-bit 3072 and 1536 outputs. Offset cancellation technique was used in DACs, and low-resistivity plated-copper was used for bus interconnections and bumps. We successfully demonstrated display operations for XGA TFT-LCDs equipped with the panel-sized drivers.
We think we have clearly shown possibility for the panel-sized drivers using system-on-glass technology. For realizing drivers with practical widths, however, such fabrication equipments may be needed as photo-steppers/aligners and dry etchers that delineate finer TFTs on large glass substrates, and device reliability has to be also confirmed in detail.

Acknowledgments

The author thanks all the people who are related to this panel-sized driver research activities. The author especially thanks Mr. D. Sasaki, Mr. O. Ishibashi, Mr. T. Toba, Mr. S. Noda, Mr. J. Yanase, Mr. H. Takaoka, Mr. Y. Satoh, Mr. Y. Kamon, Mr. H. Tsuchi, Dr. H. Okumura, Dr. H. Kanoh, Mr. M. Iriguchi, Mr. K. Kimura, Mr. J. Ishii, and Mr. H. Imai as co-authors for the related papers.

References

1. D. Sasaki et al., "A panel-sized TFT-LCD scan driver," ISSCC2005 Dig. Tech. Papers, p. 556.
2. S. Noda et al., "Flip chip assembly of panel-sized TFT-LCD scan drivers with electroless-planed bumps using anisotropic conductive films," SID Dig., p. 1788, May 2005.
3. O. Ishibashi et al., "Panel-Sized TFT-LCD Column Driver," ISSCC2006 Dig. Tech. Papers, p. 176.
4. M. Valdinoci, et.al., "Floating Body Effects in Polysilicon Thin-Film Transistors," IEEE Trans. Electron Devices, 44, No. 12, p. 2234, Dec. 1997.
5. A. Wei, et.al., "Minimizing floating-body-induced threshold voltage variation in partially depleted SOI CMOS," IEEE Electron Device Lett., 17, no. 8, p. 391, August 1996.
6. UCB BSIM GROUP, "BSIMSOI3.1 MOSFET MODEL Users' Manual", 2003

10

Technology trend of AMLCDs for mobile application

Y. Yamamoto [a], A. T. Voutsas [b]

[a] Sharp Corporation, Tenri, Nara, 632-8567, Japan
[b] Sharp Laboratories of America, Camas, WA, 98607, USA

Novel devices to serve the needs of ubiquitous network require powerful input-output terminals with exceptional characteristics. In response to these requirements, we have developed high performance TFTs and "System LCD" using CG Silicon® technology. In addition, we have demonstrated advanced system integration technology for mobile application such as sensors, the fusion of "audio" with "visual" and fine picture quality with good image control. And we have been developing Flexible Displays, and a new Reflective LCD technology for achieving super low power consumption.

Introduction

Our everyday life teaches us that we are experiencing a fundamental shift to an information-driven society. Starting with the internet, the information network – so called "ubiquitous network" – has expanded into all aspects of our life at an unprecedented speed and scale. Mobile phone is one of the leading examples of a ubiquitous information tool. We can use it as a telephone and, at the same time, as a terminal to connect to the information network and obtain data through an internet site or via e-mail. The capability, speed and quality of information exchange has received a substantial boost by the introduction of digital terrestrial broadcasting, which was launched in December 2003 in Japan and will be phased in several world capitals and major cities by 2006.

An important aspect of the ubiquitous era is the fusion of information technology and audio-visual technology. We believe that the unification of content, being character, image or voice data (in a mobile environment) is poised to become a key technology. Methods to integrate IT and AV will become necessary, in addition to the constant need for low power consumption, compactness, high-resolution, and high quality of displayed information. To meet these needs we have been developing an array of display technologies, under the umbrella of "System LCD".

CG Silicon® Technology

One of the key technologies for the realization of "System LCD" is high performance p-Si TFT [1-5]. We have developed novel crystallization technology, so-called "CG Silicon® technology", for high performance TFT, in collaboration with Semiconductor Energy Laboratory Co. Ltd. By this technology, poly-silicon thin film is formed by solid phase crystallization (SPC) using an appropriate metal catalyst to expedite crystal formation. In the CG Silicon® case, atomic arrangement at the grain boundaries of crystal domains is evident, whereas in conventional poly-Si continuity of the crystal structure at grain boundaries is lacking. As a result, the electron field-effect mobility can be as high as $300 cm^2/Vs$ for CG Silicon® TFT. Most importantly, however, CG Silicon® technology offers a substantially more stable process, which is an indispensable advantage for mass production.

---

*CG Silicon® is a trademark of SHARP CORPORATION

*CG Silicon® technology was developed in collaboration with Semiconductor Energy Laboratory Co. Ltd.

System on Panel and Mobile Applications

One key impact of CG Silicon® technology on TFT characteristics is current driving ability, which is approximately two times higher, compared with conventional p-Si TFTs. Using such high TFT performance, we have developed a new functional display, "Full−Functional system panel" [6]. This is a full-scale integrated LCD which combines a CPU, an audio circuit, an image-processing circuit, the image data program ROMs, audio ROMs, various kinds of RAMs, an electronic power supply circuit, a clock generation circuit and the LCD itself on the same glass substrate (Figure 1).

Substrate Micrograph

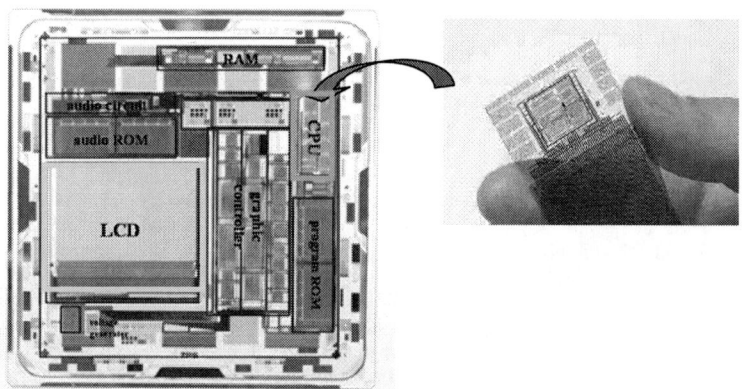

Fully-Functional panel was jointly developed with Semiconductor Energy Laboratory.

Figure 1. Full-Functional system panels (TFT substrate)

This panel was produced using dual-thicknesses gate insulator (for low and high voltage operation) and Gate-Overlapped LDD structure (GOLDD) for TFTs.

From the viewpoint of a human-friendly interface, sensor technology is an essential element for system LCD. TFT-LCD with ambient light sensor was proposed for back light control system (Figure 2) [7].

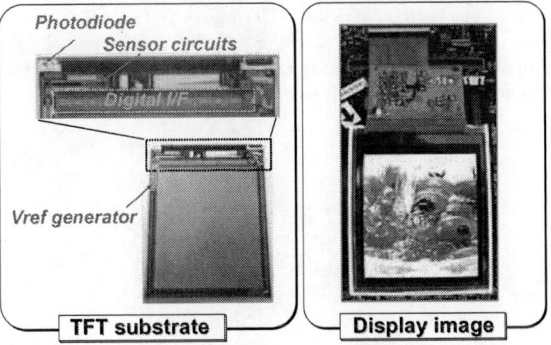

Figure 2. TFT panel and display image of Monolithic Ambient-sensor system LCD

An ambient light sensor system comprising of a lateral photo-diode and analog processing circuits was integrated directly onto the display substrate (Figure 3).

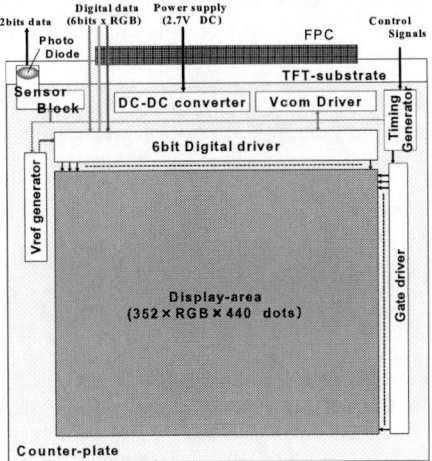

Figure 3. Panel Block Diagrams of ambient light sensor system

Intelligent control of the LCD backlight level, in response to ambient lighting conditions, leads to low power consumption and high reliability, at low cost. Furthermore, the basic components of the smart back-light technology – i.e. photo diode and analog circuit technology – are applicable to other sensing elements, such as finger-print identification system. Audio-Visual (AV) panel is defined as a display that combines audio system on LCD. Such AV panel has a 12 bit D/A converter, which enables it to play 48 kHz PCM digital audio recorded on DVD. By using an oscillating component, audio output power can be performed in the LCD simplex that carries this audio circuit.

To realize high performance audio amplifier, analog circuit technology with linear electric characteristics and low noise sound quality are important factors. Figure 4 shows the measurement data of the analog circuit.

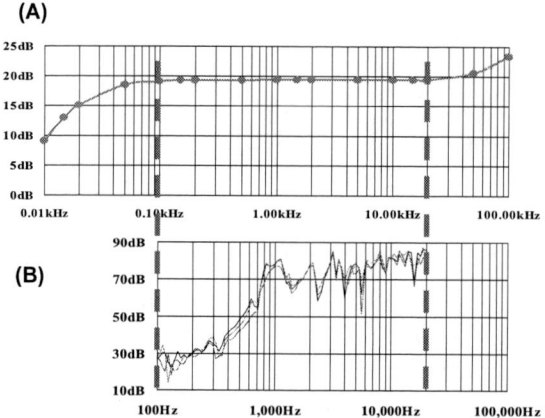

Figure 4. Frequency characteristics of the analog circuit
（A）Characteristics of Signal Amplifier shows flat (100Hz – 20 kHz)
(B) Acoustic characteristics of Panel speaker >70dB

These data prove that performance is very good not only for audio amplifier but for other System-LCD applications, as well.

Viewing Angle Control

From a LCD's image point of view, there is an increasing need to provide portable devices, like mobile phones and notebook PCs, with the capability to "veil" the display so that the screen content cannot be seen by other people in the immediate area. We have developed a new LCD that adds a switching liquid crystal material overlaid on an ordinary TFT-LCD so that light is prevented from going to the left or right, thus turning a wide viewing angle cone to a narrow viewing angle cone. As a corollary to this technology, we have developed another variant, which uses a parallax barrier superimposed on an ordinary TFT LCD. The LCD sends the light from the backlight into right and left directions (Figure 5), making it possible to show different information and visual content on the same screen at the same time.

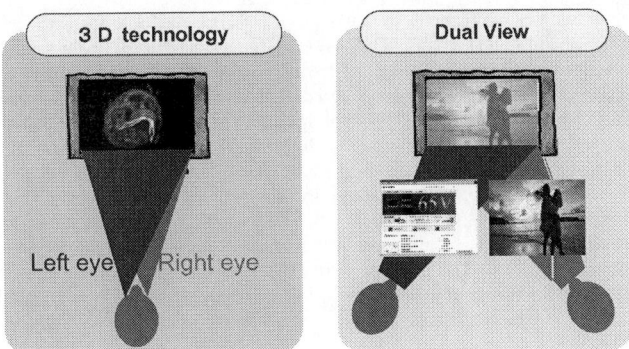

Figure 5. Dual View Technology

Controlling the viewing angle in this way allows the information or visual content to be tailored to multiple users viewing the same screen. For example, one user can view the display as a PC screen for browsing the Internet or for editing video shot using a digital camera while at the same time another user watches video content such as a movie or a TV broadcast. Digital terrestrial broadcasting allows us to receive TV programs with clear picture quality even at a street corner.

Super Reflective Color LCDs

Super low power consumption technologies are strongly required for mobile LCD applications. For example, electronic paper is attracting increasing attention. Moreover, internet viewer and mobile TV are expected to be important information tools for our life.

E-book and e-dictionary are key applications for e-paper technology. For these applications, bright display image and wide viewing angle are indispensable. In addition, for mobile information tool or TVs, Moving and Color image are essential. Nowadays newspaper and magazines are full of color contents. This means that there is an equivalent need to display color content on electronic paper. We developed new super reflective LCD technology in response to these requirements. A conceptual image is shown in figure 6.

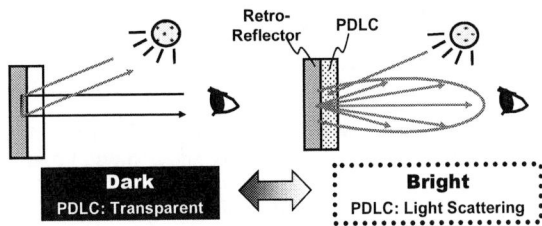

Figure 6. Physics of Super reflective color LCD

There are two key components in this display. One is Retro-reflector, and the other is Polymer Dispersed Liquid Crystal. PDLC is set on the retro-reflector as shown in figure 6. The retro-reflector reflects incoming light back to the light source and the observer. PDLC is another type of liquid crystal material. PDLC is switched between a "transparent state" and a "light scattering state" by applying electric field. In the case that PDLC is transparent, the display is in "dark" state. No light enters into the observer's eyes, and the observer perceives that as dark state. In the case that PDLC scatters incoming light, the display is in "bright" state. PDLC scatters incoming lights and the retro-reflectivity is broken, thereby allowing reflected light to reach the observer. It is bright state.

Retro-reflector must have high retro-reflectance. Retro-reflector is composed of a replicating unit structure. The pitch of retro-reflector units has to be smaller than the display pixel pitch. It was found that a corner cube array can be a very good retro-reflector. Based on this structure, we have been developing a new MCCA for retro-reflector (Figure 7).

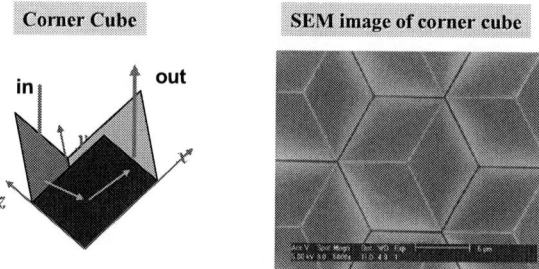

Figure 7. Retro-reflector (MCCA)

Low driving voltage and fast response are important for realizing low power consumption and displaying moving images. High transparency of PDLC is important factor for dark state. But for bright state, strong scattering is not necessary for this mode; only adequate scattering to break retro-reflectance is needed. Usually there is a trade-off relationship between driving voltage and scattering ability. In this new reflective display, the lack of strong scattering requirement enables the simultaneous realization of low driving voltage and fast response. For example, we have realized 50% reflectivity, 5V driving voltage and 30ms response time. These properties are very attractive, compared with conventional reflective mode displays or electrophoretic displays.

Flexible Display

Plastic Display

Flexible displays are one of the promising displays for mobile application. [8] [9] In the case of the plastic displays, there are some advantages against conventional displays using glass substrates, that is, light weight, thinness and durability. In most of recent mobile devices, the image quality is rapidly improved, so it is a vital requirement to a plastic display for a cellular phone that its display performance is comparable with conventional displays using glass substrates. We have developed the following key technologies.

1. Plastic substrate with high Tg, low CTE (Coefficient of Thermal Expansion) and high transparency
2. TFT process to control plastic substrate deformation
3. Low temperature process for transmissive TFT-LCDs

Plastic substrate. In order to realize a transmissive plastic TFT-LCD, a transparent plastic substrate with high thermal stability, low thermal expansion and high transparency is newly developed. For this substrate, slightly lower transparency is observed at 400-450 nm, however, no perceived difference was noted by the user. The color of the plastic substrate is clear, and no change occurs, even after annealing process.

Substrate deformation. Figure 8 shows the substrate deformation of the plastic substrate in TFT process under the process conditions of A, B. As a reference, the substrate deformation of a conventional glass substrate under process condition A is also shown. In process A, the inorganic base coat could reduce the substrate deformation to about ±70ppm In process B, the substrate deformation is reduced to ±15ppm, and it is comparable with the deformation of a conventional glass substrate.

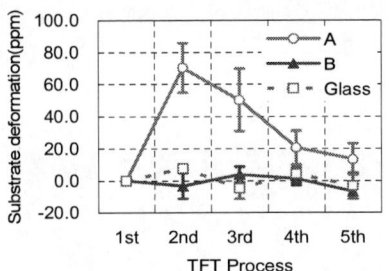

Figure 8 Substrate deformations in TFT process

Low temperature TFT process. One of the key technologies to fabricate TFT array on a plastic substrate is PECVD process at low temperature. We developed new PECVD process for low temperature deposition. The TFT characteristic fabricated on the transparent plastic substrate is shown in Figure 9. The mobility, the threshold voltage, the off current and the on/off ratio of the low temperature a-Si TFT on plastic substrate are 0.5cm2/Vsec, 0.7V, 1pA and 10^6, respectively.

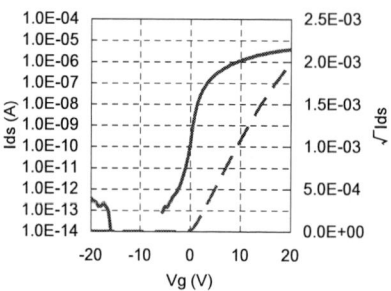

Figure 9 Vg-Ids characteristic of LT a-Si TFT on a plastic substrate

In addition, the reliability of the low temperature (LT) a-Si TFT is also investigated, and it is confirmed that there is no practical degradation in the LT a-Si TFT on the plastic Substrate [10].

Display Performance. We have fabricated a transmissive plastic TFT-LCD with high definition by using newly developed plastic substrate and TFT process (Fig. 10). The panel thickness of 0.32mm, thinner than a half of the conventional glass panel, could be realized by our newly developed technology. In manufacturing, thinner plastic substrates of less than 0.1mm can be applied.

Figure 10 Transmissive plastic TFT-LCD

Thin Metal Foil Display

Polymer substrates have been the basic contender for flexible display applications, and have received a growing research interest in the past few years. However, that is not to say that there are no drawbacks, the most important of which are their reduced compatibility with standard CMOS processing, the resulting low device stability, and large device features. Although NMOS devices with high mobility are feasible, the lack of high temperature steps compromises device stability and requires expensive low temperature processes and materials.

From that point of view, an all-inorganic, thin metal foil (TMF) backplane substrate represents an excellent alternative to polymers or hybrid backplanes (inorganic substrates with polymer/organic coatings) for use as flexible systems (Fig. 11). TMF substrate offers superior chemical resistance in a number of environments compared to plastics, and is readily compatible with med-range temperature processing, which is to say compatible with conventional display fabrication temperatures. Furthermore, the greater dimensional stability that metal foil substrates offer allows for the implementation of circuit designs with minimum feature size of 1μm or less. High performance, high speed circuits are then feasible as gate lengths are significantly reduced. Our preliminary research on this field validates poly-silicon TFT metal foil technology as a suitable platform for the fabrication of large area systems onto flexible substrates. The displays presented here meet or exceed performance requirements targeting the circuit design and implementation challenges of flexible large area electronics [11].

Poly-Silicon TFT on TMF. Poly-Si TFT devices and circuits in this work were fabricated on 150 mm flexible stainless steel type 304 substrates that were 100 micron thick. Stand alone devices of various geometries were fabricated and tested, and their electrical characteristics investigated. The high quality poly-silicon microstructure obtained with the laser annealing process used in this research make it possible to achieve some of the highest mobility recorded to date (to our best knowledge) on metal foils: over 300 cm^2/Vs for NMOS TFT and 150 cm^2/Vs for PMOS TFT. These devices show at least seven orders of magnitude difference between I_{DS} (ON) and I_{DS} (OFF), making it also one of the highest values compared to previously published results. Moreover, due to the superior dimensional stability of TMF devices with channel as short as 1μm and as narrow as 2μm have been successfully fabricated with proximity printing. The combination of all these assets; high mobility, high stability and small design rules makes these devices suitable for high performance digital and analog large area circuit applications; in particular, high resolution, low power, AMPLED displays.

Figure 11. AMPLED backplane on metal foil

Display Components. We have fabricated a display array demonstrator consisting of 480 by 640 pixels with pitch of 105 μm by 110 μm forming a full VGA display with a 3¼ inch diagonal, and a 70 μm by 74 μm emissive area, which results in 45% of aperture ratio. The standard 2 TFT pixel circuit is implemented with PMOS transistors (Fig. 12)

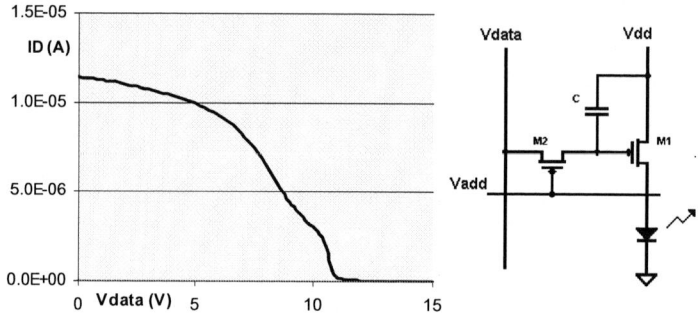

Figure 12. Pixel IV curve (left) and equivalent circuit (right)

Integrated column drivers were implemented with Static Shift Registers that were selected based on stability, reliability and yield from a variety of tested designs. The operation speed of the S/R ranges from a few Hz to several MHz, which puts them well within the operational range for VGA column driving speeds of approximately 50kHz. In order to increase yield, a half bit topology was implemented with small current buffers. The integrated shift register functionality was demonstrated with the AMPLED fabricated on silicon wafers.

Fabrication of Polymer Light-Emitting Diodes. Due to the opaque nature of the stainless steel substrates, the top-emission OLED's is adopted. Instead of small-molecular OLED, polymeric OLED (PLED) is applied in this study. There are several basic requirements needed to meet for a top-emission OLED in order to achieve the high efficiency of illumination, and they are the reflective anode, the low work-function and transparent cathode, and a smooth anode surface to prevent shorts. In our case, the anode is gold, which provides with 5.1 eV of work-function and a very smooth surface. On top of the anode, a conductive hole-transporting polymer, PEDOT, was spin-coated to make the good metal/polymer junction and smooth the anode surface even further. Sequentially, a light-emitting polymer, poly-phenylenevinylene (PPV), was spun. A thin layer of calcium with 2.8 eV of work-function was then thermally evaporated and deposited onto PPV, being capped by a thick transparent ITO cathode. This combinatorial Ca/ITO cathode gives the greatest transmittance of 85% at wavelength of 550 nm, where the peak of PLED's electroluminescence locates, so that the light output was maximized. The light turn-on voltage (V_L) of the PLED devices is 4.0 V and they become visible at 5.0 V under ambient lighting conditions (Fig. 13). For the 70 μm × 74 μm pixels in the present design, the current consumption is around 2.7 μA/pixel, which is well within the capabilities of the driving transistors.

Figure 13. AMPLED display in operation (left) and a close-up of an AMOLED pixel group (right).

New Technologies for Next Generation Information Tools

To realize the high performance system-on-panel for mobile applications, high frequency TFT circuits, and low-voltage driving performance will be essential factors. To achieve these targets, TFT characteristics such as high mobility, low threshold voltage and low sub threshold voltage are demanded. We have been developing new TFT structure and process for high performance TFT, which consist of fine lithography and thin Gate Insulator. We have fabricated an amplifier circuit using 1 micron design rule and 30nm Gate Insulator for the performance evaluation of advanced TFTs. Based on this circuit we have concluded that such transistors can be driven at very high frequency (over 1GHz) with low applied voltage.

Supported by high performance CG Silicon® TFT technology, we have been developing new technology for DCDC, timing generator, and other functional circuits. For next generation information tools, ultra high speed interface circuits, memory circuits and high performance logic circuit will be realized by new high performance TFTs. Moreover, new functional device and circuits such as sensor (light sensor, touch panel, thermo sensor, finger print sensor etc.) and analog amplifier circuits (audio amp. etc.) will be fabricated monolithically on glass.

We have also investigated new classes of display devices based on flexible array technology developed either on plastic or thin-metal-foil substrate. Different applications are envisioned by the combination of plastic and LCD or TMF and OLED technologies. By the fusion of these new technologies, personal information tools, which include high performance, multi functional display that all people can use and enjoy will come into fruition in the near future.

Conclusion

With CG Silicon® technology, we have been making a remarkable progress on "system on glass" technology. We have developed an array of new technologies for mobile display, such as "viewing angle control technology" and "flexible display". We are confident that these technologies, which have been developed or are currently under development, will enable a revolution in the field of new ubiquitous information tools.

References

[1] A. Imaya, "CG Silicon Technology and its Application", AM-LCD 2003 Digest, 1-4, 2003.

[2] H. Ohtani et al., "A 60-in. HDTV Rear-Projector with Continuous-Grain-Silicon Technology", SID'98 Digest, 467-470, 1998.

[3] T. Takayama et al., "Continuous Grain Silicon Technology and Its Applications for Active Matrix Display" AM-LCD 2000 Digest, 25-28, 2000.

[4] H. Washio et al., "TFT-LCDs with Monolithic Multi-Drivers for High Performance Video and Low-Power Text Modes", SID'01 Digest, 276-279, 2001.

[5] K. Maeda et al., "Multi-Resolution for Low Power Mobile AMLCD", SID'02 Digest, 794-797, 2002.

[6] T. Ikeda et al., "Full-Functional System Liquid Crystal Display Using CG-Silicon Technology", SID'04 Digest, 860-863, 2004

[7] K. Maeda et al., "The System-LCD with Monolithic Ambient-Light Sensor System", SID'05 Digest, 356-359, 2005

[8] Y. Okada et al, "A 4-inch Reflective Color TFT-LCD Using a Plastic Substrate", SID'02 Digest, 315, (2002)

[9] B. S. Kim et al, "Developments of Transmissive a-Si TFT-LCD using Low Temperature Processes on Plastic Substrate", SID'04 Digest, 19, (2004)

[10] H. Nishiki et al, "Reliability of Low Temperature a-Si TFTs for Plastic TFT-LCD", IDW'03, 307, (2003)

[11] M. Troccoli et al., "AMOLED TFT Pixel Circuitry for Flexible Displays on Metal Foils," Proc. MRS Spring 2003, H3.7., (2003)

ECS Transactions, 3 (8) 23-33 (2006)
10.1149/1.2356331, copyright The Electrochemical Society

Optical Feedback for AMOLED Pixel Circuits

Steve Deane[a], David Fish[a], Nigel Young[a],
Andrew Steer[a], Nicola Bramante[a], David George[a], Andrea Giraldo[b],
Herbert Lifka[b], and Wouter Oepts[b]

[a] Philips Research Laboratories, Cross Oak Lane, Redhill, Surrey, UK
[b] Philips Research Laboratories, High Tech. Campus, Eindhoven, Netherlands

Optical feedback pixel circuits for a-Si:H and LTPS technologies
will be presented. The circuits enable correction of threshold
voltage drift of the drive TFT and degradation of the OLED. In the
a-Si:H case this is achieved with a standard a-Si:H process and for
LTPS an A-Si NIP photodiode is integrated. Operation, technology
and measurements will be presented.

Introduction

OLED displays are attractive because they are thin, have perfect viewing angle, fast
response, and high contrast. For displays of any significant size and complexity, an active
matrix is needed to maintain good image quality and low power consumption. For large
displays, such as the TV market it is highly desirable to make this active matrix from a-Si
TFTs, due to the higher cost of large LTPS displays. A number of groups have used a-
Si:H technology for AMOLED displays and have shown impressive demonstrators (1,2).
Use of a-Si technology leads to several problems, such as the instability of the drive TFT
threshold voltage with time, and the large size of the drive TFT due to the low mobility
(compared to LTPS).

Correction schemes for the a-Si threshold voltage drift have been employed in many
cases (2,3). However, the accuracy of the correction and hence the overall display
lifetime that these can provide is questionable. The low a-Si TFT mobility leads to large
drive TFTs, with correspondingly large parasitic capacitances. These parasitic
capacitances generally lead to a corruption of the threshold voltage correction, due to the
limited size of designed capacitor available in the pixel circuit.

There are further difficulties with a-Si:H TFTs in that they are quite poor current
sources when the channel length is short (a requirement necessary to deliver a suitable
current to the OLED). This can give rise to image cross-talk. An example of this is
illustrated in figure 1. Here we see a typical a-Si TFT output characteristic, and the lack
of saturation is highly visible. This lack of TFT saturation means that power line voltage
drops of less than a volt will lead to image cross talk, and this constraint will make large
array design unfeasible.

The Need for Optical Feedback

A further significant issue, common to all OLED displays, is that the OLED material
itself ages. This is illustrated in figure 2. Often material lifetimes are quoted to a 50%
brightness drop, and at lower brightness than the application really requires. This 50 %
brightness drop would be acceptable in a uniform application such as a large area
backlight, or other lighting application, but is completely unacceptable on a pixel-pixel
basis in a display. Experience has shown that a burn-in of 2% is visible to most customers,

and is considered unacceptable. This occurs in typically 10-100 hours for present materials. Unfortunately, such burn in becomes more likely, due to the presence of channel logos, icon user interfaces, news tickers, and the display of media of varying aspect ratios on a single display. These artifacts then become highly visible when the image changes to be more uniform.

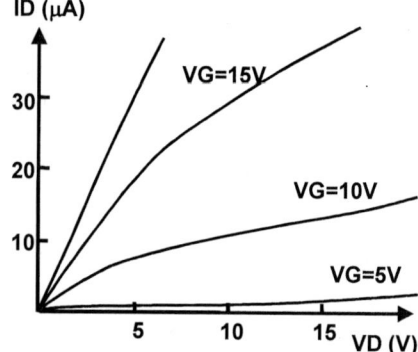

Figure 1. A-Si:H short channel TFT output characteristic, showing a lack of good saturation behaviour.

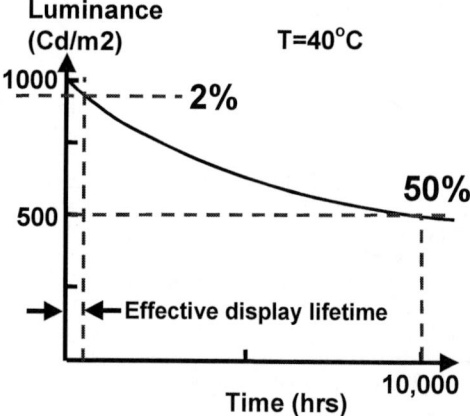

Figure 2. Typical degradation of OLED material under constant current drive. The lifetime until visible burn-in is much shorter than the lifetime to half brightness.

These problems lead us to propose the use of optical feedback to correct for these changes (4,5). The optical feedback can be performed either in-pixel, or externally. External approaches require custom external sensing and correction circuits, adding cost and restricting display design freedom, while in-pixel approaches can use standard external drive circuits. The general principle of in-pixel optical feedback is shown in figure 3. The data is addressed onto a pixel capacitor by an address TFT, much as in a conventional AMLCD. This voltage is then compared to a reference, and the outcome of this comparison is used to drive the drive TFT on or off. When the drive TFT is on, the OLED emits light, some of which is captured by a photosensor. This photosensor current is used to charge the pixel capacitor until its voltage reaches the reference, and turns the drive TFT and OLED off.

It is worth noting that this optical feedback operation makes no assumption about the OLED efficiency, the drive TFT threshold voltage, or indeed the drive TFT being an accurate current source. Differences in these factors lead to different duration of the light emission, with the circuit emitting until the total integrated light output is correct.

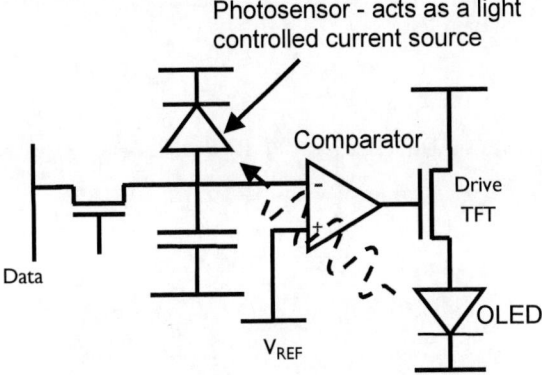

Figure 3. General principle of an optical feedback pixel circuit.

Optical Feedback in a-Si

We have therefore applied optical feedback to an a-Si:H AMOLED display to compensate for both burn-in mechanisms i.e. the drive TFT threshold voltage drift and the OLED degradation. An advantage of a-Si:H is that standard TFTs show good photosensitivity at all visible wavelengths and thus can be used as photo-detectors at no extra process cost.

An example of a practical a-Si:H optical feedback pixel circuit is shown in figure 4. Here a TFT is used as a photosensor, and the comparator is formed of two TFTs and a capacitor. In operation, T_R charges the gate of the drive TFT during the address phase, and the gate is discharged when the data capacitor reaches the threshold voltage of T_S. During the addressing phase the power line is brought low, so that the anode of the OLED is held at a known voltage by the drive TFT while the pixel data is loaded.

A process cross-section is shown in Figure 5. Here we show that the TFTs and photosensors can be fabricated is a standard a-Si TFT process. The OLED emits light through the ITO anode, and some of this light falls on the photosensor TFT, resulting in a photocurrent.

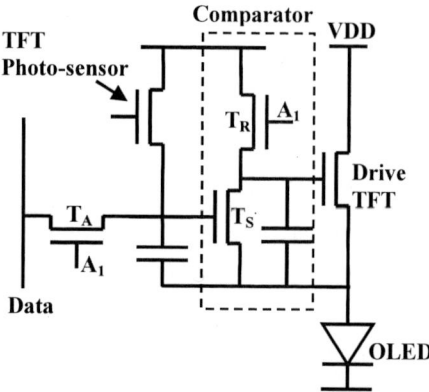

Figure 4. Example of practical A-Si:H optical feedback pixel circuit.

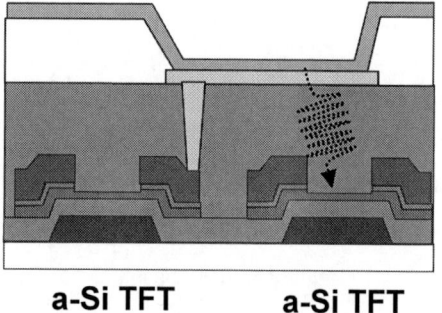

Figure 5. Cross-section of A-Si:H optical feedback pixel circuit in a standard TFT process, showing drive TFT and photosensor.

Measured results from this circuit are shown in figures 6 and 7. Figure 6 shows individual luminance measurements for undegraded and degraded pixels. We can see that the while the OLED has been substantially degraded, giving a lower initial brightness, the circuit responds by lengthening the emission pulse, so that nearly the same total amount of light is emitted. The circuit similarly corrects for drive TFT drift, and cross-talk due to the poor saturation characteristics of short channel a-Si TFTs. Figure 7 shows an image from a small test display using the pixel circuit. The left most picture shows the checkerboard image displayed to cause a burn-in problem. In the middle picture, the display is attempting to show a plain grey field, but the correction circuit has been disabled. The image shows a burn-in pattern that would be unacceptable to a customer. In the rightmost image, the correction circuit is enabled, and the checkerboard burn-in pattern is no longer visible.

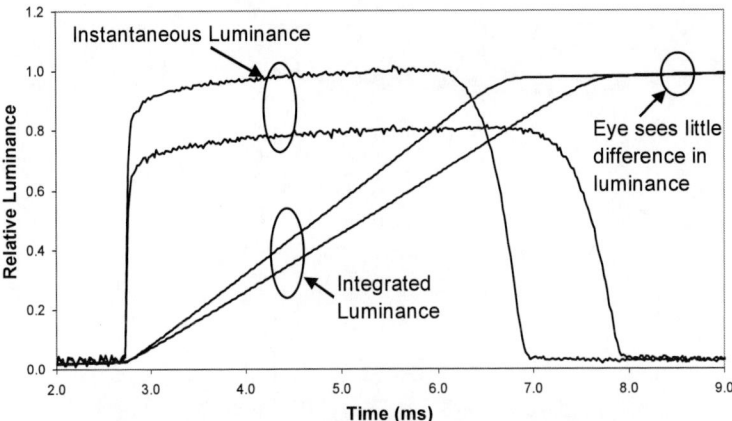

Figure 6. Measured results of A-Si optical feedback circuits correcting for burn-in.

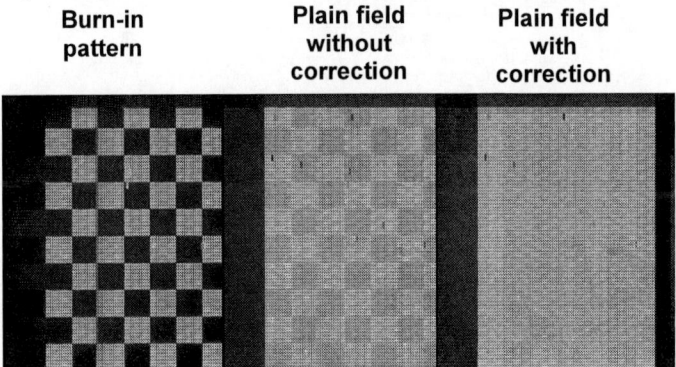

Figure 7. A-Si:H optical feedback display showing burn-in pattern, the burnt-in image when displaying a plain field with the correction disabled, and the result of enabling the correction.

Other areas of potential concern are the stability and uniformity of the photosensors. It is well known that a-Si can suffer from a light induced instability called the Staebler-Wronski effect (6) (SWE). However, it is also known that the effect is actually driven by bipolar recombination (the light only serves to generate the carriers), so the SWE can be suppressed by a bias that sweeps out the photocarriers (7,8). This bias is exactly what we desire to create a photocurrent source, so when measured under the correct operating conditions, photo-TFTs are stable as shown experimentally in figure 8. Figure 9 shows the measured uniformity of a display made with the optical feedback circuit. We see very good uniformity, which means that the uniformity of the photo-sensors, the uniformity of the optical coupling, and the uniformity of the T_S threshold voltage are all good.

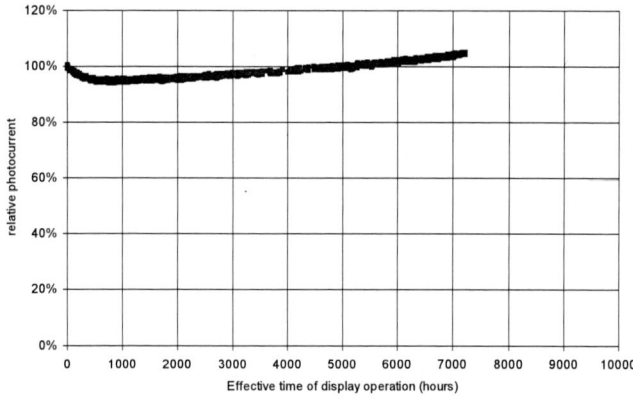

Figure 8. Measured photo-stability of a-Si TFT photosensor, showing good stability.

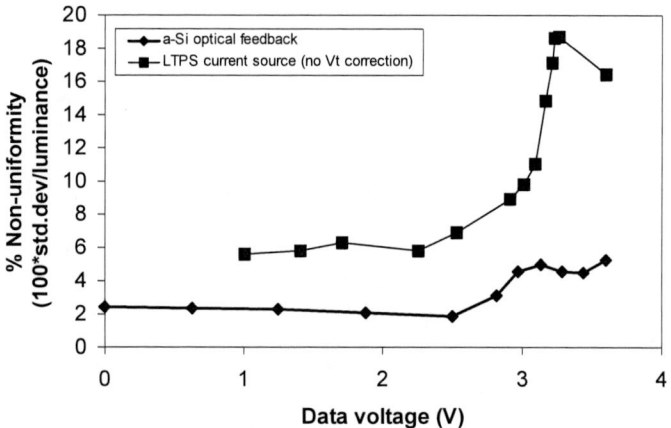

Figure 9. A-Si optical feedback circuit uniformity measurements, compared to an LTPS current source circuit (without Vt correction). Higher data voltages correspond to a darker image. This result shows that the photosensors, and their coupling to the OLEDS is uniform.

Optical Feedback in LTPS

Similarly, we can implement optical feedback in LTPS. The higher TFT mobility allows better apertures for smaller pixel sizes, and the usual arguments for reduced cost / footprint by integration of row and column drivers for small to medium sized displays apply equally to OLED as to LCD.

Figure 10 shows an example of an optical feedback circuit in LTPS, where the operation is similar to the a-Si circuit of figure 4. Notable differences are that both n-type and p-type TFTs can be used, which allows some parasitic effects to be reduced. An extra TFT has been inserted between the drive TFT and the OLED to allow the power line to be DC. Many variants are possible, and an all-PMOS version could be interesting for medium sized panels to reduce the LTPS cost.

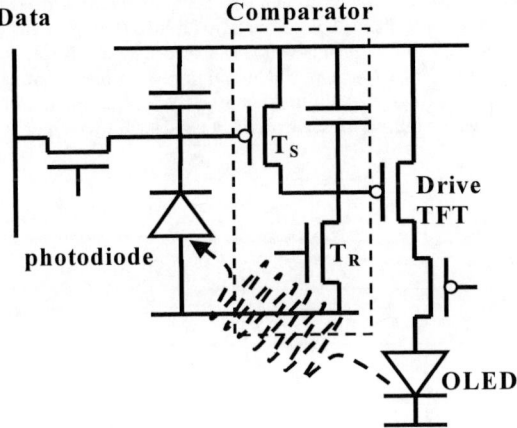

Figure 10. Example of LTPS Optical Feedback Circuit.

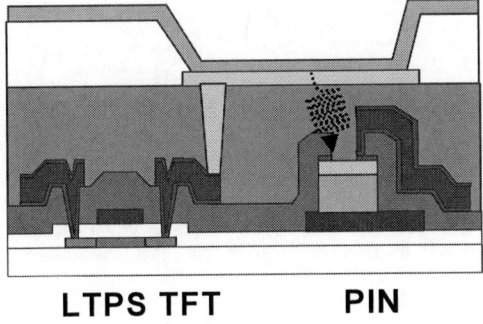

Figure 11. LTPS process cross section, showing a-Si PIN photosensor integrated with a single additional mask step.

An interesting benefit of LTPS optical feedback is that there is no need for the drive TFT to act as a current source. While this means that not only do we not need to worry about saturation in the output characteristics (like for a-Si), but also in LTPS we can use the drive TFT just as a switch, so that we reduce the power loss in the drive circuitry. This can result in up to a 2x increase in overall power efficiency compared to circuits which rely on a current source TFT. This is a significant benefit, particularly for mobile devices.

In the LTPS case, polysilicon is not so good as a photosensor, with poor absorption particularly in the red end of the spectrum. One solution to this is to integrate an a-Si photodiode into the LTPS process. This provides a high efficiency, broad spectrum photosensor. While this may sound expensive, actually it can be integrated in a standard LTPS process with only one extra mask step. An example process cross section is shown in figure 11. The PIN diode shares bottom metal with the LTPS gate electrode, and top metal with the column metal, so the only extra mask step is the diode stack.

As previously, the photosensor is operated with a reverse bias to sweep out carriers, so we expect no issue with photostability due to the SWE. The results of life testing are shown in figure 12, where indeed we see extremely good stability with light soaking.

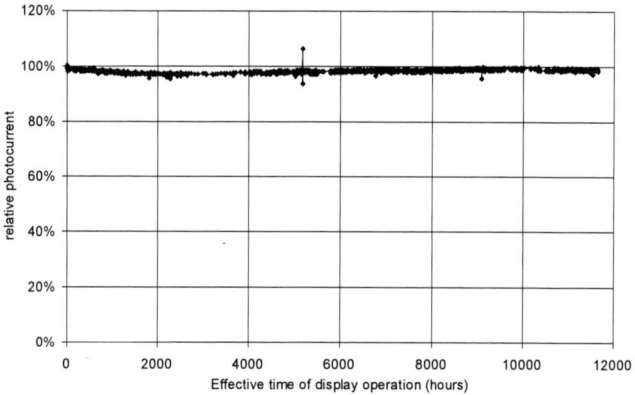

Figure 12. Measured photo-stability of a-Si PIN diode photosensor, showing good stability.

Measured operation of the pixel circuit of figure 10 is shown in figures 13 and 14. Figure 13 shows the emission duty cycle lengthening as the emission brightness falls to ~40% of the initial value. Figure 14 shows the corresponding data for integrated luminance, showing that all traces converge on the same total light output. Only as the emission pulse fills the available frame time does a significant deviation begin to occur.

The performance of the LTPS circuits can be enhanced even further (5) by the addition of an additional transistor, labeled T_F in figure 15. This has the effect of applying feedback to the circuit, so that when the turn off threshold is reached, T_F effectively adds to the photocurrent, and turns the circuit off very rapidly. This circuit is compared to than of figure 10 in figure 16, showing measured results for both circuits. Note that we use very short pulse lengths to highlight the differences in the turn-off time of the two circuits. This faster turn off leads to improved correction of all artifacts.

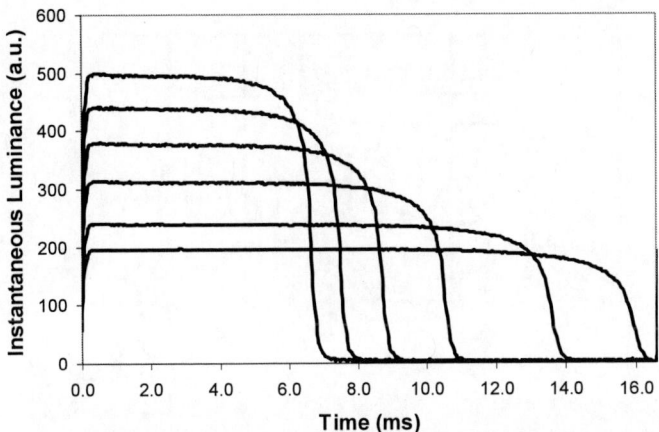

Figure 13. Measured instantaneous luminance of LTPS optical feedback circuit of figure 10 correcting for burn-in.

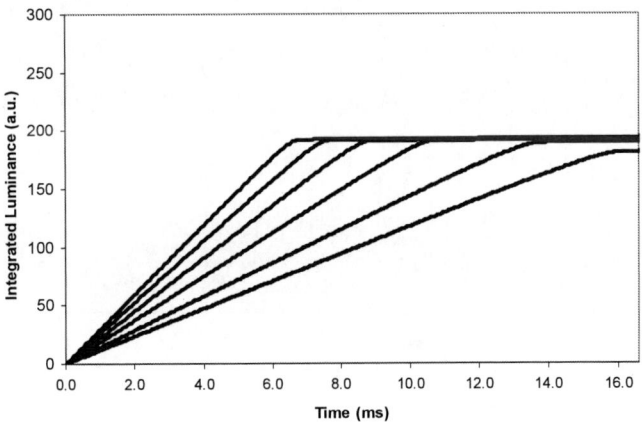

Figure 14. Measured integrated luminance of LTPS optical feedback circuit of figure 10 correcting for burn-in.

Data

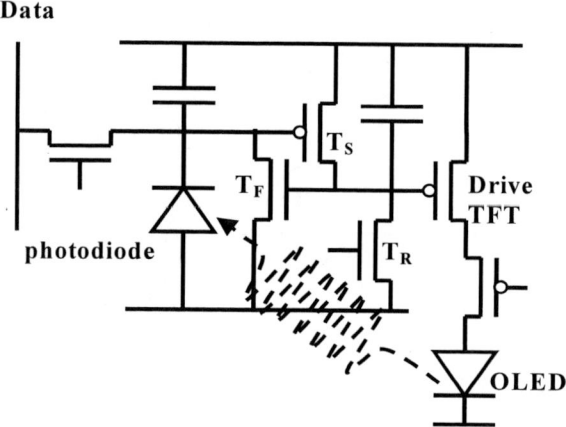

Figure 15. Fast Snap-Off Optical Feedback Circuit.

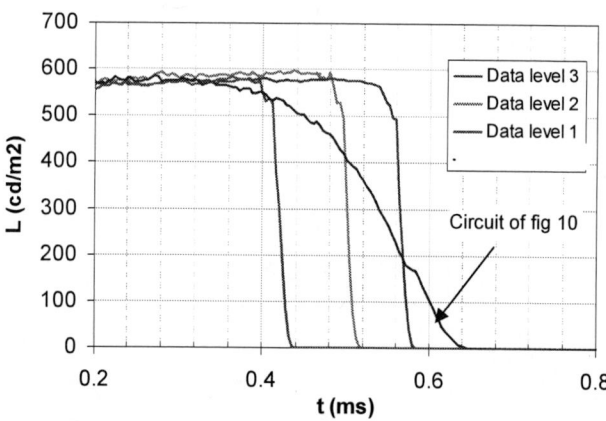

Figure 16. Experimental results showing the rapid turn-off capability of the new pixel circuit.

Conclusions

We have shown how optical feedback correction can be added to both a-Si and LTPS AMOLED displays. This provides not only significantly improved lifetime to OLED burn-in, but also a host of other benefits. In a-Si, drive TFT threshold voltage shift and poor output saturation effects are compensated, with no increase in process complexity. In LTPS, significant power consumption savings can be achieved, and only one mask step is added to the standard process.

References

1. T. Tsujimura et al, "A 20-inch OLED display Driven by Super-Amorphous-Silicon Technology", *SID03*, 4.1, p6 (2003).
2. T. Shirasaki, et al "Full Colour Polymer AMOLED using Ink-Jet and A-Si TFT technologies", *SID04* 57.4, p1516 (2004).
3. J.L. Sanford et al, "VT Compensation Performance of Voltage Data AMOLED Pixel Circuits", *IDRC 2003*, 3.1, p38 (2003).
4. D.A. Fish et al, "A comparison of pixel circuits for Active matrix Polymer Organic LED displays", *SID02*, 32.1, p968 (2002).
5. D.A. Fish et al, "Improved optical feedback for AMOLED differential ageing correction", *SID04*, 35.2, p1120 (2004).
6. D.L. Staebler and C.R. Wronski, *Appl. Phys Lett.* **31**, 292 (1977).
7. I. Sakata, Y. Hayashi, H. Karasawa, and M. Yamanaka, *Solid State Commun*, 45, 1055 (1983).
8. M. Stutzmann, W.B. Jackson, and C.C. Tsai, *Phys Rev. B*, **32**, 23 (1985).

34

ECS Transactions, 3 (8) 35-41 (2006)
10.1149/1.2356332, copyright The Electrochemical Society

High Reliability and Performance Poly-Si TFTs for System in Displays

M. Hatano [a], H.Hamamura [a], M. Matsumura [a], Y. Toyota [a], M. Tai [a],
M. Ohkura [b], and T. Miyazawa [b]

[a] Hitachi, Ltd., Central Research Laboratory,
1-280, Higashi-koigakubo, Kokubunji, Tokyo 185-8601, Japan
[b] Hitachi Displays, Ltd., 3300 Hayano, Mobara, Chiba 297-8622, Japan

> Narrow frame size IPS-mode LCD with high resolution system-in-display have been developed utilizing hybrid laser crystallization technology: SELAX are applied to low-power, high-speed circuits, and conventional ELC are used for the high-voltage circuits. The TFT degradation phenomena under dynamic stress are clarified, and the high-immunity TFT structure and high-quality gate insulator are proposed.

Introduction

"System-in-displays", based on polycrystalline-silicon thin film transistors (LTPS TFTs), have several advantages: a high resolution, fewer connections, simplified modules, high reliability, and added valuable functions. It is necessary to integrate high-performance TFTs for high-speed with low-voltage circuits, and high-voltage endurable pixel TFTs on the same substrate. To meet this requirement, we have developed a selectively enlarging laser-crystallization (SELAX) technique (1, 2), which can transfer an excimer-laser-crystallized (ELC) poly-Si to a large-grained flat poly-Si film at "selective regions".

In this paper, the SELAX technologies and high reliability and performance TFTs are described. We also discuss the degradation phenomena of LTPS TFTs under dynamic stress and highly stress- immune TFTs.

System in Displays Technologies

Low temperature processes (450°C) (3) are applied to fabricate for both ELC and SELAX TFTs. First, 50-nm-thick amorphous Si films on glass substrates are irradiated with an excimer laser in order to poly-crystallize them. Selected areas of the ELC poly-Si layers are then irradiated with the pulse duration controlled solid-state laser in order to enlarge the grains (SELAX process). Next, 100-nm-thick gate-oxide layers are deposited by PECVD with a TEOS gas source. After formation of the gate electrodes and source/drain regions, the samples are annealed for hydrogenation. Maximum process temperature is 450°C.

SELAX TFTs have high current drivability and low sub-threshold swing and are suitable for low-voltage, high-frequency operation. The dependence of the propagation delay time on the power-supply voltage, measured by 21 stages CMOS ring oscillators, is shown in Fig. 1. The delay time is shorter for SELAX than for ELC. The difference of the propagation delays between SELAX and ELC becomes larger as power-supply voltage decreases, because of the lower sub-threshold swing of SELAX TFTs.

35

In n-channel TFTs, drain-avalanche-hot-carrier (DAHC) stress is found as the worst stress condition. Figure 2 shows the dependence of power supply voltage, V_{dd} on lifetime under DAHC stress of n-channel TFTs. We define the lifetime at on-resistance is degraded 10%. ELC TFTs have longer lifetime than SELAX TFTs at the same V_{dd}. This is because the stress current of the SELAX TFT is larger than that of the ELC TFT. Therefore, ELC TFTs are used for high-voltage endurable TFTs because of their better reliability at high voltage. However, the lifetime increases exponentially with the V_{dd}. Reducing the V_{dd} should produce SELAX TFTs with a higher reliability.

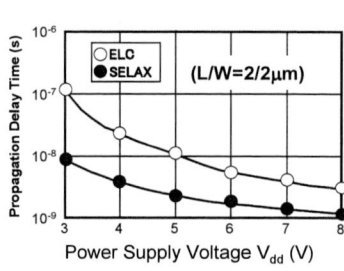

Fig.1 Propagation delay time of
SELAX and ELC TFTs.

Fig.2 Dependence of power supply voltage, V_{dd}
on lifetime under DAHC stress of n-channel TFTs.

The SELAX process can be applied to panel fabrication to realize high resolution and high performance system-in-displays. SELAX laser irradiation is carried out only on the region where the source driver circuit is formed as depicted in Fig. 3. SELAX TFTs are applied to low-power, high-speed circuits, and conventional ELC TFTs are used for the high-voltage circuits. Since the operation voltage of liquid crystal is not readily reduced, high-voltage operation of device is inevitable for pixel TFTs and TFTs directly connected to the pixels.

Thus, SELAX technology is successfully applied to the fabrication of 5-in. WXGA LCD system-in-display as shown in Fig.4. A dynamic ratio less shift register circuit using single channel TFTs is adopted in both the gate and source drivers (4). A resolution of over 300 ppi is developed by applying 4-μm layout rule, and the frame width is reduced by more than 50 %. The display using mobile IPS-mode LCD has ultimate image quality.

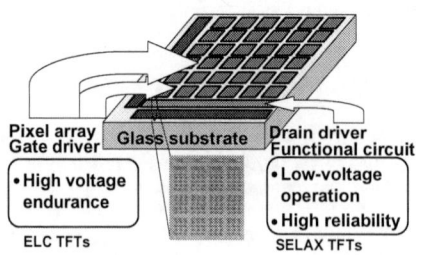

Fig.3 Schematic diagram of system in displays.

Fig.4 Display image of 5-in. IPS-mode WXGA LCD.

Degradation Mechanism in LTPS TFTs under Dynamic Stress

Reliability of the LTPS is a critical issue and the degradation properties of n- and p-channel TFTs under dc and dynamic stress were investigated (5,6).

(1) n-channel TFTs

Figure 5 is for comparison of degradation under dc and ac stress of TFTs with the single-drain structure (SD-TFT). The dc stress is DAHC stress and the ac (dynamic) stress waveform used here is shown in Fig. 7. A pulse train is applied to the gate as stress while dc stress is applied to the drain. The hatched region represents the range of gate voltages around corresponding to the maximally effective DAHC stress condition. In the case of ac stress, pronounced device degradation is observed though the period in the region of effective DAHC stress time is kept very short (2% of the period over which acstress is applied). It is clear that degradation is worse under ac stress than under dc stress. The shift in threshold voltage (ΔVth) is also shown in Fig. 5; Vth is defined as the value of Vg at which Id equals 1 pA. The ΔVth due to ac stress remains negative over the 20 s of stress application, whereas the value turns positive after only 0.2 s of dc stress. This means that more hot-holes are injected under ac stress than under dc stress. This injection of hot holes is regarded as a significant factor in the greater severity of the degradation caused by ac stress.

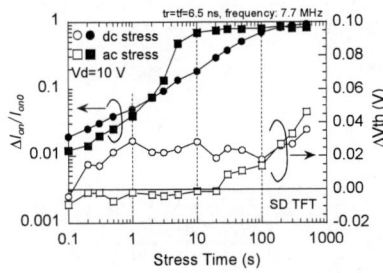

Fig. 5 Comparison of degradation under dc and ac stress.

Fig.6 Waveform of pulse applied to the gate.

To compare the severity of degradation by DAHC stress applied when the TFT is turned on and turned off, stress pulses shown in Fig. 7 are applied. Figure 8 shows the dependence of degradation in the I_{on} characteristic of SD and LDD TFTs on period of stress. The x axis has been converted to show the period in the range of maximally effective DAHC stress. This figure also shows the dc hot-carrier degradation caused by exposure to maximally effective dc DAHC stress over the same period. Degradation for the SD TFT ($\bullet$) proceeds more rapidly than it (x) does under exposure to the stress pulse of Fig. 7(a); that is, the effect is increased while the TFT is turned off. When the TFT is turned off, a transient flow of holes towards the poly-Si/SiO$_2$ interface is generated to restore the charge equilibrium associated with –2 V. These results further affirm that the injection of hot holes greatly contributes the more rapid degradation under ac stress. For the LDD TFT, on the other hand, there is no difference between the degradation of properties under the stress pulse of Fig. 7(a), the stress pulse of Fig. 7(b), and DAHC stress. Degradation of the LDD TFT's characteristics is not accelerated by ac stress.

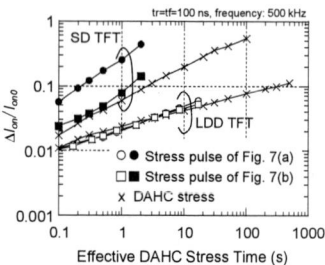

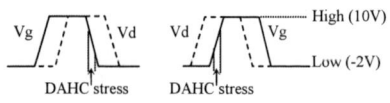

(a) Gate pulse leading drain pulse (b) Drain pulse leading gate pulse

Fig. 7 Waveform of stress gate and drain pulses.

Fig. 8 Degradation of TFTs when subjected to the stress pulses shown in Fig. 7.

(2) p-channel TFTs

Figure 9 shows the dynamic stress waveform. A pulse train is applied to the drain while dc stress is applied to the gate. This stress condition is chosen so that electrons and holes are injected into the gate oxide alternately. That is, the TFTs are biased in on state (hole injection) and subthreshold state (electron injection) when the drain pulse is respectively at high level and low level as shown in Fig. 9. Note that the bias condition between the source and drain in the on state is opposite to that in the subthreshold state. gm degradation (gm/gm_0) under exposure to this dynamic stress is shown in Fig. 10. Pronounced gm degradation and its temperature dependence are clearly observed.

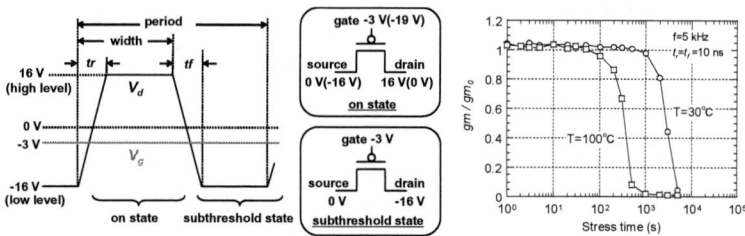

Fig.9 Waveform of stress pulse and bias.   **Fig.10 gm degradation under dynamic stress.**

Figure 11 shows the transfer characteristics after applying dynamic stress for 5000 s. This figure shows that I_d degradation in the case of reverse measurement is more severe than that in the case of forward measurement at V_d of -5 V. This means that trap states are generated near the drain junction.

To investigate the generation of trap states caused by repetition between electron injection and hole injection, gm degradation is evaluated by repeating two kinds of dc stress alternately, i.e., repetition of 1 s of dc on-state stress and 1 s of dc subthreshold-state stress.   Figure 12 shows gm degradation and temperature dependence under exposure to the alternate dc stresses. Comparing Figs. 10 and 12 shows that the pronounced gm degradation and the temperature dependence are reproduced by the alternate dc stresses. Figure 12 also shows gm degradation under exposure to the respective dc stresses. It is clear that gm does not degrade either on-state stress or the subthreshold-state stress. These results affirm that the pronounced gm degradation under dynamic stress is caused by the trap states produced by electron-hole recombination (7).

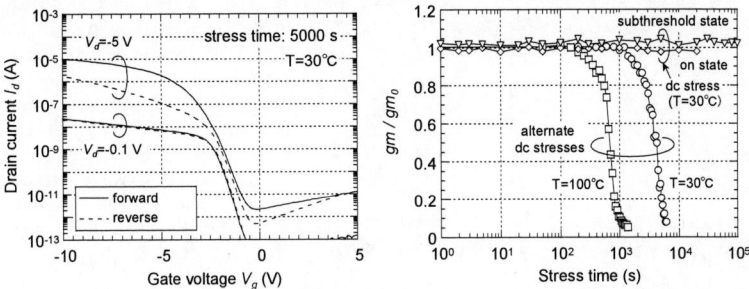

Fig.11 Transfer characteristics after dynamic stress. Fig.12 gm degradation under alternate DC stress.

Highly Reliable LTPS TFTs

To improve the reliability, we proposed the high-immunity TFT structure and high-quality gate insulator.

(1) High-immunity LTPS TFT structure to DAHC
We developed a gate-overlapped TFT (GOLD) (8). In this TFT, the gate electrode overlaps the LDD region and suppresses the effect of hot carriers injected into the LDD region as shown in Fig.13.

Gate electrode overlaps LDD

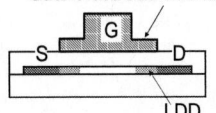

Fig.13 GOLD TFT structure.

Figure 14(a) shows the Ion degradation rate of this TFT under dynamic stress in comparison with that of the LDD TFT. The degradation rate under dynamic stress is lower than that of the LDD TFT at all frequencies of the Vg pulse.
Figure 14(b) shows the TFT driving ability at low operating voltage. The GOLD TFT has the same on current as the LDD TFT at half Vg. This means that the driving performance of GOLD is much higher than the LDD TFT. These results indicate that the GOLD TFT is well suited for high-speed low-power circuits.

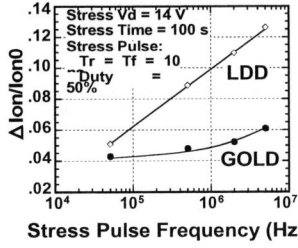

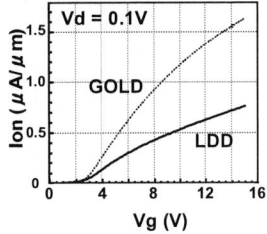

Fig.14(a) Ion degradation under dynamic. stress.   Fig.14(b) TFT driving ability.

Figure 15 shows why the Ion degradation due to hot carriers is suppressed in a GOLD TFT. We calculate the Ion degradations caused by increased LDD resistance, which is a

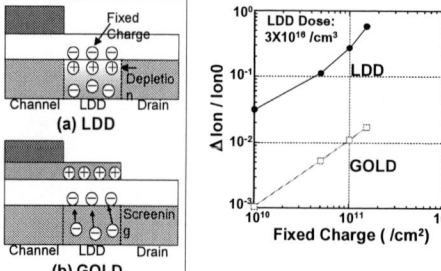

function of electron trap density at the gate-SiO2 interface. We found that the Ion degradation of the GOLD TFT is more than one order of magnitude lower than that of the LDD TFT. This is because the fixed charges due to the trapped electrons are screened by the overlapped gate electrode, that is, the so-called gate-electrode screening.

Fig.15 Suppression of hot-carrier degradation in GOLD.

(2) High-quality gate insulator

In order to improve the quality for the gate insulator, we proposed a dual layered structure, i.e., high quality interfacial layer and conventional PECVD SiO2. As the interfacial layer, a new

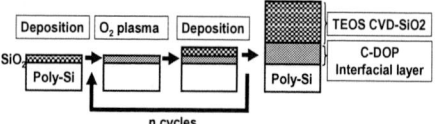

Fig.16 Highly reliable gate insulator formation using C-DOP process.

process C-DOP (Cyclic Deposition with O_2 Plasma treatment), in which very thin SiO2 deposition and O2-pasma treatment were repeated, was applied (Fig.16)(9).

The absorption spectra suggest that the amount of residual OH in the C- DOP formed SiO2 is as low as that in thermal oxide (Fig.17(a)), and the Si-O-Si network of the C-DOP is as dense as that of thermal oxide (Fig.17(b)).

By applying the C-DOP to TFTs, it is confirmed that the C-DOP formed interfacial layer is effective to improve the DAHC immunity of n-channel TFTs. The time to 10% reduction of the drain current can be prolonged one order of magnitude. gm degradation of the p-channel TFTs under AC stress are suppressed as shown in figure 18.

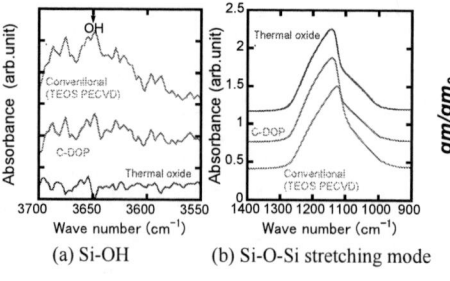

(a) Si-OH   (b) Si-O-Si stretching mode

Fig.17 P-polarized ATR-FTIR spectra.

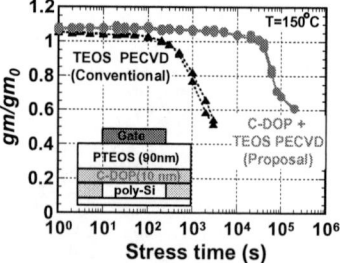

Fig.18 gm degradation of p-channel TFTs under dynamic stress.

Conclusions

SELAX technology, which is capable of forming poly-Si layers with large grain size at selected region, was applied to the fabrication of "system-in-displays". High current drivability and low sub-threshold swing TFTs were obtained using large grain poly-Si layer. Moreover, the TFT degradation phenomena under dynamic stress are clarified for n-channel and p-channel TFTs, and the high-immunity TFT structure and high-quality gate insulator were proposed. Based on these features, IPS-mode LCD panels with resolution of over 300 ppi and with narrow frames were successfully realized. These technologies have potential to be able to add valuable functions on both LCD and OLED panels.

Acknowledgments

We are grateful to the staff of the D1 pilot line (Hitachi Displays, Ltd.) for important advice on TFT technologies, and to Drs.Hongo and Yazaki for valuable production technologies.

References

1. M. Hatano et al., SID02 Tech. Dig., p 158 (2002).
2. M. Tai et al., IEEE Trans. Electro. Dev., 51, p934 (2004).
3. M. Ohkura et al., SID02 Tech Dig., p 146 (2002).
4. T. Miyazawa et al., SID05 Tech. Dig., p 1050(2005).
5. Y. Uraoka et al., IEEE Trans. Electron Devices, vol. 51, p 28 (2004).
6. Y. Toyota et al., IEEE Trans. Electron Devices, vol. 51, p 927 (2004).
7. S. K. Lai, J. Appl. Phys., vol. 54, p 2540 (1983).
8. T.Itoga et al., IDW'01 Tech. Dig., p.1733 (2001).
9. H. Hamamura et al., IDW'06 Tech. Dig., p 987 (2005).

42

CHAPTER 2
DEVICES AND RELIABILITY

Design of Ultra-high Performance Si TFTs

Genshiro Kawachi

Advanced LCD Technologies Development Center Co. Ltd (ALTEDEC)
292 Yoshida-cho, Totsuka-ku, Yokohama, Kanagawa 244-0817 Japan

Design considerations for making use of ultra-high performance Si TFTs for future display systems are presented. Careful choice of the thickness and impurity concentration of the TFT channel is important to maximize the breakdown voltage. In the sub-micron regime, a lightly doped drain (LDD) structure is needed to suppress drain-induced phenomena. Layout design considerations based on the knowledge of the channel location dependence of device parameters and alignment method of TFT channels to the grain pattern are important to achieve high performance, while avoiding the undesirable effects of grain-boundary defects.

Introduction

Ultra-high performance Si thin-film transistors (TFTs) are key components for enabling large-scale integration of functional circuits in displays. There are a number of issues that need to be overcome for enhancing device performance. High-quality poly-Si films with very large grains, whose dimensions are larger than the TFT channel length, are essential for increasing carrier mobility. The phase-modulated excimer laser annealing (PMELA) technique has been proven to be an adequate solution for these issues (1).

Using a lateral re-crystallization technique such as the PMELA, one can drastically reduce the number of grain-boundary defects in the channel. The device characteristics then become closer to those of SOI MOSFETs. This is both beneficial and detrimental for display applications. An increase in the carrier mobility is an obvious benefit of the improved Si films. On the other hand, the floating-body effect (FBE), which is prominent in high-quality Si films having longer carrier life times, yields undesirable effects, including lowering of the source-drain breakdown voltages (V_{BD}) or the threshold voltage (V_T) roll-off at large drain voltages (V_D), on device performance.

The present paper discusses the issues involved in designing high-performance Si TFTs for use in high-speed digital and/or analog circuits on glass substrates.

Device fabrication

Co-planar-type poly-Si TFTs with channel lengths ranging from 1 to 3 μm were fabricated on 200 nm-thick poly-Si films with large grains grown by the phase-modulated excimer laser annealing (PMELA) method (1). A stacked oxide film consisting of a 2 nm-thick microwave-plasma oxide layer and a 28 nm-thick VHF plasma CVD oxide layer was used as the gate insulator (2). The maximum processing temperature was 600 °C during impurity activation. Details of the device fabrication process are reported elsewhere (3). The TFT channels are placed such that the current direction is parallel to the lateral growth direction. Therefore, the majority of carriers are free from scattering at grain boundaries during their travel through the channel. The V_T values of these TFTs were measured at various drain voltages (V_D) ranging from 0.1 V to 5.1 **V**. When measuring the body current induced by impact ionization, 4-terminal TFTs with a body terminal were used. MOSFETs fabricated on SOI wafer using the same process sequence were characterized as the reference sample. Such a comparison is very useful to distinguish the effects of grain-boundary defects from those associates with device scaling.

Channel design

The first decision we have to make is whether we are dealing with a partially depleted (PD) device or a fully depleted (FD) device. The PD MOSFETs can basically be regarded as the same device as the bulk MOSFETs except for their dynamic behavior. Fixing the body potential by attaching a body terminal, problem arising from FBE can almost be avoided. However, introduction of a body terminal causes vast complexity in the layout design, and furthermore, expertise is needed in circuit design to deal with complex dynamic behavior due to the re-distribution of excess holes in the neutral body (4). The FD devices, on the other hand, exhibit no such dynamic behavior since there is no neutral region in the channel. In addition, we expect a steeper sub-threshold slope for FD devices compared to PD devices, and thus, FD is favorable for poly-Si TFTs for which the sub-threshold slope tends to be gentle due to the existence of grain-boundary traps. A major drawback of FD devices is their relatively small V_{BD}. This is due to

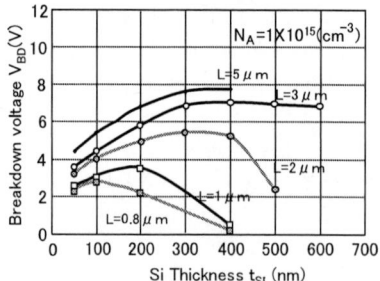

Fig. 1 Simulated source-drain breakdown voltage (V_{BD}) as a function of Si thickness for non-LDD TFTs with various gate lengths.

strong parasitic bipolar effects in FD MOSFETs. Considering that the high-performance TFTs should coexist with the liquid-crystal driver circuits that require 10 **V** or more, careful channel design is needed to maximize **V$_{BD}$**.

Figure 1 shows **V$_{BD}$** values of non-LDD TFTs obtained by 2-dimensional device simulation (DESSIS) as a function of the Si thickness **tsi**. **V$_{BD}$** is defined as the drain voltage for the onset of single transistor latch. Short channel TFTs with **L**=1 μm or less exhibit maximum values of **V$_{BD}$** at an optimum **tsi**, which depends on **L**. This **V$_{BD}$** peak occurs as a consequence of the competition between the increasing impact ionization rate for the thinner Si layers and the punch-through near the back interface for thick Si layers.

The calculation was extended to various combinations of **t$_{si}$** and the channel doping concentration, **N$_A$**. Figure 2 shows a **V$_{BD}$** contour plot of 1 μm n-channel TFTs with respect to **t$_{si}$** and **N$_A$**. The design space for the threshold voltage, **V$_T$**, which is set to 0.6±0.1 **V**, is also indicated. The maximum **V$_{BD}$** of 3.6 V was obtained around **t$_{si}$** =200nm and **N$_A$**=1×10^{16} cm^{-3}. With decreasing **t$_{si}$**, the optimum doping concentration that satisfies the **V$_T$** target shifts to larger values. This leads to an increase in the internal field strength thereby decreasing **V$_{BD}$** to 2.4 **V**. The decreasing trend of **V$_{BD}$** with decreasing Si thickness has been confirmed experimentally. The supply voltage for the CMOS circuits is constrained by the value of **V$_{BD}$**. According to the above results, circuit operation at 5 **V** seems to be difficult and 3.3 **V** is barely acceptable using non-LDD devices. When the channel length decreases to the sub-micron regime, **V$_{BD}$** of non-LDD devices decreases below 3 **V**. LDD is then required in such cases.

In LTPS-TFTs, very thin Si films around 50 nm are commonly used, mainly due to two reasons; for reducing the leakage current of the pixel TFTs and due to the limitation of the excimer laser power for re-crystallization. The present results show that 50 nm is not necessarily the optimum thickness when considering integration of high-speed

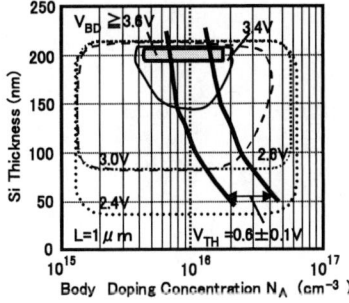

Fig. 2 Optimum channel design area for 1 μm N-type SOI MOSFETs as a function of body doping concentration and Si thickness.

peripheral TFT circuits. The optimum Si thickness should be chosen carefully by taking these properties into account.

Threshold voltage

The threshold voltage, V_T, of FD MOSFETs with the floating body is not a static parameter but is a dynamic one that varies depending on the operating conditions. The important parameter that governs V_T of FD devices is the body potential. In short channel devices, the drain-induced body potential modulation is the most vital problem that causes substantial V_T roll-off at elevated V_D. Figure 3 shows the V_D dependence of V_T for TFTs with various L values. The results for SOI MOSFETs with the same dimensions are also indicated for comparison. In TFTs, V_T decreases almost linearly with increasing V_D. TFTs exhibit a large V_T roll-off even at $L=3$ μm, whereas SOI MOSFETs have a constant V_T up to 5 V. A large V_D dependence results in a large standby current at the operating voltage. The decrease in V_T at small V_D can be attributed to body potential elevation caused by excess holes originating from the drain junction leakage (5). The catastrophic drop in V_T for a 1 μm-long device at a critical V_D corresponds to the onset of the single-transistor latch caused by the parasitic bipolar action. In addition, TFTs occasionally exhibit asymmetric V_D-dependence of V_T under source-drain swapping. These phenomena can be modeled by taking the V_D-dependent drain leakage current, the parasitic bipolar gain, β, and avalanche multiplication at the drain junction, as indicated in Fig. 4 (6). According to this model, V_D dependence of V_T can be expressed by:

$$V_T = V_{T0} \gamma \, n \, k_B \, T \, / \, q \, \ln \{(\beta + 1)(I_{HOLE}) \, / \, Ijo \; + 1 \} , \qquad [1]$$

where q is the elemental charge, k_B is the Boltzmann factor, T is the temperature, V_{T0} is the unperturbed threshold voltage, β is the bipolar current gain, Ijo and n are the reverse saturation current and the ideality factor of the source junction, respectively, γ is the

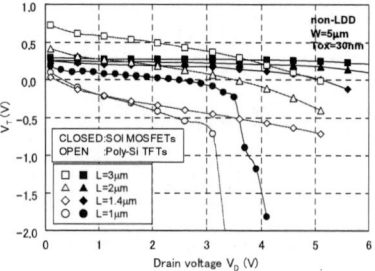

Fig. 3 Threshold voltage (V_T) as a function of drain voltage for non-LDD TFTs with various gate lengths. (6)

substrate-bias coefficient, and I_{HOLE} is the hole current flowing into the source. The quantities β, Ijo, n, and I_{HOLE} can be obtained using 4 terminal TFTs with a p+ body terminal.

Figure 5 indicates V_D dependence of V_T for 1 μm-long 4 terminal TFTs. The results of analytical calculation obtained using eq. (1) are also plotted in Fig. 5. The observed results, including asymmetric V_D-dependence of V_T under source-drain swapping, can be satisfactorily explained by the model described. The asymmetry under source-drain swapping arises from a difference in β and the junction leakage current between the source and drain junctions. Such differences in β and in the junction leakage current are supposed to be originating from the local variation in the grain boundary structure in the channel. Grain-boundary defects near the drain junction induce a V_D-dependent junction leakage current, thereby enhancing the V_T roll-off. The defects near the source-body junction, on the other hand, reduce the lifetimes of excess holes resulting in reduced β and an increase in V_{BD}.

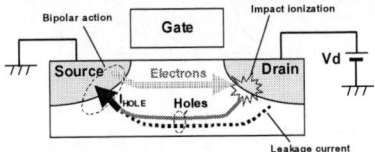

Fig. 4 Carrier dynamics model of TFTs under a large V_D. (6)

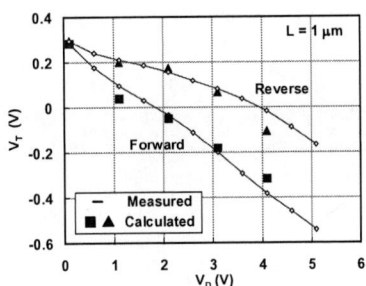

Fig .5 V_D dependence of V_T for a TFT with L = 1μm. Results obtained by source-drain swapping are plotted. (6)

Besides the drain-induced phenomena described above, variation in V_T under the triode condition is also a major concern. When the grain area becomes comparable to the channel area, device-to-device variation of these quantities can be quite large. Control of the channel location in the grains and crystal orientation of each grain are then very important. The former problem can be solved by using the PMELA technique, which involves providing a phase-shifter that can create alignment patterns for photolithography onto the Si film during the crystallization step, as described later. The latter problem is more difficult to overcome. Further extensive effort will be needed to obtain an orientation-controlled large mono-crystalline grain. At the present, the use of a poly-Si film with a long and narrow grain structure could be a practical solution for suppressing fluctuations in device characteristics (3). Figure 6 shows the relationship between V_T dispersion and channel width for 1 μm-long TFTs fabricated on a poly-Si film with long and narrow grains. Increasing the channel area effectively reduces the V_T fluctuation similar to typical MOSFETs on crystalline Si.

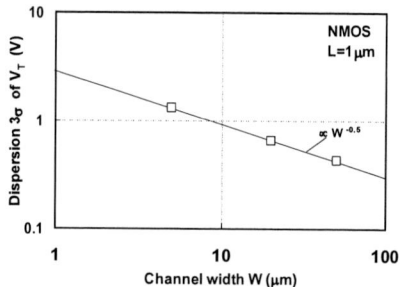

Fig. 6 Dispersion of V_T for a 1 μm-long TFT fabricated on long and narrow grains as a function of channel width.(3)

Mobility

Since poly-Si films fabricated by the PMELA method have an inherent inhomogeneous grain structure, the TFT performance shows a strong dependence on the channel location. Knowing the channel location dependence of the TFT parameter is of primary importance for considering the layout design. Figure 7 shows the field effect mobility, μ_{FE}, of 1 μm-long and 2 μm-wide n-channel TFTs, as a function of the channel location. The poly-Si film used for this device has 2.5 μm-long grains as indicated in the upper part of the figure. μ_{FE} larger than 400 cm^2/V·s can be obtained in the range of X = 0.7 μm to 1.8 μm and X = 3.2 μm to 4.1 μm, wherein large grains were grown. Indeed, it has been confirmed that TFTs with μ_{FE} greater than 600 cm^2/V·s have almost a single grain in the channel. On the other hand, TFTs located close to the grain boundaries, around X = 0 μm, 2.5 μm and 5 μm, show low μ_{FE} below 200 cm^2/V·s, indicating that carrier transport is affected by the grain boundaries.

Figure 8 compares the universal plots of three TFTs, indicated in Fig. 7 as **SD, GB** and **LG**, and SOI MOSFETs. The effective mobility μ_{eff} is plotted as a function of the effective vertical field E_{eff}. The TFT located on the large grains, denoted as **LG**, shows a large peak value of μ_{eff} comparable to the SOI MOSFET. For the high-E_{eff} regions above 10^5 V/cm, μ_{eff} of SOI MOSFET decreases with increasing E_{eff} as $\mu_{eff} \propto E_{eff}^{-0.34}$, indicating that the dominant carrier scattering mechanism is acoustic phonon scattering. The TFT (**LG**), on the other hand, shows a somewhat stronger dependence on E_{eff}, as $\mu_{eff} \propto E_{eff}^{-0.77}$. This suggests that μ_{eff} of this TFT is suffered by the surface roughness scattering. The TFTs including the grain boundaries, denoted as **GB** and **SD**, show completely different dependence on E_{eff}, since carrier transport is mostly dominated by grain boundary scattering in these devices.

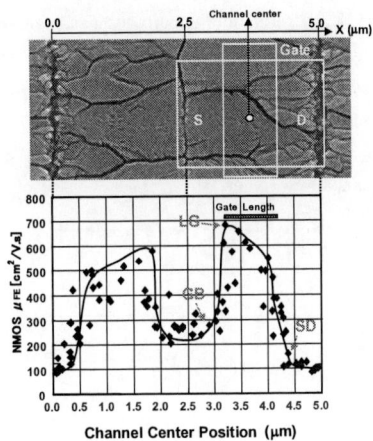

Fig. 7 Field effect mobility μ_{FE} of 1 μm-long n-channel TFTs as a function of channel location.(3)

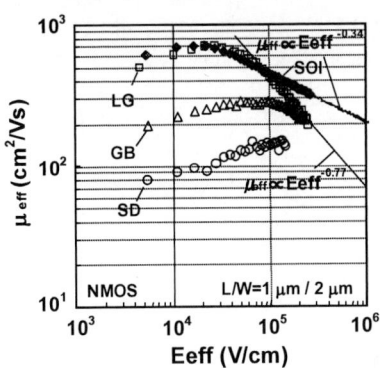

Fig. 8 Effective mobility μ_{eff} of TFTs, denoted as **LG**, **GB**, and **SD** in Fig. 7, and SOI MOSFET as a function of effective vertical field E_{eff}. (3)

LDD design

In addition to improvement in carrier mobility, a scaling down of the gate length is required for improving the circuit performance. As already discussed, a simple scaling down of non-LDD devices to the sub-micron regime will fail because of the strong drain-induced effects. LDD seems to be the most natural solution for this problem. LDD, however, adds a large series resistance to the intrinsic channel resistance, thereby reducing the drive current. Careful design of the LDD parameters, i.e., length and doping profiles, is needed to minimize the degradation of the current derivability, while maintaining the breakdown voltage.

Figure 9 shows simulated V_{BD} as a function of impurity concentration of the LDD n- region for 0.5 μm-long TFTs having various LDD lengths d_L. To allow for operation at a supply voltage of 3.3V, d_L of at least 0.2 μm is needed. For devices with d_L of 0.2 to 0.3 μm, the maximum V_{BD} is obtained at an impurity concentration of 6×10^{16} cm^{-3}. This, however, is not the optimum concentration since the on-current is too low at an N_{LDD} of 6×10^{16} cm^{-3}. The compromise for N_{LDD} is obtained at a somewhat larger N_{LDD} value of around 1×10^{17} cm^{-3}.

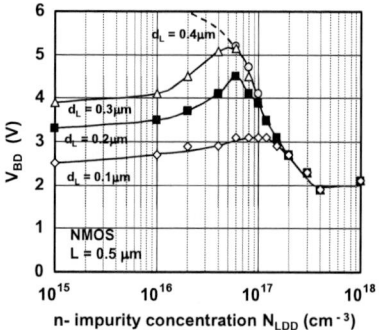

Fig. 9 V_{BD} as a function of impurity concentration for an LDD n- layer.

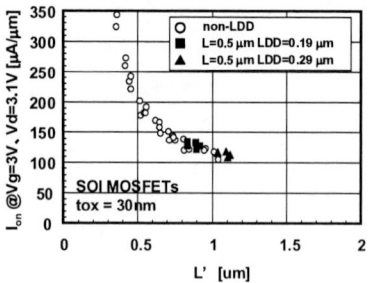

Fig. 10 I_{on} as a function of L', which is the sum of the gate length and double the LDD length.

Figure 10 plots the on-current measured using SOI MOSFETs as a function of L' which is the sum of the gate length L and double the LDD length, $2 \times d_L$. It is clear from Fig. 10 that the on-current of 0.5 μm-long TFTs with 0.2 μm- and 0.3 μm-long LDD are almost equal to that of 0.8 μm and 1.0 μm-long non-LDD TFTs. Even so, efforts for scaling down to the sub-micron regime are still worth carrying out, since the delay time of CMOS logic gates is expected to be shorter in 0.5 μm LDD TFTs because of the small gate capacitance. This is an important benefit of device scaling on the insulating substrates wherein the parasitic capacitance is negligibly small compared to the gate capacitance.

So far, the discussion has been focusing on the n-channel devices, since FBE has lesser impact on device performance in p-channel TFTs. However, assuming that a supply voltage of 3.3 V, LDD is required for p-channel TFTs when the gate length is reduced to 0.5 μm. Similar design considerations on the LDD parameters are then needed.

Channel location control

As discussed previously, precise control of channel location against the elongated grain patterns is essential to obtain good performance. In the PMELA method, this can be done by preparing a specially designed alignment pattern area in a part of the phase modulator. The alignment patterns, consisting of islands of non-crystallized a-Si embedded in the poly-Si matrix or vice-versa, could be detected utilizing the difference in reflectively for a He-Ne laser probe light between the a-Si and poly-Si areas. An alignment error of ±0.2 μm, which was tolerable to form 1 μm-long TFTs within 4 μm-long grains, could be achieved using this method (7).

Figure 11 shows the transfer curves of 1 μm-long CMOS TFTs that are aligned to the grains. The variations in V_T are ±0.32 V and ±0.34 V for NMOS and PMOS, respectively, which are sufficiently small for a 3.3 V operation of CMOS circuits.

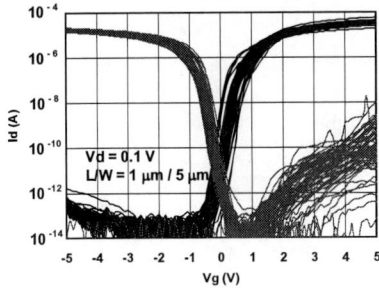

Fig. 11 Transfer characteristics of a 1 μm-long CMOS TFT with channel location control.(3)

Figure 12 shows the propagation delay time, τ_{pd}, of a CMOS inverter deduced from the oscillation frequency of a 21-stage ring oscillator as a function of the gate length **L**. For comparison, data for bulk MOSFETs, taken from previous publications (8), are also plotted. The value of τ_{pd} for TFTs was almost comparable to that of the bulk MOSFETs with the same gate length even though the static characteristics of the transistors are not necessarily comparable. Furthermore, it is interesting to note that τ_{pd} of TFT was proportional to L^2, while that of bulk MOSFETs was proportional to **L**. These results are due to the small junction capacitance of the SOI structure and the small parasitic capacitance associated with the interconnect wiring. Further improvements in circuit performance are expected by proper device scaling and by improvements in the material properties.

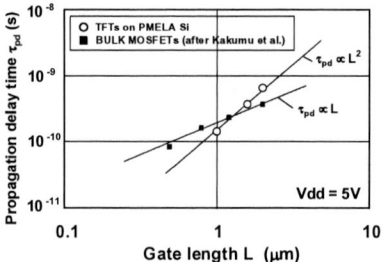

Fig. 12 Propagation delay time, τ_{pd}, of a CMOS inverter as a function of gate length.

Summary

The design considerations for making use of ultra-high performance Si TFTs for future display systems are presented. In the sub-micron regime, a LDD structure is needed to suppress the drain-induced phenomena. Layout design considerations based on the knowledge of the channel location dependence of device parameters and alignment method of TFT channels to the grain pattern are important to achieve high performance, while avoiding undesirable effects of grain-boundary defects. High-speed digital circuits with a delay time less than 100 ps are expected to be realized by combining these technologies.

Acknowledgments

The author acknowledges to Drs. M. Matsumura, H. Abe, T. Mizutani and Mr. H. Kobayashi and other colleagues of ALTEDEC for simulating discussions and technical support. The present study was supported by NEDO and METI.

References

1. M.Jyumonji, Y.Kimura, Y.Taniguchi, M.Hiramatu, H.Ogawa and M.Matsumura., *Jpn.J.Appl.Phys.* **43**, 739 (2004).

2. M. Goto, K. Azuma, T. Okamoto and Y. Nakata, *Jpn. J. Appl. Phys.* **42**, 7033 (2003).

3. Y. Nakazaki, G. Kawachi, H. Ogawa, M. Jyumonji, M. Hiramatsu, K. Azuma, T. Warabisako, and M. Matsumura, *Jpn. J. Appl. Phys.* 45, 1489 (2006).

4. K. Bernstein and N. J. Rohrer, *SOI Circuit Design Concept,* Kluwer Academic Publishers, Boston (2000) p. 95.

5. J. Tihanyi and H. Schlötterer, *Solid-State Electron.* **18**, 309 (1975).

6. G. Kawachi, S. Tsuboi, T. Okada, M. Mitani, and M. Matsumura, *Dig. of 13th International Workshop on Active-Matrix Flatpanel Displays and Devices*, 55 (2006).

7. M. Matsumura, *Dig. of 12th International Workshop on Active-Matrix Liquid-Crystal Displays,* 1 (2005).

8. M. Kakumu, N. Kinugasa and K. Hashimoto, *IEEE Trans Electron Devices* **37**, 1334 (1990).

ECS Transactions, 3 (8) 57-62 (2006)
10.1149/1.2356334, copyright The Electrochemical Society

Electrical Hysteresis Behavior of Low Temperature Polycrystalline Silicon Thin Film Transistors

Dae-Hyun Nam, Hong-Koo Lee, Sang-Hoon Jung, Tae-Joon Ahn, Chang-Yeon Kim, Chang-Dong Kim and In-Jae Chung

LG.PHILIPS LCD R&D Center, 533 Hogye-dong, Dongan-Gu, Anyang, KOREA

Image sticking of AMOLED is caused by the property variation of driving transistors. In this paper, we investigated the hysteresis behavior of p-type p-Si thin film transistors as driving device for AMOLED. As the temperature increase, it was observed that hysteresis phenomenon was suppressed. This can be explained by that the trapping and de-trapping rate of carriers is much faster in high temperature than in room temperature.

Introduction

Active Matrix Organic Light Emitting Diode (AMOLED) displays have been considered as the most promising candidate for a post Active Matrix Liquid Crystal Display (AMLCD) due to higher brightness, faster response time, wider viewing angle, and thinner thickness.[1] In order to commercialize AMOLED displays, not only lifetime of organic EL materials but also quality of transistor should be improved. Among the properties required for driving transistor of AMOLED, the most important factor is uniformity and reliability of device. a-Si TFTs are superior in uniformity but poor in reliability and laser crystallized p-Si TFTs have no issues in reliability but suffer for the non-uniformity problem. Recoverable image sticking is another important issue in AMOLED. Due to the hysteresis behavior of driving transistor of AMOLED, image sticking is observed for a few minutes.[2] To eliminate the image sticking, various compensation pixel designs have been proposed.[3,4] But complex structure of compensated pixel limits the aperture ratio of pixel. During light emitting of organic EL, heat generation is occurred. In case of full white operation, the temperature of panel surface reaches ~100℃. This heat causes degradation of transistor. In this study, we fabricated p-type p-Si TFTs and investigated the hysteresis properties of devices in high temperature.

Experimental

Top-gate p-type p-Si TFTs were made by using low temperature process. First, a buffer oxide layer and an a-Si were deposited by PECVD on a glass substrate. The silicon film was crystallized by SPC (Solid Phase Crystallization) method. After the active patterning, a SiO_2 layer as a gate insulator was deposited by PECVD and a gate metal (Mo/AlNd) was deposited by sputtering. Self-aligned p^+ ion doping and activation process were performed to form the source and drain junctions. Then, SiN_x/SiO2 interlayer was deposited by PECVD. After contact holes were formed, data bus metal (Mo/AlNd) was formed. In order to study and characterize the hysteresis and reliability properties of p-type p-Si TFTs, we extracted the current-voltage (I-V) characteristics and the device parameters by using the Keithley 4200 system. The channel width and length of typical TFTs used in this experiment were 5 μm and 20 μm, respectively.

57

Results and Discussion

Figure 1 (a) shows the chessboard pattern composed of black and white to evaluate the image sticking of AMOLED. After the chessboard pattern was displayed a few minutes, a middle gray signal was inputted. The initial middle gray image was displayed image sticking as shown in figure 1 (b). It is remarkable that the brighter regions of the image sticking were the area of the previous black pattern and the darker regions were the area of the previous white pattern. Figure 1 (c) explains the mechanism of image sticking by transfer curve. As shown in figure 1 (c), hysteresis in transfer curve can be observed with the sweep direction of gate voltage. In figure 1 (a), the gate voltages of black and white are V_{BLACK} and V_{WHITE}, and the drain currents are I_{BLACK} and I_{WHITE} respectively. And then the V_{GRAY} is applied to display the middle gray, previous black patterns move to I_{GRAY1} and previous white patterns go to I_{GRAY2}. The current difference (I_{GRAY1}-I_{GRAY2}) makes the luminance difference and this phenomenon is called as the image sticking.

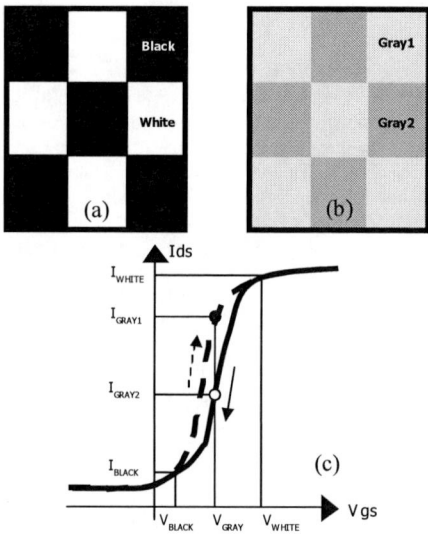

Figure 1. Image sticking and hysteresis. (a) initial image of chessboard pattern, (b) image sticking, and (c) transfer curve that show device hysteresis.

It is well known that the image sticking caused by hysteresis disappears after a few minutes. This can be described by the current changes with time. As shown in the figure 2, after the voltages which corresponded to the white level (0.84 μA) current and black level (0.10 μA) current were applied for 120 s, the gray level (0.25 μA) voltage was inputted. In the early stage of voltage change, drain current of white to gray showed lower than 0.25 μA, and that of black to gray showed higher than 0.25 μA. But, the difference of current level decreased with time and current converged into 0.25 μA about 30 s.

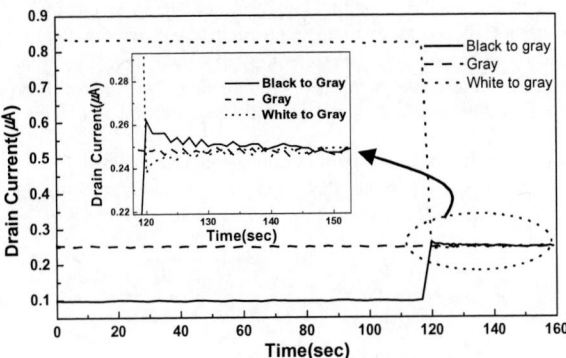

Figure 2. Current difference and convergence with time.

The temperature of panel surface during full white luminance of AMOLED is near ~100℃. It is important to examine the nature of transistor at high temperature. Figure 3 shows the hysteresis behavior of transfer curve with different temperatures. The image sticking caused by hysteresis can be estimated by the threshold voltage variation (ΔV_{th}) with the gate voltage sweep direction.

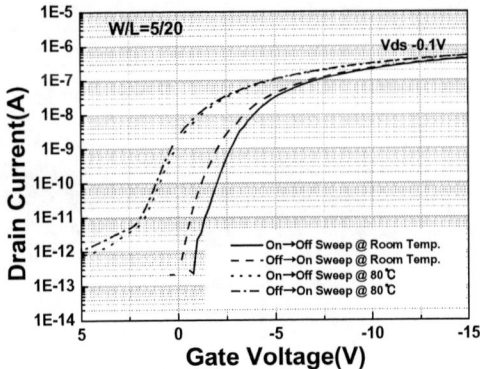

Figure 3. The comparison of hysteresis behavior in case of room temperature and 80℃.

The mechanism of the threshold variation can be described by the trapping of carriers in trap sites of the bulk oxide and the interface of gate oxide/p-Si. In case of p-type devices, trapping and de-trapping of the holes occurs with the applied voltage. If the voltage for a high current is applied, trapping of the holes occurs and the threshold voltage would increase and current decrease, consequently. On the contrary, if the voltage for a low current is applied, the trapped charge would be de-trapped and the current increase. In this experiment, the reduction of threshold voltage variation at high temperature can be explained by that the trapping and de-trapping rate of carriers is much faster in high temperature than in room temperature. Another estimation of the image sticking caused by hysteresis can be expressed by the drain current variation (ΔI_{ds}) with the gate voltage sweep direction. Because the luminance of organic EL is proportional to drain current, larger drain current variation results in serious image sticking problem.

Figure 4 (a) and (b) show the drain current variations in case of room temperature and 80 ℃, respectively. As same the threshold voltage variation, the higher the temperature, the lower the drain current variation.

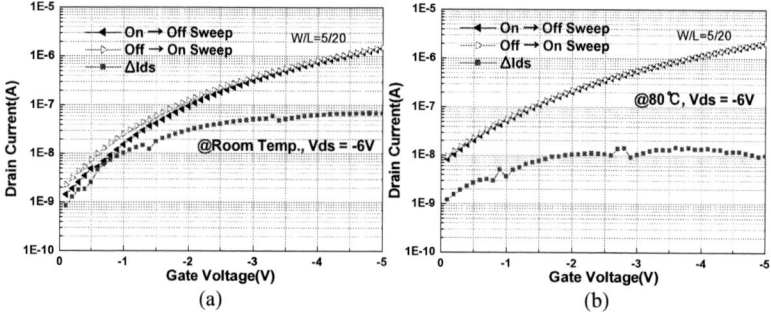

(a)   (b)

Figure 4. Drain current variations caused by hysteresis (a) room temperature, and (b) 80℃.

As the temperature increase, the variations of threshold voltage and drain current are suppressed. It is shown in figure 5. The threshold voltage variation shown in figure 5 (a) of room temperature is 0.48 V and that of 80℃ is 0.24 V. The drain current variation shown in figure 5 (b) of room temperature is 50 nA and that of 80℃ is 10 nA. It is expected that the image sticking could be suppressed at high temperature.

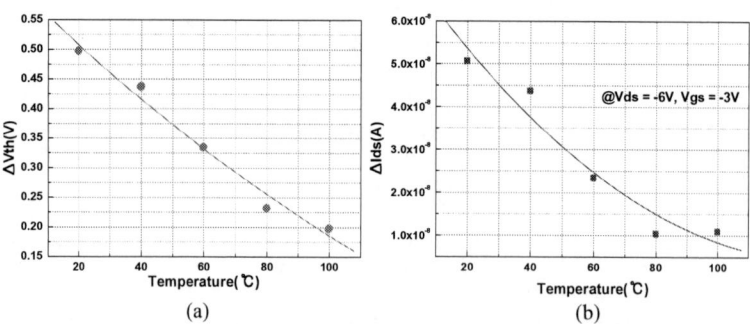

(a)   (b)

Figure 5. Hysteresis behaviors with temperature (a) threshold voltage variation, and (b) drain current variation.

As varying the holding time of each points of voltage sweeping, room temperature hysteresis behavior was characterized. Figure 6 (a) and (b) show the transfer curves of holding time of 0 s and 60 s, respectively. Regardless of gate voltage sweep directions, transfer curve of holding time of 60 s is almost coincident.

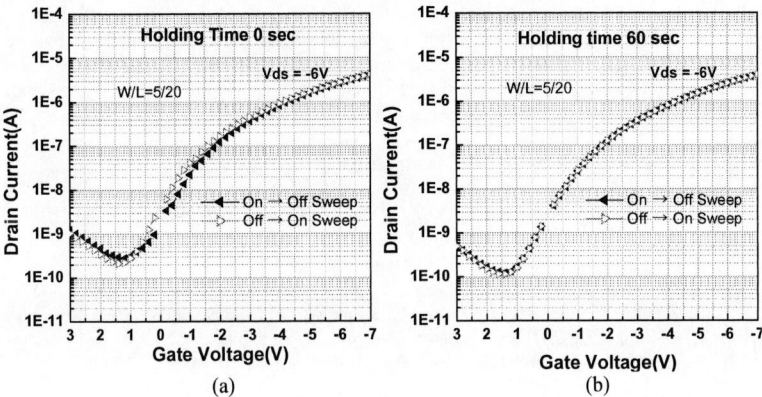

Figure 6. Hysteresis of transfer curve (a) holding time 0 s, and (b) holding time 60 s.

Figure 7 shows the drain current variations with holding time of each points of voltage sweeping at room temperature. Decrease of drain current variation (ΔI_{ds}) is observed in the holding time of 10 ms ~ 30 s. After the 30 s, decrease of drain current variation is saturated. If there is enough time to reach the equilibrium state of tapping and de-trapping of carriers, the variation of threshold voltage and drain current caused by hysteresis can be negligible.

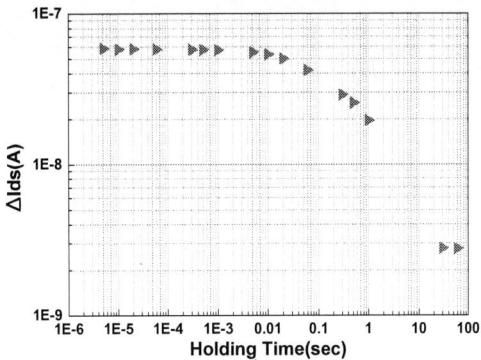

Figure 7. Drain current variations (ΔI_{ds}) caused by hysteresis with holding time.

Conclusion

In this paper, we investigated the hysteresis and reliability properties of solid phase crystallized p-type p-Si TFTs for AMOLED application. In the case of emission of full white, transistors for driving AMOLED are heated near ~ 100℃. As the temperature increase, the behavior of transistors was changed. It was observed that hysteresis phenomenon was suppressed at the high temperature. This can be explained by that the trapping and de-trapping rate of carriers in trap sites is much faster in high temperature. Consequently, it could be expected that the image sticking would be reducible in high temperature such as high luminance condition.

References

1. G. Gu and Stephen R. Forrest, *IEEE J. Sel. Top. Quantum Electron.*, **4**, 83 (1998).
2. B. K. Kim, O. Kim, H. J. Chung, J. W. Chang, and Y. M. Ha, *Jpn. J. Appl. Phys.*, **43**, 482 (2004).
3. M. Kimura, I. Yudasaka, S. Kanbe, H. Kobayashi, H. Kiguchi, S. Seki, S. Miyashita, T. Shimoda, T. Ozawa, K. Kitawada, T. Nakazawa, W. Miyazawa, and H. Ohshima, *IEEE Trans Electron Dev.*, **46,** 2282 (1999).
4. Y. He, R. Hattory, and J. Kanicki, *Jpn. J. Appl. Phys.*, **40**, 1199 (2001).

ECS Transactions, 3 (8) 63-67 (2006)
10.1149/1.2356335, copyright The Electrochemical Society

Analysis of the Hump Characteristics in Poly-Si Thin Film Transistor

Soopool Kim, Jae Young Oh, Joon Young Yang, Myoung Su Yang, In Jae Chung

LG.Philips LCD R&D center, 533 Hogae-dong, Dongan-gu, Anyang-shi, Kyounggi-do, 431-080, Korea(Republic)

> To reduce the hump characteristic which is one of the critical
> issues in poly-Si TFT, we focus on crowding of gate fringing field
> at channel edge and suggest new modified structure of channel
> edge. Using dry etching process, poly-Si layer of channel edge is
> partially removed and the steep profile of channel edge is made
> into a step profile to decrease the number of carriers from active
> area. The edge step active (ESA) structure we proposed has a
> positive effect on removing the hump characteristics.

Introduction

The major advantage of Poly-Silicon thin-film transistors (poly-Si TFTs) is its high mobility and on-current which is better than that of amorphous silicon (a-Si) TFTs for active matrix liquid crystal display (AMLCD). However, appearance of the hump characteristics is one of the critical issues in poly-Si TFT. There are two main factors by which the hump characteristics are explained. One of the factors is the inverse narrow-channel effect explained by crowding of gate fringing field at the corner of channel edge [1] and the other is non-activated phosphorus dopant in channel due to insufficient thermal activation treatment.

In this work, we focus on crowding of gate fringing field at channel edge and suggest new modified structure of channel edge to reduce the hump characteristics in poly-Si TFTs using dry etching process. The steep profile of channel edge, as shown in Figure 1 (b), induces crowding of gate fringing field and this phenomenon increases corner parasitic effect, which leads to an abnormal current [2]-[3].

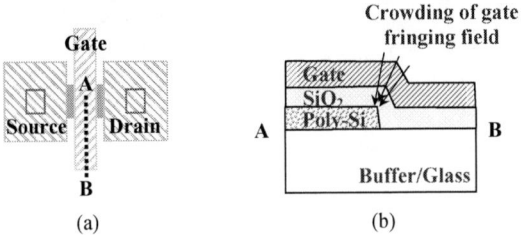

Figure 1. Scheme of TFT structure (a) TFT device structure, (b) cross section of the edge of channel.

63

Experiment

In order to reduce the abnormal current, poly-Si layer of channel edge is partially removed and the steep profile of channel edge is made into a step profile to decrease the number of carriers from active area. Using dry etching process in poly-Si layer, the step profile can be formed effectively. Figure 2 shows dry etching process flow of modifying the edge shape profile of channel. First, poly-Si layer is fully patterned by plasma etching using photolithography and then remaining PR (Photo Resist) is partially removed using ashing process to reveal the edge of patterned poly-Si. Finally, the partial etching makes each side of edges of active area into the step profile. Figure 3 shows SEM (Scanning Electron Microscope) images of the edge step profile of active area and a cross sectional image after deposition of gate insulator SiO_2 and gate metal. SiO_2 and gate metal are covered with the edge step profile of channel.

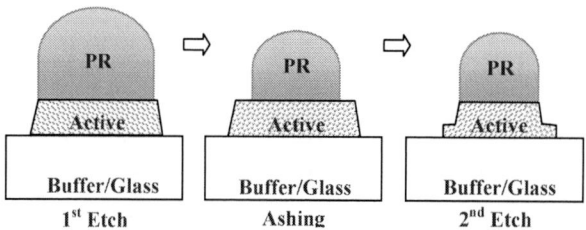

Figure 2. Scheme of dry etching process of forming edge step structure of channel.

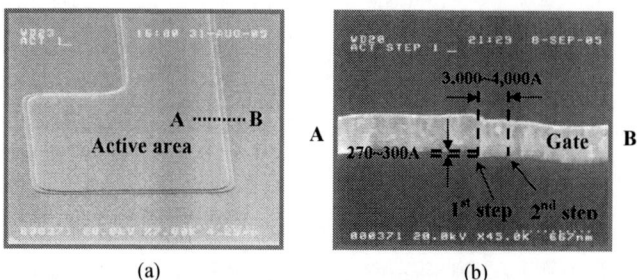

Figure 3. SEM image of (a) active area with edge step, (b) cross section after deposition of gate insulator and gate metal.

Result and Discussion

We achieved good electrical characteristics with W/L=4/4 $\mu m/\mu m$ p-channel and n-channel TFT device with a thickness of 50nm poly-Si. The thickness of the gate insulator is 75nm and boron(B) ions are doped with dose of 3×10^{15} cm^2 at the acceleration voltage (Vacc) of 50KeV for p-channel and phosphorus(P) ions are doped with dose of 2.5×10^{15} cm^2 at the Vacc of 60KeV for n-cl   ˙ TFT. To reduce high electric field in the fringe of drain junction of poly-Si TFT, LDD (Lightly Doped Drain) structure has been adopted. P ions are doped with dose of $3 \times 10^{13} cm^2$ at the acceleration of 60KeV at LDD region.

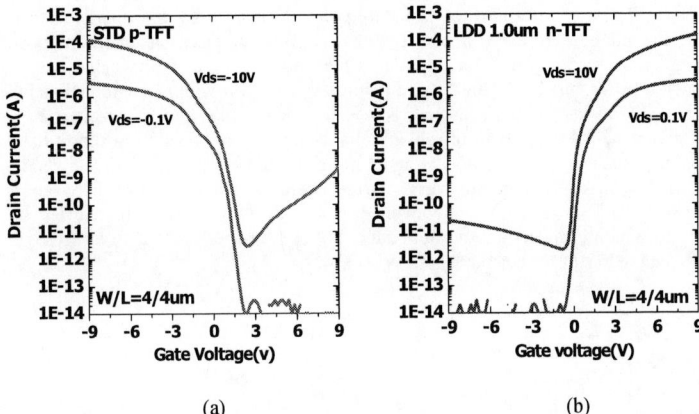

(a)   (b)

Figure 4. Conventional edge structure TFT device transfer characteristics where drain voltage is ±0.1V and ±9.0V with hump characteristics (a) p-channel TFT, (b) 1 μm LDD n-channel TFT.

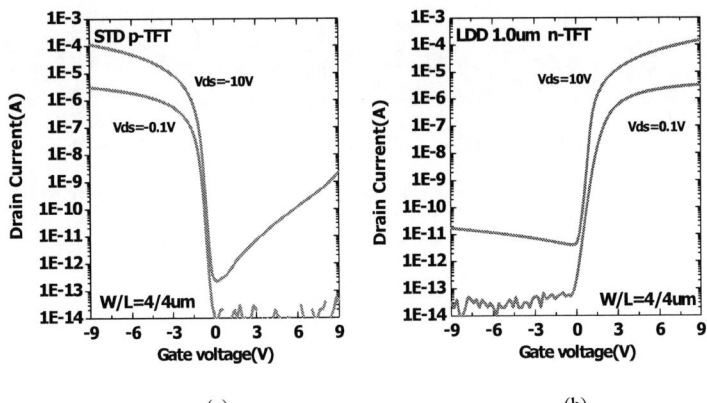

(a)   (b)

Figure 5. The ESA structure TFT device transfer characteristics where drain voltage is ±0.1V and ±9.0V (a) p-channel TFT, (b) 1 μm LDD n- channel TFT.

Figure 4 shows initial transfer characteristics of n-channel TFT and p-channel TFT with a conventional channel edge structure. The hump characteristic is observed in TFT device with the conventional edge structure, as shown in Figure 4 (a), (b) due to crowding of gate fringing field. However, it is not observed in TFT device with the ESA structure we proposed, as shown in Figure 5 (a), (b).

The TFT device with the ESA structure is tested under high drain current stress to analyze the reliability of the TFT device. Figure 6-(a) shows electrical characteristic after the high drain current stress is applied. When stress of gate voltage exceeds -13V (100sec), transfer curve is shifted and this result is common with conventional samples. Figure 6-(b) shows comparison between TFT device with the ESA structure and conventional TFT device about threshold voltage change. This result also shows that the step edge structure we proposed is more reliable than compared with conventional structure TFT devices. For Hot carrier stress, there are no differences between TFT device with the ESA structure and conventional TFT device. In case of hot carrier stress, it has affected on channel current not in edges but in interface, so the ESA structure dose not affected on results of hot carrier stress.

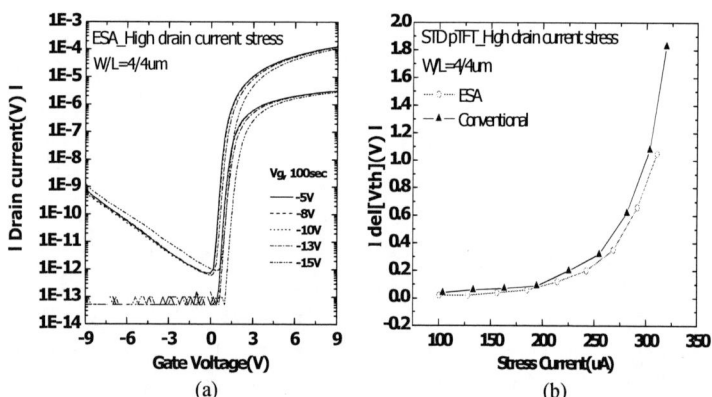

(a)   (b)

Figure 6. The ESA structure p-channel TFT device high drain current stress characteristics where gate voltage varied from -8V to -16V, 100sec (a) transfer characteristic after stress, (b) threshold voltage shift value through stress current between conventional edge structure and step edge structured of p-channel TFT device.

Conclusion

In this paper, we analyzed the hump characteristic in poly-Si TFTs which is dependent on gate fringing field at steep active area edge. By removing the steep edge and transforming into the ESA structure using the dry etching process, the abnormal current due to the crowding of gate fringe field is reduced and it makes removes the hump characteristic is not in TFT transfer characteristics in both n- and p-channel TFTs.

References

1. N. Shigo, S. Fukuda, T. Wada, K. Hieda, *Jpn J. Appl Phys.*, **39**, p.2136 (2000).
2. P. Sallagoity, M. Ada-Hanifi, M. Paoli, and M. Haond, *IEEE Trans. Electron Device.*, **43**, 1903 (1996).
3. N. Shigo, S. Fukuda, T. Wada, K. Hieda, *IEEE Trans. Electron Device.*, **35**, p.945 (1988).

68

Analysis of Characteristics in Poly-Si Thin Film Transistor Crystallized by a New Alignment SLS Process

K. S. Hwang, J. Y. Yang, T. J. Ahn, Y. J. Kim, M. S. Yang and I. J. Chung

LG. Philips LCD R&D Center, 533, Hogae-dong, Dongan-gu,
Anyang-shi, Kyongki-do, 431-080, Korea
Phone : +82-31-450-7494   E-mail : funnyvibe@lgphilips-lcd.com

ABSTRACT

In the conventional SLS crystallization method, we just perform a basic pre-align on the substrate stage and apply a whole substrate area scanning. Therefore, each TFT has different grain boundary(GB) location in channel region. The number of grain boundaries in channel also varies from one to two, which can cause non-uniform TFT characteristics and image quality deterioration of the panel. In this paper, we present work that has been carried out using the SLS process to control grain boundary(GB) location in TFT channel region and it is possible to locate the GB at the same location in the channel region of each TFT. We fabricated TFT by applying a new alignment SLS process and compared the TFT characteristics between a normal SLS method and the grain boundary location controlled SLS method.

INTRODUCTION

Polycrystalline Silicon thin-film transistors(poly-Si TFTs) have been very attractive for the integration of driver circuit into microelectronic applications such as active matrix liquid crystal displays(AMLCDs) or organic light-emitting diodes(OLEDs). In order to improve the TFT device performances, various crystallization schemes have been proposed. Among them, SLS(Sequential Lateral Solidification) technique has attracted a considerable attention due to its excellent throughput and high field-effect mobility.[1, 2, 3, 4, 5]

However, using the SLS technique to manufacture OLED devices, grain boundary(GB) location in channel region of SLS process varies between the TFTs and it results an image quality deterioration of the panel. Therefore grain boundary location control in channel region is an important factor to improve the uniformity and performance of the TFTs, which enhances the image quality of the OLED panel.

In this work, we carried out a grain boundary location controlled SLS method by applying a new key recognition alignment crystallization and it is possible to locate the GB at the same location in the channel region of each TFT.

EXPERIMENTS

Fabrication I

Top gate n-channel poly-Si TFTs were fabricated on SiO_2/ $SiNx$ (300 nm/100 nm) which was stacked as buffer layer on NEG glass substrate. A 50 nm-thick amorphous silicon was deposited on the buffer layer by plasma enhanced chemical vapor deposition (PECVD) as the precursor to laser crystallization. Precursor was dehydrogenated in furnace at 430℃. Then, alignment process keys were formed on an amorphous silicon layer by photo-lithography at each corners of the glass substrate. This would be the first photo-lithography layer and following process steps were aligned with these keys. After that, SLS crystallization (2-shot process) were carried out.

Key alignment SLS process

In the normal SLS crystallization method, we just perform a basic substrate pre-align on the stage and apply a whole substrate area scan. Therefore, after we complete all the TFT fabrication processes, each TFT has different GB location in channel region. The number of grain boundaries in the channel also varies from one to two, which can cause non-uniform TFT characteristics and image quality deterioration of the panel. This is because the tilt between the grain boundary and the following gate line cannot be aligned well when applying the normal SLS crystallization method. In this work, we carried out a grain boundary location controlled SLS method by applying a new key recognition alignment crystallization and it is possible to locate the GB at the same location in the channel region of each TFT.

Figure 1. shows a concept of a new key alignment SLS process. It was preformed as follows: 1] Glass is loaded onto a substrate stage. 2] Laser diode sensors detect loaded glass edges and calculate the right stage angle. 3] Stage rotates slightly to the calculated angle.(paralleled to substrate scan direction) 4] CDD cameras search alignment process keys on the glass substrate and system calculates the angle and perform a slight rotation to a correct scanning angle. 5] Whole substrate area laser scanning is performed. This new method of key alignment SLS process enables grain boundary to be aligned more finely to following gate lines so it can be advantageous to have more uniform grain boundary location. If we apply an offset value from the line scanning start position, we can control the exact location of the grain boundary.

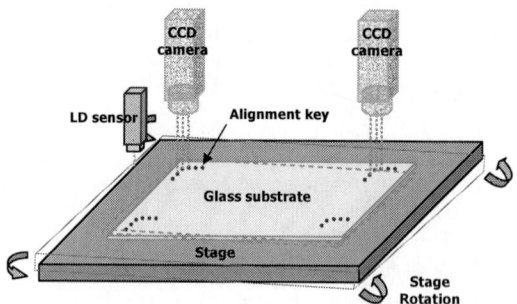

Figure 1. Conceptual image of new key alignment SLS process a) At first, LD sensors search the glass edge and stage rotates primarily b) And then, CCD cameras calculate the correct scan angle and 2nd stage rotation is performed. c) After that, substrate area laser scan begins.

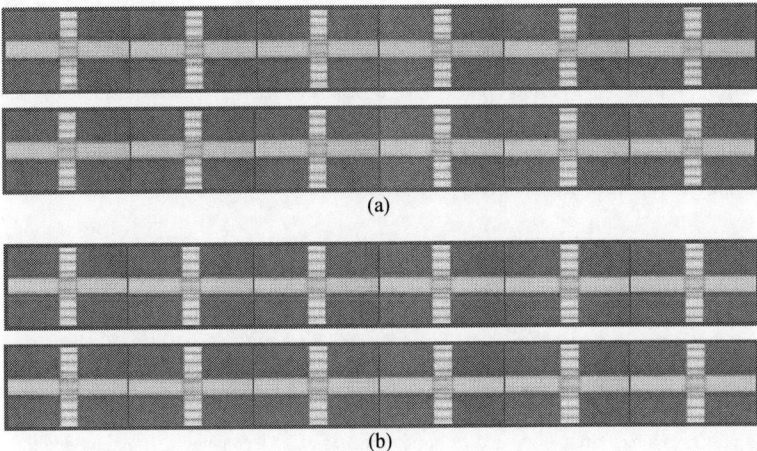

(a)

(b)

Figure 2. Optical TFT images (a) There are changes in grain boundary locations within channel region when normal SLS crystallization method was applied and (b) There is no change in grain boundary locations within channel region when a new alignment SLS crystallization method was applied (In case of channel W/L=4 μm/4 μm).

Figure 2. shows continuous optical images of grain boundary within TFT channel region on the same gate line. As shown in the optical image, there is a change in the grain boundary location within channel region in case of normal SLS crystallization method (a) however, when we apply the grain boundary location controlled SLS method there is no change in grain boundary location within channel region on the same gate lines (b). Figure 3. shows a tilted view of SEM image of TFT channel region with two grain boundaries located. The change in the GB location can affect TFT performances. Using the grain boundary location controlled SLS method, we have crystallized and fabricated poly-Si-based TFTs with only one grain boundary located at the same location in channel region on a same gate line.

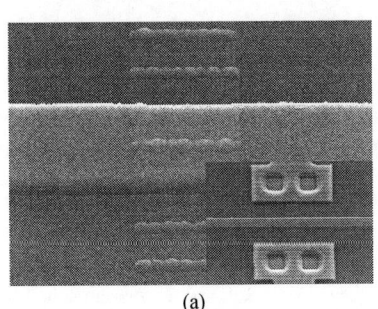 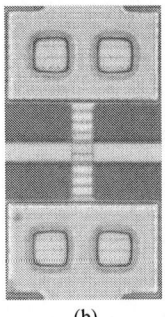

(a)   (b)

Figure 3. (a) Tilted view SEM image of TFT channel region of one grain boundary located and (b) Optical image of TFT chan ⋅ ion

Fabrication II

After laser crystallization, active layer was patterned and gate insulator(SiO₂) of 75nm and gate metal(Mo) of 300nm were deposited and patterned. Doping process was carried out in the source and drain region. Source and drain doping concentration was $3x10^{15}$cm-2 for n-type TFT. LDD (Lightly Doped Drain) structure was adopted to reduce the high electric field in the fringe of the drain junction of the poly-Si TFT. Additional nitride layer was deposited for hydrogen passivation. Source and drain metal(Al) of 500nm and ITO were deposited and patterned.

Results and Discussion

The transfer characteristics of TFTs were measured at room temperature with V_{ds}=0.1 V and V_{ds}=9.0 V, where the gate voltage(V_g) varied from -9.0 to +9.0 V. We used the 1.0 μm LDD (Lightly Doped Drain) n-type TFT(channel width(W)/length(L)= 4 μm/4 μm) in TFT characteristics measurement.

Figure 4. shows Id-Vg transfer characteristics for TFTs crystallized with normal SLS process crystallization method and its optical images of TFT channel (a) and applying the grain boundary location controlled SLS method and its optical images of TFT channel (b). As shown in the transfer curves(figure 4.), when we applied the grain boundary location controlled SLS method, the TFT characteristic showed more uniform performances than the TFT with normal crystallization process. In order to explain these results, the GB trap density (N_t) was calculated from the slope of $\ln(I_d/V_g)$ vs. $(1/V_g)$ for each TFT based on Levinson's model.[6] As for the threshold voltage (V_{th}) is defined at a fixed drain current at V_{ds}=0.1V.

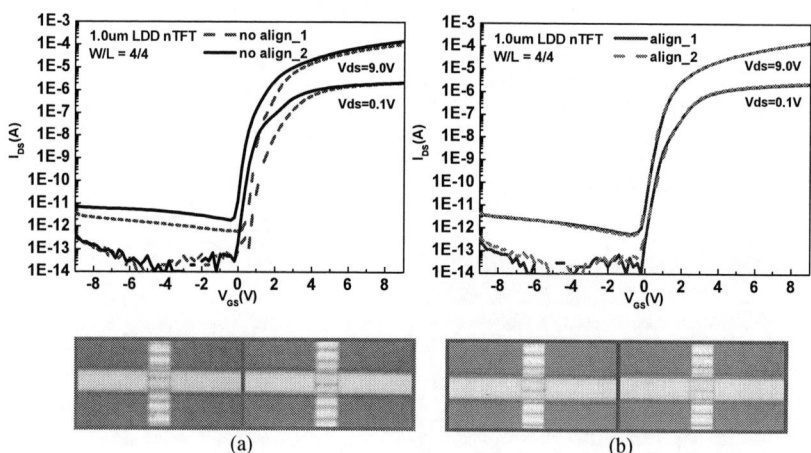

(a)   (b)

Figure 4. 1.0 μm LDD (Lightly Doped Drain) n-type TFT device transfer characteristics and its optical images of TFT channel (a) when normal SLS crystallization method was applied and (b) when a new alignment SLS crystallization method was applied

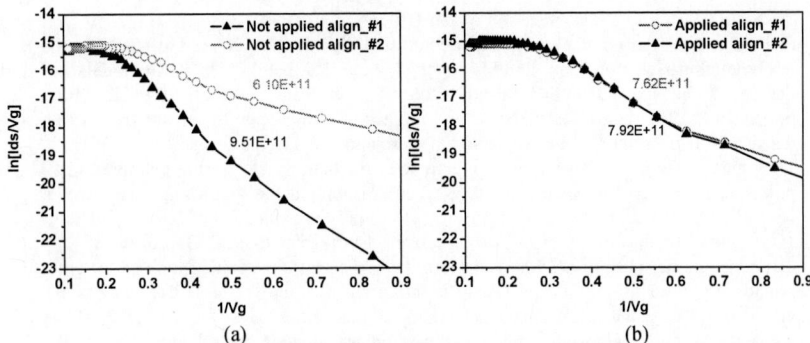

Figure 5. Comparison of the GB trap density (N_t) which was calculated from the slope of $\ln(I_d/V_g)$ vs. $(1/V_g)$ (a) when normal SLS crystallization method was applied and (b) when a new alignment SLS crystallization method was applied

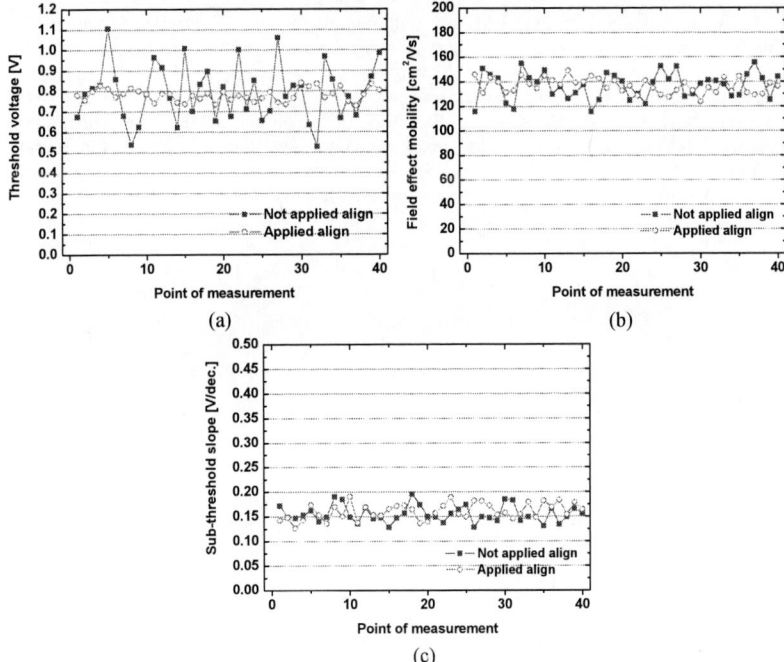

Figure 6. Results of 1.0 μm LDD n-type TFT parameters fluctuation of a normal SLS method and the grain boundary location controlled SLS method, which were measured in 40 points in series. (a) Threshold voltage (b) field-effect mobility uniformity and (c) sub-threshold slope

Figure 5. shows the comparison of the GB trap density (N_t) between a normal SLS method and the grain boundary location controlled SLS method. In case of applying the grain boundary location controlled SLS method, we found that the trap density in GB was decreased and showed more uniform values. The uniform GB location by the grain boundary location controlled SLS method might affect these improvements of GB trap density (N_t) uniformity and was resulted in more uniform TFT performances.

Figure 6. shows the results of TFT parameter fluctuation which were measured in 40 points in series. As shown in Figure 5, when we applied the grain boundary location controlled SLS method, the Vth uniformity was improved from 35% to 7% and the field-effect mobility uniformity was also improved from 15% to 9%. But there was no difference in the sub-threshold slope uniformity. We came to the conclusion that the grain boundary control in the channel region is important to improve the uniformity of TFT characteristics. As a result, we have accomplished more uniform TFT characteristics than conventional poly TFT by applying the grain boundary location controlled SLS method.

Conclusion

We fabricated a poly-Si TFT using a grain boundary location controlled SLS method by applying a new key recognition alignment crystallization. In this way, it is possible to locate the grain boundary at the same location in the channel region of each TFT. When we applied the grain boundary location controlled SLS method, the TFT characteristic showed more uniform performances than the TFT with normal SLS crystallization process. The Vth uniformity was improved from 35% to 7% and the field-effect mobility uniformity was also improved from 15% to 9%. In conclusion, we have accomplished more uniform TFT characteristics than conventional poly TFT by applying the grain boundary location controlled SLS method. Also, it can be said that control of grain boundary location is an important factor for the image quality improvement of the panel.

References

[1] P Mark A. Crowder, *Member, IEEE*, A. Tolis Voutsas, Steven R. Droes, Masao Moriguchi, and Yasuhiro Mitani
IEEE Transactions on ELECTRON DEVICES, VOL. 51, NO. 4, APRIL 2004
[2] Y.J.Kim, K.S.Hwang, Y.I.Park, S.W.Lee, H.C.Kang, K.M.Lim, C.D.Kim, I.J.Chung, AMLCD 05, 249(2005)
[3] L. Mariucci_, A. Pecora, R. Carluccio, G. Fortunato
Thin Solid Films 383Ž2001.39_44
[4] G. A. ARMSTRONG, S. UPPAL, S. D. BROTHERTON AND J. R. AYRES
Jpn. J. Appl. Phys. Vol. 37(1998) pp.1721-1726
[5] A. Bonfiglietti, A. Valletta, P. Gaucci, L. Mariucci, and G. Fortunato, S. D. Brotherton
JOURNAL OF APPLIED PHYSICS **98**, 033702 (2005)
[6] J. Levinson, F. R. shepherd, P. J. Scanlon, W. D. Westwood, G. Este, and M. Rider, J. Appl. Phys. 53, 1193(1982).

ECS Transactions, 3 (8) 75-80 (2006)
10.1149/1.2356337, copyright The Electrochemical Society

Front and Back Channel Properties of Asymmetrical Double-Gate Polysilicon TFTs

F. V. Farmakis[a], D. N. Kouvatsos[a], A. T. Voutsas[b], D. C. Moschou[a],
G. P. Kontogiannopoulos[a], and G. J. Papaioannou[c]

[a] Institute of Microelectronics, NCSR Demokritos, Agia Paraskevi 15310, Greece
[b] LCD Process Technology Laboratory, Sharp Labs of America, 5700 NM Pacific Rim
Boulevard, Camas, Washington 98607, USA
[c] Department of Physics, University of Athens, Athens 15784, Greece

> Polycrystalline silicon TFTs with different front- and back-gate
> lengths are investigated In addition, the laser annealing process
> yields high quality directional grains that enable us to orient TFT
> channels parallel or perpendicular to the grain boundaries. It is
> demonstrated that double-gate TFTs are fully depleted and
> therefore back interface properties exert critical influence to the
> overall TFT electrical performance. More specifically, it is
> demonstrated that the back interface contains a large number of
> defects, resulting in a deterioration of carrier conductance.

Introduction

Polycrystalline silicon thin film transistors (TFTs), fabricated at low temperatures, are
essential for flat panel display applications. In particular, high resolution and small form
factor displays with advanced input / output functions can only be realized by high
performance TFTs fabricated at low temperatures (1). Breakthroughs have recently
occurred in the polysilicon crystallization process using various excimer laser anneal
(ELA) methods (2-4), particularly with the advent of the sequential lateral solidification
technique (SLS) (5). Such advances have greatly improved TFT performance, reaching
the point of allowing the use of TFTs for the monolithic integration of circuits on display
substrates; thus, silicon on insulator (SOI) and system on glass applications are facilitated.
The SLS ELA technique yields polysilicon films with an excellent intragrain material
quality and grains with very elongated or engineered shapes. In the case of films with
elongated grains, it is possible to locate the TFT channels so as to be either parallel or
perpendicular to the grain boundaries.
Such monolithic integration in circuits demands absolute control of electrical TFT
parameters, such as turn-on voltage, subthreshold swing and transconductance. Towards
this direction, double-gate (DG) TFTs have been developed that enable electrical
parameter control through the application of an electric field on the opposite gate.
However, even though DG fully depleted SOI MOSFETs have been extensively studied,
DG TFTs lack such an investigation. In this paper, we attempt to clarify the electrical
properties of DG TFTs with asymmetrical top and bottom gates in terms of length.

Experimental

TFTs were fabricated in 50 nm thick polysilicon films, formed at low temperature by
ELA crystallization of a-Si, using the SLS technique (4,5). This process results in
polysilicon films with grains much longer than the TFT channel lengths, separated by
roughly parallel boundaries. Moreover, the polysilicon channel was lightly doped with

boron resulting in inversion-mode n-type devices. The TFTs were directed either parallel or perpendicular to the elongated direction, as shown in Figure 1. The double gate (DG) device structure consisted of self-aligned front gates, with a 30nm thick PECVD SiO$_2$ gate dielectric, and non-self-aligned back gates, with a 50nm thick PECVD SiO$_2$ gate dielectric. The channel width W was 8 μm and the front and back gate lengths L$_t$ and L$_b$ were 1.6 μm and 0.5-1.3 μm. In addition, top and bottom gate devices were fabricated to serve as references. The devices were characterized using a semiconductor analyzer HP4140B and a Keithley 230 voltage source, with the transfer characteristics and the parameters extracted in the linear regime (V$_D$=0.1 V) for both front and back channel operation.

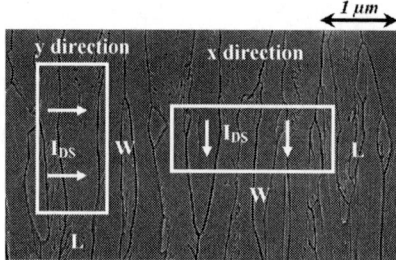

Figure 1. SEM micrograph for directional polycrystalline silicon. TFT y-direction and x direction representation is schematically shown.

Discussion

Before discussing our data it is important to note that for front operation L=L$_t$ and for back operation L=L$_b$. Since a gate mask with L=L$_t$ was used for source and drain doping, back operation TFT may be simulated as that of a TFT with L=L$_b$ connected to series resistances that may be modulated by front-gate biasing. We apply the terms "front operation" and "back operation" to indicate electrical measurements and parameter extraction having the opposite gate at constant bias.

Dependence on opposite gate biasing

Figure 2 shows the turn-on voltage dependence as a function of the opposite gate biasing for front and back operation. In both cases, we observe an almost linear behavior that reflects the polycrystalline silicon fully depleted condition, already expected from the thickness of the active layer (50 nm). Moreover, it is demonstrated that there is no effect of the TFT orientation on the turn-on voltage variation, leading to the conclusion that the grain boundary defect density can be neglected in the following analysis. It is noted that the turn-on voltage variation is more pronounced for back operation and this cannot solely be explained by the difference in gate oxides (t$_{ox,b}$/t$_{ox,f}$=50/30). This behavior also reflects the interface trap density difference in both interfaces (N$_{ss,f}$ and N$_{ss,b}$). For fully depleted devices (6), we can estimate N$_{ss,f}$ and N$_{ss,b}$ for devices with L$_b$=L$_t$ by using the following equation for front operation:

$$\Delta V_{ON,f} = -\frac{C_{ox,b}C_{Si}}{C_{ox,f}(C_{Si} + C_{ox,b} + C_{ss,b})}V_{G,b} \qquad [1]$$

where $C_{ox,f}$ and $C_{ox,b}$ represent the gate capacitances of front and back interfaces respectively, C_{Si} the silicon (in our case polysilicon) capacitance and $C_{ss,b}$ the back interface capacitance due to surface states. An analogous equation is applicable for back operation. Applying the previous equation to our data and taking into account that $C_{ss,b}=qN_{ss,b}$, we evaluate the back interface surface state density for our devices; it has values ranging from $3\text{-}6\times10^{11}$ cm^{-2} eV^{-1}.

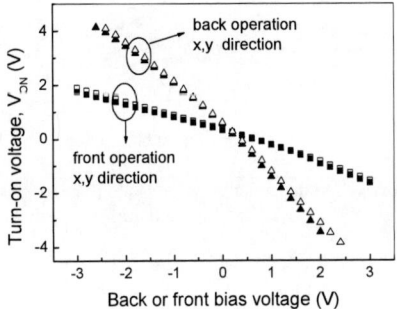

Figure 2. Turn-on voltage dependence on back or front gate bias voltage in the case of a double-gate TFT (W/L$_b$/L$_t$=8/0.8/1.6 μm).

However, in the case of back operation, the transfer characteristics (not shown here) indicate that the drain current is most efficiently controlled by the front gate and thus it is assumed that carriers flow mostly through the front polysilicon/SiO$_2$ interface. This is further supported by the high surface state density of the back interface, as evaluated above. Therefore, an equation analogous to [1] cannot be applied for back operation.

Figure 3 shows the subthreshold swing dependence on opposite gate biasing (back or front gate biasing for front or back gate operation, respectively). In the case of front operation, the subthreshold swing takes its minimum value when the back interface accumulates (holes) at around $V_{G,b}=-1V$ and it remains constant for lower $V_{G,b}$ values, indicating that the drain current flows mainly along the front polysilicon/SiO$_2$ interface. For positive back gate bias, the subthreshold swing is drastically increased, indicating that the inversion layer becomes thicker, as shown in the following equation (7):

$$S = \frac{kT}{q} \cdot \ln 10 \cdot \left[1 + \frac{q^2 (D_{bulk} t_{inv} + N_{ss})}{C_{ox}} \right] \qquad [2]$$

where kT is the thermal energy, q is the electron charge, D_{bulk} is the mean bulk density (eV^{-1} cm^{-3}), t_{inv} is the inversion layer thickness and N_{ss} is the surface state density in the front gate interface (eV^{-1} cm^{-2}).

In addition, the grain orientation in the device channel affects the subthreshold swing value, but not its variation with back gate bias. That is, the difference in the S value for x-oriented or y-oriented TFTs remains constant for any given $V_{G,b}$. From this observation and from equation [2] it can be inferred that the variation of t_{inv} with $V_{G,b}$ is the same for both orientations.

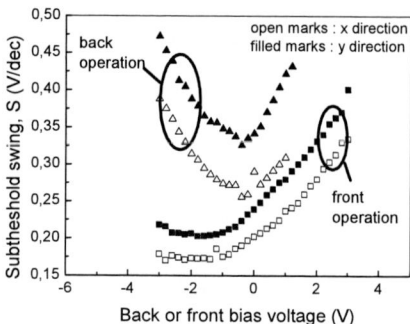

Figure 3. Subthreshold swing dependence on opposite (back or front) gate bias voltage in the case of a double-gate TFT ($W/L_b/L_f$=8/0.8/1.6 μm).

The back operation subthreshold swing dependence on front gate bias exhibits a U-shape behavior. It is noted that for accumulated front interface ($V_{G,f}$<0), the subthreshold swing is increased, demonstrating a high back interface state density, as the drain current is mainly controlled by back gate biasing. On the contrary, for inverted front interface ($V_{G,f}$>0) the subthreshold swing is also increased. Considering that in this case the current mainly flows through the front interface, one would expect a subthreshold swing improvement. However, it has to be noted that the subthreshold swing is measured from the back transfer characteristic, giving rise to an "apparent" value. This appears more evident by taking into account equation [2], where C_{ox} denotes the back oxide capacitance.

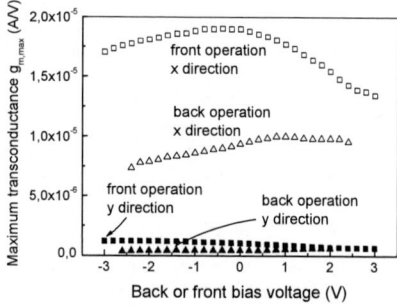

Figure 4. Maximum transconductance dependence on opposite (back or front) gate bias voltage in the case of a double-gate TFT ($W/L_b/L_f$=8/0.8/1.6 μm).

Figure 4 shows the maximum transconductance $g_{m,max}$ as a function of opposite gate biasing for front and back channel operation. In the case of front channel operation, $g_{m,max}$ is more than a decade higher in x direction than in y direction TFTs, indicating the important role of the direction of grain   laries in device transconductance (8).

Moreover, a clear decrease in the transconductance is observed as the bottom interface reaches accumulation ($V_{G,b}<0$). This may be attributed to the increase of the electric field in the inversion region that leads to an enhanced carrier scattering in the front polysilicon/SiO$_2$ interface. In addition, we also observe a severe decrease of the transconductance as the bottom interface is inverted ($V_{G,b}>0$). In this situation, the current flows not only in the front interface but also deeper in the polysilicon bulk and/or the back interface and is thus affected by additional trapping mechanisms. Finally, it is also shown, that in the case of back operation, device transconductance is low and it also exhibits the same behavior regarding the grain orientation as mentioned for front operation. This further supports that back interface contains a large number of defects.

Dependence on back over front gate ratio

In order to further investigate carrier conductance through DG-TFTs, we performed electrical measurements on DG-TFTs with various L_b/L_t ratios, as well as on single bottom and front gate devices. Figure 5 shows the extracted turn-on voltage as a function of L_b/L_t ratio. Top gate (TG) and bottom gate (BG) TFT data are also added for comparison. It is observed that for front operation the turn-on voltage remains unchanged; top gate TFT exhibit the same turn-on voltage with DG-TFTs with $L_b/L_t<0.5$. In the case of longer bottom lengths ($L_b/L_t>0.5$), the turn-on voltage appears to increase almost linearly with L_b.

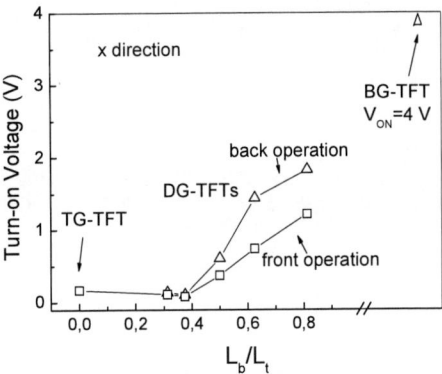

Figure 5. Turn-on voltage as a function of bottom and top channel length ratio for back and front operation. Top-gate and bottom-gate device data are also shown.

In the case of back operation, the turn-on voltage for DG-TFTs with $L_b/L_t<0.5$ exhibits the same behavior than for front operation. In addition, for longer bottom gates, it seems that back interface influence becomes more pronounced. Taking into account the large value of V_{ON} for BG-TFT, it is evident that the back interface defect density pins the Fermi level. Therefore, the inference that back operation in DG-TFTs is mainly due to the front gate potential is supported.

Conclusion

In this paper we investigated asymmetrical double gate TFTs for various bottom over top gate length ratios. Analysis of the electrical measurements functioning for front and back operation demonstrated that back interface contains a large number of defects that drastically modifies device electrical properties. Moreover, we suggested that for back operation drain current flows not only through the back interface but also from the bulk and/or the front interface. Finally, it was shown that grain orientation along the device channel influences mainly the transconductance and has insignificant effects on turn-on voltage.

Acknowledgments

The authors acknowledge financial support through the research project PENED 3ED550, administered by the Greek General Secretariat for Research and Technology.

References

1. T. Matsuo and T. Muramatsu, *Proc. SID Intern. Symposium*, 856, (2004).
2. S. D. Brotherton, D. J. McCulloch, J. B. Clegg, and J. P. Gowers, *IEEE Trans. Electr Dev.*, **40**, 407 (1993).
3. K. M. Chang, Y. H. Chung, G. M. Lin, C. G. Deng and J. H. Lin, *IEEE Electron Dev. Lett.*, **22**, 475 (2001).
4. A. T. Voutsas, *IEEE Trans. Electron Dev.*, **50**, 1494 (2003).
5. R. S. Sposili and J. S. Im, *Appl. Phys. Lett.*, **69**, 2864 (1996).
6. F. Balestra, M. Benachir, J. Brini and G. Ghibaudo, *IEEE Trans. Electron Dev.*, **37**, 2303, (1990).
7. M. Miyasaka and J. Stoemenos, *J.Appl.Phys.*, **86**, 5556, (1999).
8. A. T. Voutsas, *Applied Surface Science* **208-209**, 250-262, (2003).

ECS Transactions, 3 (8) 81-85 (2006)
10.1149/1.2356338, copyright The Electrochemical Society

Threshold-Voltage Instability of Single-Crystal Si Thin-Film Transistors Fabricated on Plastic Substrate

H.-C. Yuan[a], Z. Ma[a,*], M. G. Lagally[b], and G. K. Celler[c]

[a] Department of Electrical and Computer Engineering,
[b] Department of Material Science and Engineering,
University of Wisconsin-Madison, Madison, Wisconsin 53706, USA
[c] Soitec USA, 2 Centennial Dr, Peabody, MA 01960, USA
* mazq@engr.wisc.edu

Single-crystal Si thin films are transferred onto plastic host substrate after selectively removing the buried oxide (BOX) layer on the silicon-on-insulator (SOI) substrate. A layer of SU-8 epoxy is used as gate dielectric for the thin-film transistors (TFTs) fabricated on the transferred Si thin films. Here we report the first observation of threshold voltage (V_t) instability on these n-type, single-crystal Si TFTs on flexible plastic substrate. It is observed that V_t shifts to higher (lower) value under high positive (negative) gate-bias stressing. The logarithmic time-dependence of the V_t shift suggests that the instability is attributed to charge trapping in the gate dielectric layer.

Introduction

Amorphous-Si (α-Si) TFTs, together with other active devices such as high-temperature poly-Si TFTs and low-temperature poly-Si TFTs, have been widely used for the backplane of active matrix displays (1). The V_t instability of the α-Si TFTs has been extensively studied and a summary can be found in Ref. 2. For α-Si TFTs, the V_t shift is generally due to two distinct mechanisms: charge trapping in the gate dielectric layer and the creation of metastable dangling bond state in the α-Si. Going beyond the rigid substrates, flexible electronics enable the possibility of realizing flexible displays, electronic tags, and other applications that desire electronics to be conformal on the surface of the substance. While the TFTs made on α-Si, poly-Si and organic semiconductors on plastic substrates offer limited carrier mobility, very high electron mobility has been demonstrated on single-crystal Si TFTs on flexible substrate (3-5). In this report, we present the fabrication of single-crystal Si TFTs on plastic substrate and focus on the characterizations of V_t instability of these devices. We observe, for the first time, the V_t shift under gate bias stressing of single-crystal Si TFTs and identify that the instability of V_t is attributed to charge trapping in the gate dielectric layer.

Device Fabrication

The starting material is (001) SOI substrate with 200 nm, lightly-doped p-type Si template on top of a 200 nm BOX layer. Si template is first patterned into 20 μm wide strips with 5 μm spacing in between. BOX layer is then selectively removed by concentrated (49%) HF, which leaves the "released" Si strips settling down on the Si handling substrate, as show in Fig. 1(a). Indium-tin-oxide (ITO) coated, 175-μm thick polyethylene terephthalate (PET) is chosen : plastic host substrate. A layer of SU-8

81

(Microchem Corp.) epoxy is spun on the plastic substrate. Then the "released" Si strips are printed face-to-face against the adhesive epoxy. Figure 1(b) shows the 20 μm wide Si strips that have been transferred onto the plastic host substrate with the spatial position almost unchanged as they were on the SOI substrate. After the printing transfer, the SU-8 layer is fully cross-linked by exposing it under UV light and a subsequent baking step. This layer serves as the gate dielectric for the TFTs whereas the conductive ITO coating acts as the gate electrode. Finally, metal pads consisting of 30 nm Ti and 60 nm Au are formed on the Si strips as source and drain electrodes. The separation between the two metal pads determines the channel length of the TFTs. Figure 1(c) and 1(d) show the top view microscopic image of the finished TFTs and the schematic device cross-section, respectively. The SU-8 layer is around 1.8 μm after finishing the full fabrication process. Detailed fabrication has been depicted in Ref. 5.

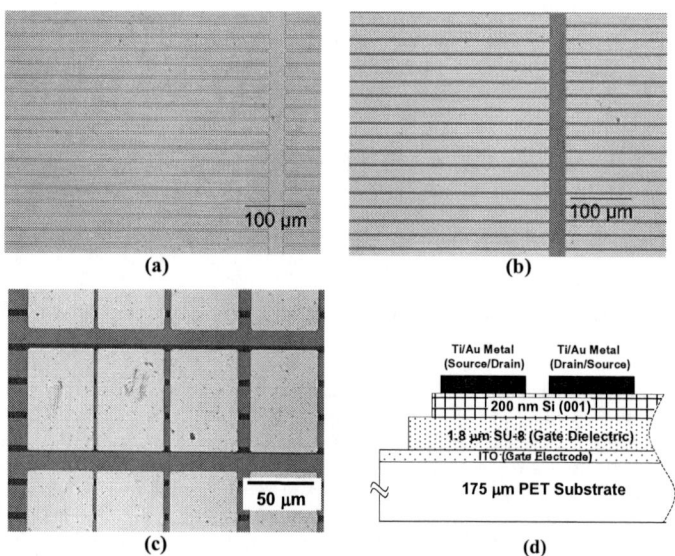

Figure 1. (a) 20 μm wide Si strips settle down on Si handling substrate after BOX was entirely removed in 49% HF. (b) Si strips transferred onto plastic host substrate by printing. (c) Finished TFTs on plastic with Ti/Au metal pads on top of the Si strips. (d) Schematic cross-section of the single-crystal Si TFT on plastic substrate.

Device Characterizations and Discussions

Output and Transfer Characteristics

The DC characteristics of the single-crystal Si TFTs are measured with an Agilent 4155B semiconductor parameter analyzer at room temperature. Figure 2 shows the typical normalized output and transfer characteristics of a single-crystal Si TFT on plastic substrate with channel length (L_g) and wi   ) of 3 and 60 μm, respectively. Due to the high Schottky barrier for holes between p⁻-Si and Ti metal, electrons are the primary conducting carriers. As a result, the TFTs demonstrate typical I-V characteristics similar to that of n-type MOSFETs (6). Well-defined linear and saturation regions are observed on the single-crystal Si TFT using Schottky barrier source/drain contacts. The sub-threshold swing of the TFT is around 5 V/decade, which is presumably due to the very thick gate dielectric layer.

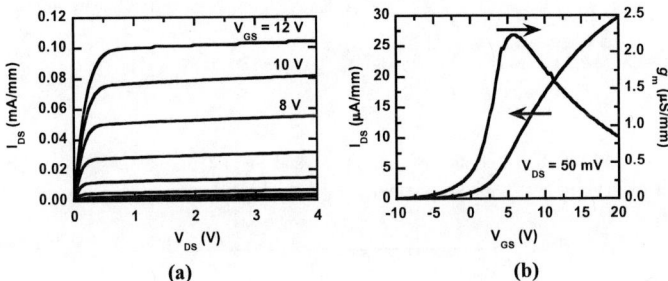

(a)   (b)

Figure 2. DC characteristics of a single-crystal Si TFT on plastic substrate with $L_g = 3$ μm, $W_g = 60$ μm. (a) Normalized I-V curves with gate voltage varied from -4 to 12 V in 2 V intervals. (b) Normalized transfer characteristics under $V_{DS} = 50$ mV.

Threshold Voltage Characteristics

The instability of threshold voltage V_t is characterized by applying a DC bias to the gate electrode while keeping both source and drain electrodes grounded. Figure 3 shows the measured transfer characteristics of the single-crystal Si TFT before and after gate stressing. Significant source-to-drain current (I_{DS}) shifts were observed after high gate bias (±20 V) stressing for 120 sec on the device. For consistency, we define the V_t to be the gate voltage at which the I_{DS} reaches 5×10^{-8} A during the sweeping of V_{GS} from the off-state to the on-state at a constant V_{DS} of 50 mV. Based on this defined rule, Fig. 3 shows that the values of V_t before any DC stressing, after +20 V DC stressing for 120 sec, and after -20 V DC stressing for another 120 sec are 0.8, 3.6, and -2.4 V, respectively. The values of V_t shift are thus 2.8 V and -3.2 V under 20 V and -20 V gate stressing, respectively. It is also observed that both I_{DS} and V_t show almost identical shifts when repeating the bias stressing cycles (both positive and negative gate bias stressing). Figure 3 shows the overlapped curves for two gate stressing cycles. The repeatability of I_{DS} under different stressing conditions suggests that the changes of I_{DS} caused by gate stressing can be fully recovered under the reversed stressing conditions (bias is under opposite polarity, with same length of duration).

To further investigate the V_t shift (ΔV_t), the device was stressed under different bias voltages, varying from -30 V to +30 V. The stressing duration for each voltage is 180 sec this time. The V_t shift (ΔV_t) results are plotted in Fig. 4. Significant values of ΔV_t are observed at gate stressing voltages of higher than 10 V and lower than -10 V. Interestingly, there is a fairly small V_t shift to the opposite signs when the stressing voltage is within ±10 V. The cause of this small but opposite shift is not clear at this moment.

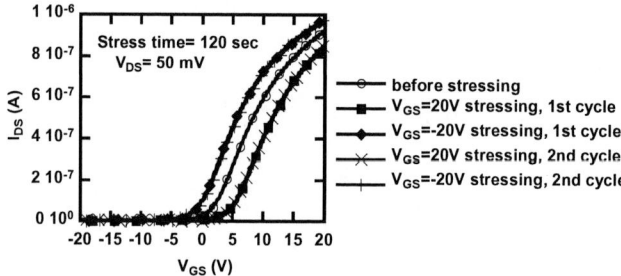

Figure 3. Transfer characteristics of a single-crystal Si TFT on plastic substrate before and after two cycles of gate bias stressing. L_g= 3 μm, W_g= 60 μm.

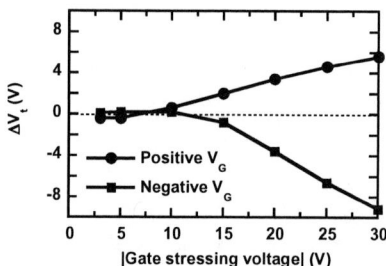

Figure 4. V_t shift (ΔV_t) of a single-crystal Si TFT on plastic substrate at different gate stressing voltage. The stressing time is 180 sec for each bias condition. L_g= 3 μm, W_g= 60 μm

The V_t shift to higher (lower) values under high positive (negative) gate bias stressing reveals that the shift is not attributed to dipole or mobile charge in the gate dielectric layer. To determine the cause of V_t shift, time dependence of the V_t shift is studied. Figure 5 shows the ΔV_t under different stressing times at a constant stressing voltage of 20 V. The logarithmic time-dependence of the ΔV_t is clearly demonstrated. It is the strong evidence that the V_t instability of the single-crystal Si TFT using SU-8 epoxy as gate dielectric layer is caused by charge trapping in the dielectric layer, since charge injection into the dielectric/insulating layer depends exponentially on the density of previously injected charge (2, 7).

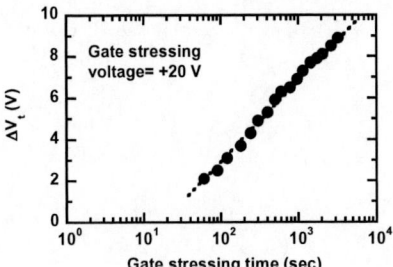

Figure 5. Time dependence of the V_t shift (ΔV_t) of a single-crystal Si TFT on plastic substrate. The gate stressing voltage is 20 V. $L_g = 3~\mu m$, $W_g = 60~\mu m$

Conclusion

V_t instability on single-crystal Si TFTs on plastic substrate is observed for the first time. With the TFT layer structure used in the current study, significant V_t shift to higher (lower) value after high ($> |10|$ V) gate bias stressing for 180 sec is observed. The logarithmic time-dependence of V_t shift is the clear evidence that the V_t instability of the single-crystal Si TFTs is attributed to the charge trapping in the gate dielectric layer.

Acknowledgments

This work is supported by NSF-MRSEC, DOE and AFOSR.

References

1. T. Suzuki, *J. Appl. Phys.*, **99**, 111101 (2006).
2. M. J. Powell, *IEEE Trans. Electron Devices*, **36**, 2753 (1989).
3. E. Menard, R. G. Nuzzo and J. A. Rogers, *Appl. Phys. Lett.*, **86**, 093507 (2005).
4. Z.-T. Zhu, E. Menard, K. Hurley, R. G. Nuzzo and J. A. Rogers, *Appl. Phys. Lett.*, **86**, 133507 (2005).
5. H.-C. Yuan, Z. Ma, M. M. Roberts, D. E. Savage and M. G. Lagally, *J. Appl. Phys.*, **100**, 013708 (2006).
6. H.-C. Yuan, G. Wang, Z. Ma, M. M. Roberts, D. E. Savage and M. G. Lagally, to appear in *Semicond. Sci. Technol.*.
7. R. H. Walden, *J. Appl. Phys.*, **43**, 1178 (1972).

ECS Transactions, 3 (8) 87-92 (2006)
10.1149/1.2356339, copyright The Electrochemical Society

The Role of Grain Boundaries on the Performance of Poly-Si TFTs

L. Michalas[a], G. J. Papaioannou[a], D. N. Kouvatsos[b] and A.T. Voutsas[c]

[a] Physics Department, National and Kapodistrian University of Athens, Athens 15784, Greece
[b] Institute of Microelectronics, NCSR "Demokritos", Ag. Paraskevi, Athens 15310, Greece
[c] LCD Process Technology Laboratory, Sharp Labs of America, Inc., Camas, WA 98607, USA

> The role of grain boundaries on the performance of polycrystalline Si thin film transistors is investigated through the temperature analysis of the transfer characteristics. The investigation is performed on devices fabricated on films that consist of long grains, separated by parallel boundaries. The transport properties are studied by employing channel orientations parallel and vertical to the grain boundaries. Finally, the data analysis reveals that the major device parameters are thermally activated.

Introduction

The possible future applications of polycrystalline silicon thin film transistors (poly-Si TFTs) in integrated circuits on display substrates, the so called systems on panel (SOP) (1), requires the comprehensive clarification of the role of grain boundaries on the device performance. This is because the presence of grain boundaries in the active area constitutes a major drawback of TFTs technology limiting the devices performance. Recently, advances in laser crystallization techniques allowed the fabrication of poly-silicon films that consist of very long crystal domains (grains), separated by roughly parallel boundaries (2). In spite of the dependence of crystal properties on the film thickness (2), these techniques allowed the fabrication of TFTs with desirable channel orientation, which were used on the systematic investigation of the influence of the grain boundaries on the electrical properties of the devices.

Regarding the device performance, although there has been a significant effort on the understanding of the effect of grain boundaries on the electrical properties of TFTs, in order to generate more accurate models and improved simulation tools (3,4), the experimental data are still limited. So, there are reports on the transfer characteristics and hot carriers degradation at room temperature (5,6) as well as isolated results on the mobility at different temperatures (2).

The aim of the present work is to obtain a better insight on the effect of the grain boundaries and the thermally activated mechanisms on TFTs electrical characteristics. The investigation has been performed on devices fabricated on films that consist of long grains separated by parallel boundaries obtained by the sequential lateral solidification (SLS) process (7). The study included the effect of temperature and grain boundaries orientation, with respect to the current flow, which made them particularly useful for the investigation of their influence on the performance of poly-Si TFTs, and to address the issue of the role of spatially localized trapping states. Finally, devices with different body layer thicknesses were employed.

Experimental

Poly-silicon TFTs were fabricated on films formed by crystallization of amorphous silicon. The initial hydrogenated a-Si film   H) were deposited by plasma enhanced chemical vapor deposition (PECVD) on quartz substrates. The resulting a-Si films were transformed to poly-silicon ones by excimer laser annealing (ELA), using the sequential lateral solidification process (SLS) (7). This process results in a polysilicon film composed of long crystal domains (grains), separated by roughly parallel grain boundaries (Fig. 1). Devices were processed on films with thickness of 50 nm and 100nm. Finally, the die included devices with channel aligned parallel (d0) and vertical (d90) to the boundaries direction.

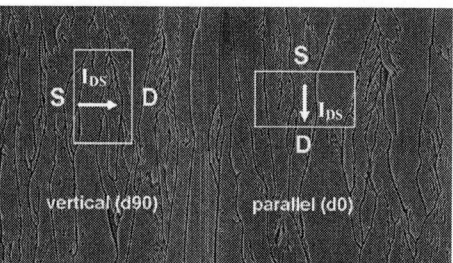

Figure 1. SEM micrograph of an SLS ELA polysilicon film and device orientation

I_{DS}-V_{GS} characteristics have been recorded in the temperature range of 150 to 440 K, under linear operation regime. The slope of a line fitted to the I_{DS}-V_{GS} curve at the point of maximum transconductance g_m, yielded to the field effect mobility and the intercept of that line to the extrapolated threshold voltage V_T. Additional results were obtained from the application, on the experimental data, of a modified (8) Levinson analysis (3), yielding at the calculation of flat band voltage V_{FB} .

Results and Discussion

A key issue parameter parameter that characterizes the TFT performance in the ON regime is the field effect mobility. In devices with parallel orientation of boundaries to the current flow higher mobility values are observed (Fig.2). This is expected since under such orientation the scattering rate on potential fluctuations, arising from grain boundaries barriers, and charged defect centers, surrounding the grain boundaries, decreases significantly. Due to scattering on distributed charged and neutral defects the temperature dependence in the high temperature region is not limited by phonon scattering ($T^{-3/2}$ to $T^{-5/2}$) although the mobility maximum is attained at about 160K to 180K, depending on the film thickness. In contrast in devices with vertical orientation the mobility is lower and the mobility maximum appears above room temperature (350K to 400K) due to scattering on these barriers. Moreover, in thinner film devices the temperature of mobility maximum is attained at higher temperatures due to the narrower grains hence the larger number of boundaries in the channel (2).

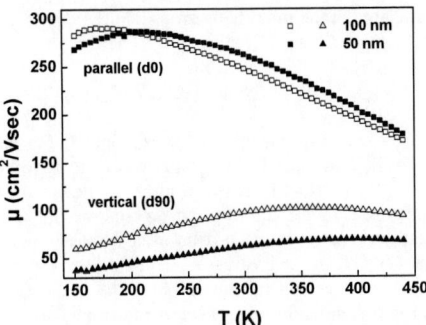

Figure 2. Field effect mobility for different film thickness and device orientation.

The Mathiessen law for mobility was applied for the calculation of temperature dependence of potential barrier height at the grain boundaries. The calculation was based on two assumptions: i) the measured mobility in parallel case (d0) is practically equal to the grain mobility μ_{gr}, and ii) the mobility term for the grain boundary area μ_{gb}, is $\mu_{gb}=\mu_{gr}*\exp(-E_B/kT)$.

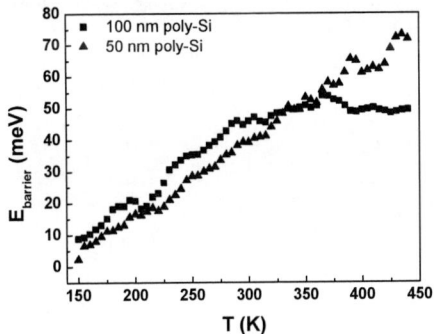

Figure 3. Grain boundaries potential barrier height vs temperature

The temperature dependence of the grain boundaries potential barrier height is presented in Fig. 3. The barrier height increases with temperature for both film thickness and in the case of 100nm thick films saturates above 350K. In order to understand this behavior we must take into account that in the present case the poly-silicon TFTs body is slightly p-type, so that the Fermi level in the body will shift upwards, towards midgap, as temperatures increases. Moreover, as will be shown bellow, the threshold voltage decreases with temperature. So, under a constant gate bias the Fermi level at the gate oxide interface will shift closer to the conduction band. Both mechanisms will lead to a larger trapped charge at the grain boundaries thus to higher potential barriers at higher temperatures. Here it must be pointed out that the barrier height is proportional to the square of the trapped charge at the grain boundaries and the device operating point (V_G-V_T) (9).

$$E_B = q^2 N_T^2 t_{ch}/8\varepsilon C_{ox}(V_G-V_T) \qquad [1]$$

where Cox is the gate oxide capacitance, t_{ch} the channel depth and N_T is the grain boundary trap density. Thus, the fact that the barrier height, in 100 nm films seems to saturate, while in 50 nm films further increases leads us to the conclusion that the density of states at grain boundaries is larger in thinner polycrystalline films.

Another important parameter for thin film transistors is the threshold voltage. Generally the threshold voltage decreases as temperature increases (10). In non-crystalline devices the major mechanism responsible for the reduction is the excitation of carriers through band gap states. The rise in temperature increases the number of free carriers that leads to channel formation at lower gate voltage (11). Moreover, the density and distribution of these defects is expected to affect the temperature dependence of the threshold voltage. This hypothesis is confirmed by the typical calculated values, which are -36 mV/C for a-Si TFTs, and -6 mV/C for poly-Si TFTs (11). Furthermore almost the same temperature dependence is obtained on devices fabricated on the same thickness film (Fig. 4).

However, higher threshold voltages are obtained on TFTs with orientation of boundaries vertical to the current flow. Even if there could exist enough free carriers, the current flow is prevented by the presence of grain boundaries potential barriers, on these devices. As gate voltage further increases, a potential barrier lowering is induced (12), allowing the carriers motion to the drain.

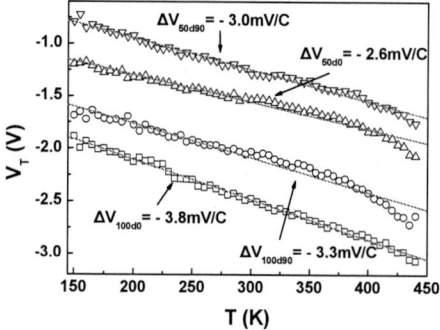

Figure 4. Threshold voltage temperature dependence for parallel (d0) and vertical (d90) orientation

In the OFF state poly-silicon TFTs suffer from high leakage currents. These currents arise from carriers generated in the drain – body contact depletion region (13).
Since Fermi level position is temperature dependent, the traps occupation changes. This is leading to flat band voltage shift. To take this shift into account, as well as the characteristic curve shift (14), in the present work, the leakage current is measured at $V_G=V_{FB}$ for each temperature.

At low drain voltage, thermal generation is proposed to be the major generation mechanism (13).

$$J_{Leak} = q \frac{n_i}{\tau_e} w \qquad [2]$$

where n_i is the intrinsic carrier concentration, w is the depletion layer width and τ_e is the effective electron generation lifetime. In polycrystalline materials the trapping, emission, and thermal generation of carriers are determined by the continuous distribution of band gap states, which consists of band tails and deep traps (15)

Since the generation mechanism is trap assisted and regional, thus the activation energy is related to the nature of the grain boundaries trap states. This is the reason that the same activation energy is calculated in devices fabricated on the same thickness films. Devices fabricated on different film thickness exhibit different activation energies (Fig. 5). This is consistent with the fact that the defect characteristics in poly-Si film, obtained from the SLS-ELA process, are affected by the film thickness (2).

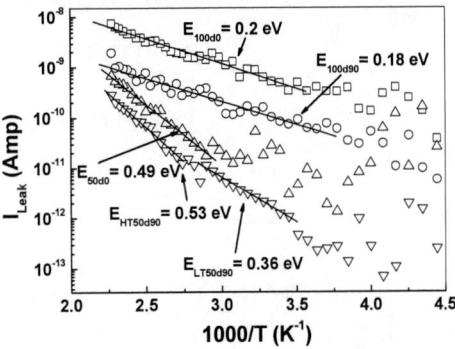

Figure 5. Leakage Current calculated at V_{FB} for parallel (d0) and vertical (d90) orientation

Regarding the leakage current magnitude, a higher current was monitored in devices with channel orientation parallel to the grain boundaries. A similar behaviour has been reported in (16) and attributed to the fact that the leakage current appears to be dominated by the overall density of sub-grain boundaries, and in this case those close to the drain terminal. Higher currents also observed on devices fabricated on thicker films, an effect attributed to the contribution of a larger material volume to carriers generation. Finally, the value of the activation energy indicates the distance of the contributing defects from the intrinsic level

Conclusions

The role of grain boundaries on the performance of poly-silicon TFTs has been investigated through the temperature analysis of the transfer characteristics. The investigation has been performed on devices fabricated on films with long grains separated by parallel boundaries, oriented parallel and vertical to the current flow. Major device parameters behaviour, was determined by the thermal activation of carriers via band gap states, introduced by the boundaries. Furthermore their presence in the active area is responsible for the formation of potential barriers that affect the transport properties. The activation energies of the thermal processes and the barriers heights were found to be related to the densities and the distribution of the traps states. The results suggest that the device electrical properties are strongly related to the polycrystalline material properties.

Acknowledgments

Finally the authors would like to acknowledge that the present work has been performed within the Greek GSRT project 3EΔ550.

References

1. N.Bavidge, M. Boero, P. Migliorato, T. Shoimoda, *Applied Physics Letters,*77 ,23, 3836 (2000)
2. A. Voutsas, *IEEE Transactions on Electron Devices*, **50**, 1494 (2003)
3. J. Levinson, F.R. Shepherd, P.J. Scanlon, W.D. Westwood, G. Este, M. Rider, *Journal of Applied Physics,* **53,** 1193 (1982)
4. Mutsumi Kimura, Satoshi Inoue, Tatsuya Shimoda, Tsukasa Eguchi, *Journal of Applied physics*, **89**, 596 (2001)
5. Y.H. Jung, J.M. Yoon, M.S. Yang, W.K. Park, H.S. Soh, Material Research Society Symposium Proceedings, 621 (2000)
6. D. Kouvatsos, L. Michalas, A. Voutsa, G.J. Papaioannou, *IOP Journalof Physics: Conference Series*, **10**, 45 (2005)
7. R.S. Sposili and J.S. Im, *Applied. Physics. Letters*, **69**, 2864 (1996)
8. R.E. Proano, R.S. Misage,D.G. Ast, *IEEE Transactions on Electron Devices,* **36**, 1915 (1989).
9. Y. Morimoto, Y. Jinno, K. Hirai, H. Ogata, T. Yamada, K. Yoneda, *Journal of Electrochemical Society,* **144**, 2495 (1997)
10. S.M. Sze, Physics of semiconductor devices, p.257, Wiley & Sons, New York (1981)
11. L.Wang, T.A. Fjeldly,B. Iniguez, H.C. Slade,M. Shur, *IEEE Transactions on Electron Devices*, **47**, 387 (2000)
12. P. Walker, H. Mizuta, S. Uno, Y. Furuta, D.G. Hasko, *IEEE Transactions on Electron Devices*, **51**, 212 (2004)
13. C.H.Kim, K.S.Sohn, J.Jang, *Journal of Applied Physics*, **81**, 8084 (1997)
14. L. Michalas, M.A. Exarchos, G.J. Papaioannou, D. Kouvatsos, A. Voutsas, 25th International conference on Microelectronics Proceedings, **2**, 597 (2006)
15. M.A. Exarchos, G.J. Papaioannou, D.N. Kouvatsos, A.T. Voutsas, *Journal of Applied Physics,* **99**, 024511 (2005)
16. A. Bonfiglietti, A. Valletta, P. Gaucci, L. Mariucci, and G. Fortunato, *Journal of Applied Physics*, **98**, 033702 (2005)

ECS Transactions, 3 (8) 93-97 (2006)
10.1149/1.2356340, copyright The Electrochemical Society

Hydrogen Passivation and Channel Capping for Threshold Voltage Shift and Off-Current Reduction in Nanocrystalline Silicon TFTs

M. R. Esmaeili Rad, A. Sazonov, and A. Nathan

Department of Electrical and Computer Engineering, University of Waterloo, Waterloo, ON, Canada, N2L 3G1

> Bottom-gate thin-film transistors were fabricated using a nanocrystalline silicon channel with amorphous silicon cap layer along with hydrogen plasma treatment on silicon nitride gate dielectric to enhance the stability and reliability of TFTs. The hydrogen plasma reduces the rate of threshold voltage shift, while amorphous silicon cap layer plays the role to reduce the TFT off-current. The TFT exhibits mobility of ~0.8 $cm^2V^{-1}s^{-1}$, threshold voltage of ~4V, on/off current ratio about 10^8 with off-current of 10^{-13}A, and subthreshold slope of 0.8 V/dec. No threshold voltage shift was observed following application of 15V gate stress for 5 hours.

Introduction

Thin-film transistors (TFTs) are the fundamental switching and driving devices in large-area active-matrix liquid crystal displays, X-ray imagers and the newly-emerging organic light emitting diode (OLED) displays (1-2). Polycrystalline silicon (polysilicon) and amorphous silicon (a-Si:H) TFTs are being considered for driving active-matrix OLED (AMOLED) display pixels. Here, they are required to supply a stable drive current to the OLED at each pixel. Polysilicon TFTs are capable of driving OLED pixels since they can render high field-effect mobility (μ_{FE}) and stable current. However, this technology is costly with a further drawback of non-uniformity of device parameters over large-area substrates. On the other hand, a-Si:H is a low-cost technology for large-area fabrication. But it suffers from low device mobility and electrical instability under prolonged gate voltages (2-4).

The solution may lie in directly-deposited nanocrystalline silicon (nc-Si) as an active layer in TFTs, since it offers higher mobility and stability compared to its a-Si counterpart (5-7). While top-gate TFTs with nc-Si active layer have shown promise to render high drive current, compatibility requirements with current display manufacturing, dictates a bottom-gate TFT structure. The basic requirements for obtaining high-performance bottom-gate nc-Si TFTs include: 1) high crystallinity nc-Si active layer, 2) minimum incubation layer at the interface with the gate dielectric, 3) low defect density and low weak bond density at the gate dielectric interface, and 4) low impurity concentration in the active layer.

The aim of the research presented here is to fabricate stable bottom-gate (BG) TFTs compatible with existing a-Si:H process facilities. We describe the fabrication and performance characteristics of TFTs employing a nc-Si/a-Si:H active bi-layer and amorphous silicon nitride (a-SiN$_x$:H) as the gate dielectric. These layers were deposited by 13.56 MHz plasma-enhanced chemical vapor deposition (PECVD) technique. We also present the steps taken, from basic material development to TFT structure optimization, toward obtaining high performance BG nc-Si TFTs.

Experimental

The undoped nc-Si, a-Si:H, phosphorous-doped (n+) nc-Si, and a-SiN$_x$:H films used in the TFT structure were deposited at 280°C using a multi-chamber PECVD system (MVSystems cluster tool). Undoped and doped nc-Si films were deposited using silane highly diluted by hydrogen to induce micro-structural change from amorphous to nc-Si silicon. In this method, we use a mixture of 1% silane in hydrogen. The crystallinity of nc-Si films was obtained from Raman spectra measured by using a He-Ne laser with a wavelength of 632.8nm. Dark conductivity measurements were performed using nc-Si test structures with Al co-planar top contacts sputtered through shadow mask. The conductivity activation energy values were obtained from temperature dependence of dark conductivity. The a-SiN$_x$:H gate insulator was deposited from a mixture of silane and ammonia. The FTIR spectra were measured by Shimadzu FTIR-8400S spectrometer in the range of 600-3600 cm^{-1} wavenumbers. The gate insulator leakage current and breakdown field were measured by Keithley 4200 semiconductor characterization system on fabricated metal-insulator-semiconductor (MIS) structures.

Figure 1 shows the cross section of the tri-layer inverted-staggered TFT structure, which consists of five lithography steps. First, 100nm-thick molybdenum (Mo) was sputtered on glass substrate and patterned to define the gate area. Subsequently, the tri-layer was deposited in one PECVD run which comprises of 300nm a-SiN$_x$:H as the gate dielectric, 50nm nc-Si/a-Si:H active bi-layer, and 300nm a-SiN$_x$:H as the channel passivation dielectric. Before deposition of the active layer, surface of the gate dielectric was treated by hydrogen plasma to passivate interface defects and reduce weak bonds at the interface with nc-Si active layer. Details of fabrication sequence have been reported elsewhere (8). TFTs with channel lengths ranging from 25μm to 200μm were fabricated while the channel width was kept constant at 100μm. After fabrication, TFTs were annealed at 175°C for two hours in air to improve the source and drain contact interface. The TFT characteristics were measured by aforementioned characterization system. To evaluate the TFT stability, a DC bias gate voltage with and without a drain voltage was applied at room temperature. Subsequently, the threshold voltage shift was obtained from retrieved transfer characteristics in saturation mode.

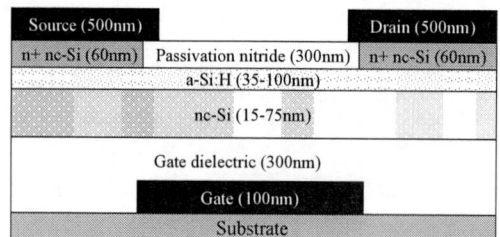

Figure 1. Cross section of the tri-layer inverted staggered TFT structure.

Results and discussion

Figure 2 shows the Raman spectrum of a 50nm-thick undoped nc-Si film deposited at the same conditions as that of TFT channel layer. The film crystallinity was evaluated by fitting the spectrum by Gaussian curves centered at $518cm^{-1}$ and $480cm^{-1}$, which corresponds to the crystalline and amorphous phase, respectively. From Figure 2, nc-Si film has a crystallinity of 63±3%. This crystallinity was obtained at low film thickness necessary for nc-Si TFT applications. The film exhibits dark conductivity of $\sim10^{-6}$S/cm and conductivity activation energy of 0.43±0.03eV.

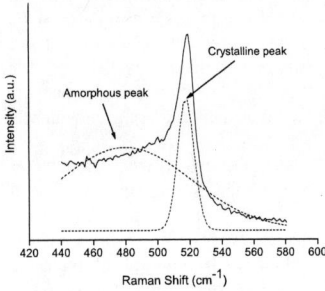

Figure 2. Raman spectrum of the 50nm-thick nc-Si film.

The infrared spectrum (FTIR) and current-voltage characteristics of a-SiN$_x$:H film are shown in Figure 3. The main absorption peaks, usually observed in PECVD-deposited a-SiN$_x$:H films, are present in the spectrum: namely, Si-N stretching mode (at 875-$890cm^{-1}$), N-H bending (at 1180-$1190cm^{-1}$) and stretching (at $3340cm^{-1}$) modes. The spectrum shows the presence of a weak peak due to N-H$_2$ bonds at $1545cm^{-1}$ and presence of a small peak due to Si-H bonds at $2100cm^{-1}$. It was shown that low concentration of Si-H bonds correlates with low leakage current in nitride films (9). As the I-V characteristics shows, the leakage current density obtained from MIS structures is $\sim10^{-9}$A/cm^2. The resistivity and breakdown field are $\sim10^{15}\Omega$cm and ~2.5MV/cm, respectively.

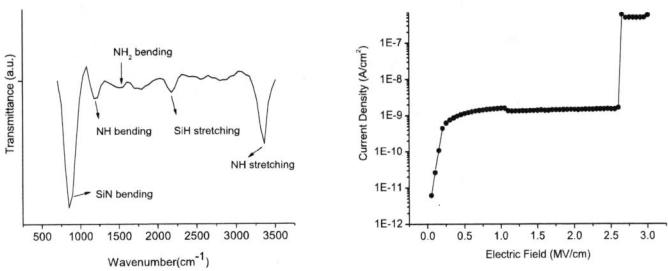

Figure 3. FTIR spectrum and I-V characteristics of a-SiN$_x$:H film used in TFT.

Figure 4 shows the transfer and output characteristics of the TFT. The TFT exhibits μ_{FE} of ~0.8 cm^{2}V^{-1}s^{-1}, threshold voltage (V_T) of ~4V, on/off current ratio about 10^8 with off-current of 10^{-13}A, and subthreshold slope of 0.8 V/dec. As the output characteristics show, there is no current crowding at low drain-source voltages and there is current saturation at high drain voltages, demonstrating a negligible effect of parasitic resistances on TFT performance. To investigate this, the total TFT resistance was obtained for different channel lengths from the linear output characteristics (10). At low drain-source voltages, our TFT resistance is modeled as two source-drain contact resistances in series with the a-Si:H parasitic resistance plus the channel resistance. The channel resistance is a function of gate voltage and channel length. Figure 5 shows dependence of the total TFT resistance versus the channel length at gate voltages of 10V and 20V. The parasitic resistance was obtained from the y-intercept of this plot. The linear fit, as shown in the figure, gives a value of ~80kΩ, which is less than 10% of the minimum channel resistance in the figure (~1MΩ). Thus, we conclude that parasitics are minimal and do not impact the TFT performance for the channel lengths considered in this work.

Figure 6 shows the TFT transfer characteristics before and after bias stress. The fabricated TFTs show stable operation in saturation regime following application of a bias stress of 15V to gate and 10V to drain terminal (V_{DS}=10V, V_{GS}=15V) for 5 hours at room temperature. As is widely believed, defect creation inside the active layer due to the breaking of strained bonds and charge trapping at the interface of the gate dielectric are the two mechanisms leading to TFT instability. As discussed, high crystallinity active layer helps to minimize density of strained bonds at the gate dielectric interface. In addition, hydrogen plasma treatment of the gate dielectric was also necessary, as the bias stress measurements showed that TFTs were not stable without hydrogen plasma treatment. Indeed, the as-deposited gate dielectric surface contains loosely-bound atoms which are considered as the source of interface defects. Presumably, hydrogen plasma removes weak bonds and newly created dangling bonds are passivated by atomic hydrogen (11).

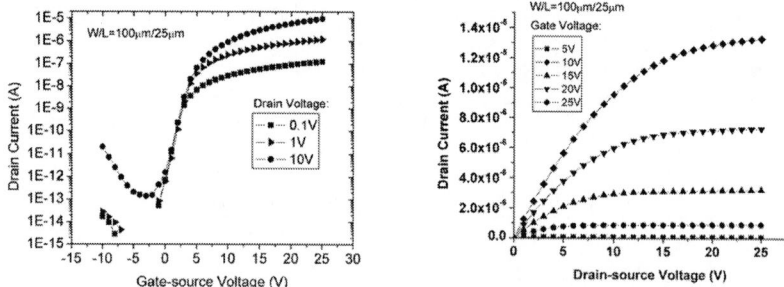

Figure 4. Transfer (left) and output (right) characteristics of TFT.

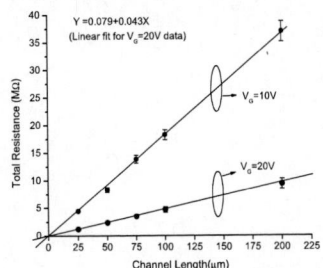
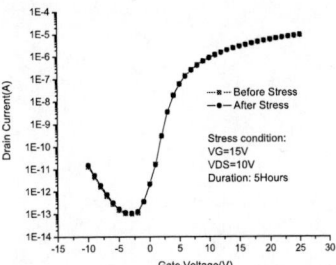

Figure 5. Total TFT resistance at different gate voltages as a function of channel length.

Figure 6. Transfer characteristics before and after bias stress for 5 hours ($V_{GS}=15V$, $V_{DS}=10V$).

It should be noted that the TFT shows good stability even at higher gate voltages. Comprehensive characterization of stability is in progress.

In conclusion, we presented bottom-gate nc-Si TFTs fabricated by RF PECVD in the sequence typical for standard a-Si:H devices. The TFT can be applied to AMOLED displays, since the pixel driver works in saturation regime and under gate voltages less than 15V. Due to compatibility with existing TFT technology, this process can be a viable candidate for the flat panel industry.

References

1. R. A. Street, *Technology and Application of Hydrogenated Amorphous Silicon,* New York: Springer-Verlag (2000).
2. A. Nathan, A. Kumar, K. Sakariya, P. Servati, S. Sambandan, and D. Striakhilev, IEEE J. Solid-State Circ., **39**, 1477 (2004).
3. M. Hack, A. Chwang, Y.-J. Tung, R. Hewitt, J. Brown, J. P. Lu, C. Shih, J. Ho, R. A. Street, L. Moro, X. Chu, T. Krajewski, N. Rutherford and Robert Visser, Mater. Res. Soc. Symp. Proc. **870E**, H3.1.1 (2005).
4. T. Tsujimura, Jpn. J. Appl. Phys. **43**, 5122 (2004).
5. C.-H. Lee, A. Sazonov and A. Nathan, Appl. Phys. Lett. **86**, 222106 (2005).
6. I-C. Cheng, S. Allen, and S. Wagner, J. of Non-Crys. Solids **338–340**, 720 (2004).
7. P. Roca i Cabarrocas, R. Brenot, P. Bulkin, R. Vanderhaghen, B. Drevillon, and I. French, J. of Appl. Phys. **86**, 7079 (1999).
8. A. Sazonov, D. Striakhilev, C.-H. Lee, and A. Nathan, Proc. IEEE, **93**, 1420 (2005).
9. K. Jin Park and G. N. Parsons, J. Vac. Sci. Technol. A **22(6)**, 2256 (2004).
10. P. V. Necliudov, M. S. Shur, D. J. Gundlach, T. N. Jackson, Solid-State Electronics **47**, 259 (2003).
11. G. Lavareda, C. Nunes de Carvalho, A. Amaral, E. Fortunato, P. Vilarinho, Materials Science and Engineering B **109**, 264 (2004).

98

CHAPTER 3

THIN FILMS DEPOSITION
AND CHARACTERIZATION

Development of ALD/PECVD reactor for high quality LTPS-TFTs insulator

Kazutoshi Murata[1], Naomasa Miyatake[1], Yasunari Mori[1], Hiroyuki Tachibana[1], Yukiharu Uraoka[2] and Takashi Fuyuki[2]

[1]Tamano Technology Center, Research & Development Hdq., Mitsui Engineering & Shipbuilding Co., LTD.,　16-1, Tamahara 3-chome, Tamano, Okayama 706-0014, Japan
[2]Materials Science, Nara Institute of Science and Technology,
8916-5, Takayama, Ikoma, Nara 630-0192, Japan

A high quality gate insulator is one of the key technologies for improving the performance of Low Temperature Poly Silicon TFTs. We propose an all-new stacked gate insulator, which is prepared with ALD technology for an interfacial layer, and a monopole antenna Plasma Enhanced Chemical Vapor Deposition (PECVD) technology for a secondary layer. The ALD/PECVD reactor, which handles glass substrates of 370 mm $\times$ 470 mm in size was developed, and its performance was evaluated. Silicon dioxide films were deposited at 400℃ with the reactor. Thickness variations of the ALD film and the PECVD film were confirmed to be less than $\pm 5\%$ and $\pm 10\%$, respectively. The stacked insulator, which was composed of the 2nm-ALD film and the 100nm-PECVD film, had an excellent Si/SiO_2 interface of 1×10^{11} cm^{-2}eV^{-1} and breakdown electrical field of 7.5 MVcm^{-1}.

Introduction

The rapid progress of a light-weight and high-resolution in a compact size Liquid Crystal Display presses on a research of a high performance of Low Temperature Poly Silicon (LTPS) TFTs. A high quality gate insulator is a one of the key technologies to improve its performance. The general method to prepare the gate insulator of the TFT is a capacitive coupling type Plasma Enhanced Chemical Vapor Deposition (PECVD) method, using Tetraethylorthosilicate (TEOS) as a source gas (1). However, the prepared SiO_2 insulators have defects caused by plasma damage in them. There are reports concerning high quality gate insulators prepared with different types of plasma sources (2,3).

Recently, an Atomic Layer Deposition (ALD) technology has attracted much attention as a new technology for a high-k gate insulator in LSI's. The ALD oxide film is deposited with alternating exposures of a source gas and an oxidant (4). The ALD film has additional features of precise thickness control and high conformity, because of the alternating gas supplies. However, the growth rate of the ALD film is not satisfactory for preparing the 50-100nm-thick gate insulator of the TFTs. We have developed a low energy monopole antenna plasma technology. The plasma is an electromagnetic wave coupled type, which works at very high frequency (VHF). The plasma does not cause severe damage to the film surface, because it is localized around the monopole antenna, and its electron temperature at the substrate position is lower.

Figure 1 shows an all-new stacked gate insulator, which is prepared with the ALD technology for an interfacial layer, and the monopole antenna PECVD technology for a secondary layer. As a fundamental study, three different SiO_2 films of an ALD-SiO_2, a PECVD-SiO_2, and a stacked ALD-SiO_2/PECVD-SiO_2 were prepared on 2-inch Si (100) wafers, and their electrical properties were evaluated. The interface trap density of the stacked SiO_2 was lower than that of PECVD-SiO_2. And the stacked film had a comparable dielectric strength to the PECVD one (5).

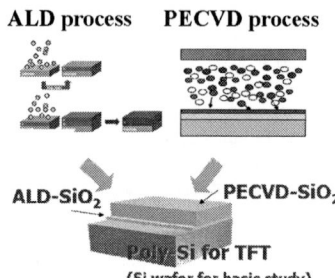

Fig.1 Concept of a new stacked insulator for LTPS-TFTs.

In this paper, we report characteristics of the developed ALD/PECVD reactor, and the quality of the SiO_2 film prepared by the reactor.

Experimental

Reactor

Figure 2 shows a schematic view of the ALD/PECVD reactor. Gases for the ALD reaction are supplied through an injector, which is mounted on a side wall of the reactor and swept out through exhaust openings mounted on the opposite side wall. The gases are supplied and exhausted in a co-flow pattern. Gases for the PECVD are fed to the substrate through a showerhead mounted on the reactor ceiling, and pumped out from an exhaust opening on the bottom plate.

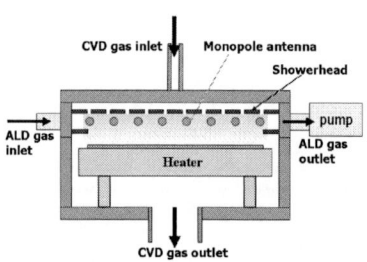

Fig.2 Schematic view of the ALD/PECVD reactor.

The specifications of the ALD/PECVD reactor are summarized in Table I . The reactor consists of a process chamber and a load-lock chamber, which can handle 370 mm× 470 mm glass substrates. The substrate is heated up to 450 ℃ by the hot plate-type heater in the reactor. The ALD interfacial layer and the PECVD secondary layer could be deposited continuously in the reactor.

The ALD gas cycle for SiO_2 deposition is composed of 4 steps, 1) Si source gas supply to make the source

Table I . Specification of the ALD/PECVD reactor.

Item	Specification
System component	Process chamber:1 Load-lock chamber : 1
Deposition formula	ALD system Monopole antenna PECVD system
Substrate size	370mm×470mm
Plasma source	Monopole antenna type 80MHz, 1.5kW
Exhaust System	Dry pump : 3 Turbo molecular pump : 3
Substrate temperature	Max. 450℃
Application	SiO_2

adsorbed to a surface of the substrate, 2) Inert gas supply to sweep out of the non-absorbed source, 3) Oxidizer supply to make the absorbed one oxidized, 4) Inert gas supply to sweep out of the excess and the reactant gases. We adopted Aminosilane and ozone as a silicon source and an oxidant. The following points are important for a large-scale ALD process: 1) constant gas supply to the reactor, 2) uniform gas distribution to the substrate, 3) rapid exhaust from the reactor. The developed reactor has a buffer tank, placed between the vaporizer and the reactor. The tank keeps the vapor pressure of the source gas constant, and the stable supply of the gas is realized. Typical amount of Aminosilane supply is 17 mg/cycle, and its variation was less than ±1% during 1000 times repetition.

One of the characteristics of the developed reactor is that the reactor has the monopole antenna plasma source. This plasma source is an electromagnetic wave coupled type, which works at the very high frequency (VHF). The length of the antenna is equal to quarter- wavelength to bring the monopole antenna into resonance. The electrical field strength is

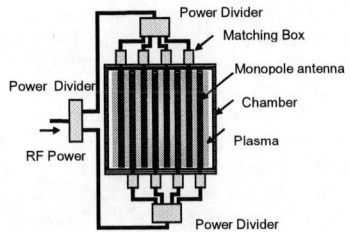

Fig.3 Monopole antenna arrangement in the reactor.

maximum at the tip of the antenna and minimum at the foot under the antenna resonance condition. The plasma density under an equilibrium condition is proportional to $\sin^2(2\pi \cdot x/4L)$, where L is the length of the monopole antenna. Figure 3 shows the monopole antenna arrangement, which is applied to the reactor. We obtained the uniformity of the plasma density in the direction along the antenna with alternating antenna arrangement. The number of the antennas is eight in the reactor as shown in Fig 3. Reflection wave is suppressed to be lower than 3% with the matching box within several seconds after ignition of the plasma.

As a fundamental study, we measured an electron temperature with a Langmuir probe to grasp the characteristic of the monopole plasma source. Two small reactors were fabricated for the study, one is with the monopole plasma source and the other is with CCP. The monopole plasma is excited under the condition of a forcing frequency of 130 MHz, an electric power of 15-30 W and a gas pressure of 27 Pa of Ar. The electron temperature in the monopole antenna plasma is shown in Fig.4 with triangle symbols. The temperature decreases with increasing distance from the antenna. On the other hand, the CCP has an almost constant temperature at the different distance as

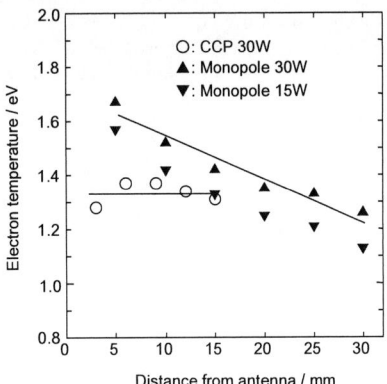

Fig.4 Electron temperature in the monopole antenna plasma and the CCP.

shown in Fig.4. The electron temperature of the monopole would be lower than that of CCP at the distance of ≥ 30 mm. Therefore the lower damage deposition can be expected in comparison with the CCP.

Film preparation

We used Si (001) wafers with resistivity of 1-10 Ω cm for measuring the electrical properties of the prepared films. The glass substrate has circular-shaped depressions on its surface for the Si wafers setting. The ALD film was deposited with alternating exposures of Aminosilane and 5% ozone. Typical exposure conditions of Aminosilane and ozone were 60 Pa $\times$ 1 s and 500 Pa $\times$ 2 s , respectively. The ALD gas cycle was repeated 200 times for the ALD film and 20 times for the ALD/PECVD film. The PECVD film was deposited with the monopole antenna plasma. Oxygen and TEOS gases were used as source gases. The substrate temperature, RF power, frequency, process pressure and antenna-substrate distance were 400℃, 1500 W, 80 MHz, 110 Pa and 55 mm, respectively.

Metal-oxide-semiconductor (MOS) capacitors were prepared to evaluate their electrical properties. As an electrode, aluminum was deposited on the SiO$_2$ films. Typical diameter of the Al was 400 μ m. Post-annealing process was conducted in a forming gas (10%H$_2$ + 90%N$_2$) at 400℃ for 30 minutes.

Results and Discussion

Thicknesses of the ALD, the PECVD and the stacked insulator were measured with an ellipsometer. The deposition rates of the ALD process were between 0.10 nm/cycle and 0.20 nm/cycle, and they varied with gas exposure conditions. A constant deposition rate of 0.1 nm/cycle was obtained with an Aminosilane exposure of 60 Pa $\times$ 1 s. The thickness of the ALD film increased linearly with increasing repetition time, and it could be controlled with a number of the ALD cycles. Figure 5 shows the ALD thickness distribution on the glass substrate and on the 6-inch Si wafer. Thickness variation was $\pm$ 5% on the 370 mm $\times$ 470 mm glass and $\pm$3% on the wafer. The source gases were fed to the substrate in horizontal direction. However, the thickness variation in the flow direction was small. Surface morphology of the ALD film prepared with 200 times

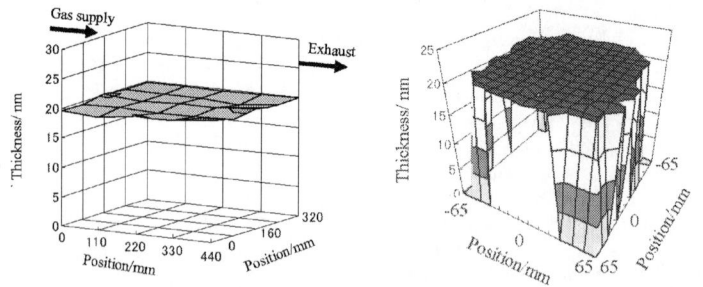

Fig.5a Thickness distribution of the ALD film on the 370mm $\times$ 470mm glass.

Fig.5b Thickness distribution of the ALD film on 6-inch Si wafer.

repetition was observed by AFM(Atomic Force Microscopy). A quite smooth surface was observed as shown in Fig.6 and its average roughness was 0.1 nm in an area of $1 \times 1 \ \mu \mathrm{m}^2$. Typical electrical breakdown field was 4 MVcm^{-1}, which was not enough for the insulator application.

The thickness of the PECVD film on the glass area was evaluated. Thickness was uniform on a line, which is parallel to the antenna, and not uniform in a line perpendicular to the antenna. However, variation on the glass was less than ±10%. Current-voltage performance of the 100nm-thick PECVD film was measured, and its electrical field at a leakage current of 1×10^{-6} Acm^{-2} was 7.5 MVcm^{-1}. The dielectric strength voltage was good enough for TFT application. Step coverage of the PECVD film was evaluated. We used 200nm-height polycrystalline Si stripes, which placed on a Si wafer with 50 nm intervals. The 17nm-thick PECVD film was deposited on the top of the Si stripes, then the thickness was 10.6 nm on the 100nm-deep side wall as shown in Fig.7. The calculated step coverage was 62%.

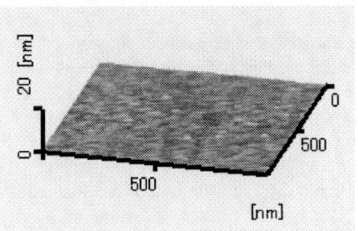

Fig.6 Surface morphology of the 20nm-thick ALD film.

Fig.7 TEM image of the 17nm-thick PECVD film deposited on the Si pattern.

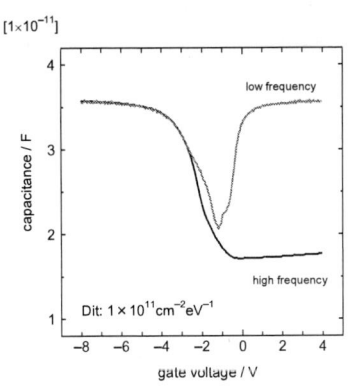

Fig.8 Capacitance-voltage performance of the ALD/PECVD stacked insulator.

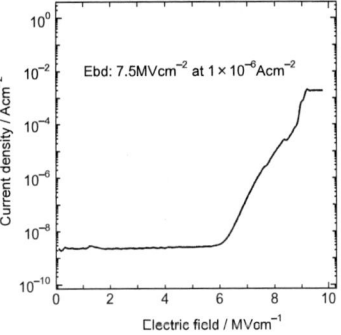

Fig.9 Current-voltage performance of the ALD/PECVD stacked insulator.

The ALD/PECVD stacked insulator was deposited with 2nm-ALD film followed by 100nm-PECVD deposition. Figure 8 shows a capacitance-voltage performance of the stacked insulator. Interface trap density was calculated by Hi-Lo method. The insulator had the interface trap density of 1×10^{11} cm^{-2}eV^{-1}, effective charge density of 4×10^{11} cm^{-2}, and a flat band voltage of -2.4 V. Figure 9 shows a current-voltage performance. The electrical field at a leakage current of 1×10^{-6} Acm^{-2} was 7.5 MVcm^{-1}. It was almost equal to that of PECVD film. The interfacial ALD film, which electrical breakdown field was 4 MVcm^{-1}, did not drop that of the stacked insulator. The stacked insulator has an excellent Si/SiO$_2$ interface and comparable dielectric breakdown strength to that of the PECVD. It is expected that the quality of the gate insulator would be improved with the newly developed reactor.

Conclusion

We developed the ALD/PECVD reactor, which handles glass substrate of 370mm $\times$ 470mm in size. The reactor has the unique plasma source of monopole antenna, which is the electromagnetic wave coupled type and operates at VHF range. The ALD film, the PECVD film and the staked ALD/PECVD SiO$_2$ film were deposited at 400℃ in the developed reactor, and the following performance of the reactor was confirmed.

1) The ALD SiO$_2$ film had a uniform thickness on the glass substrate and its variation was less than $\pm$5%. The constant deposition rate of 0.1 nm/cycle was obtained with the alternative exposures of Aminosilane and 5% ozone.

2) The PECVD film had a thickness variation of less than $\pm$10%, and its electrical field was 7.5 MVcm^{-1} at a leakage current of 1×10^{-6} Acm^{-2}.

3) The ALD/PECVD stacked insulator, which was composed of the 2nm-ALD film and the 100nm-PECVD film, had the interface trap density of 1×10^{11} cm^{-2}eV^{-1} and the electrical field of 7.5 MVcm^{-1}.

References

1. T. Takehara, T. Ide, F. Nakano, Y. Nakazaki, G. Kawachi and K. Azuma, *Proceedings of IDW'04*, p.513 (2004)
2. S. Higashi, D. Abe, Y. Hiroshima, K. Miyashita, T. Kawamura, S. Inoue and T. Shimoda, *Jpn. J. Appl. Phys.*, **41**, 3646 (2002)
3. T. Ide, A. Sasaki, T. Okamoto, K. Azuma and Y. Nakata, *Proceedings of IDW'03*, p.1677 (2003)
4. O. Sneh, M.L. Wise, A.W. Ott, L.A. Okada and S.M. George, *Surf. Sci.* **334**, 135 (1995)
5. K. Murata, N. Hattori, K. Washio, N. Miyatake, Y. Uraoka and T. Fuyuki, *Proceedings of AM-LCD 05*, p.109 (2005).

Room-temperature Sputter-deposited Gate SiO$_2$ Films
for High Quality Poly-Si TFTs

Tadashi Serikawa [†], Takamasa Miyamoto [‡], Hitoshi Ueno [‡], Yuta Sugawara [‡],
Yukiharu Uraoka [‡], and Takashi Fuyuki [‡]

[†] Osaka University, Mihogaoka 11-1, Ibaraki-shi, Osaka 567-0047, Japan

[‡] Nara Institute of Science and Technology, Takayama 8916-5, Ikoma, Nara 630-0192,
Japan

[†] E-mail: serikawa@jwri.osaka-u.ac.jp

Gate SiO$_2$ films in poly-Si TFTs were fabricated by sputtering at substrate temperature from 300℃ down to room-temperature. It was found that poly-Si TFT characteristics of mobility, threshold voltage and subthreshold slope were greatly improved by decreasing substrate temperature. Mobility of the poly-Si TFT for room-temperature was 90 cm^2/V·s, much higher than 50 cm^2/V·s for 300℃. Moreover, sputtered gate SiO$_2$ poly-Si TFTs showed high reliability against hot carrier effect even for room-temperature-deposition.

Introduction

Low temperature deposition of gate SiO$_2$ film in polycrystalline silicon thin film transistor (poly-Si TFT) is one of key issues to fabricate high quality poly-Si TFTs. Gate SiO$_2$ films usually have been deposited at substrate temperature higher than 200℃ by sputtering, PECVD, etc. (1,2,3,4). However, advanced flat panel displays, such as flexible displays essentially necessitate very low temperature fabrication of poly-Si TFTs (5,6). In this study, we have fabricated poly-Si TFTs with SiO$_2$ films sputter-deposited at substrate temperature down to room temperature and characterized them.

SiO$_2$ film Depositions and Poly-Si TFT Fabrication

SiO$_2$ films were deposited by radio-frequency, planar magnetron sputtering at various substrate temperatures. The target and sputtering gas used were high quality quartz (99.999%) and 30% oxygen-70% argon mixture (1,4).

Coplanar structure poly-Si TFTs were fabricated on glass substrate using sputter-deposited gate SiO$_2$ films. Precursor Si films 50 nm thick were deposited at 200℃ substrate temperature by radio-frequency planar-magnetron sputtering, where argon gas was used as sputtering gas (6,7). As-deposited a Si films were crystallized by furnace annealing at 600℃ for 10 hrs. After patterning the poly-Si films by plasma etching technique, 100 nm thick gate SiO$_2$ films were sputter-deposited at substrate temperatures of room-temperature, 100℃, 200℃ and 300℃. Gate electrodes were patterned, then source and drain regions were formed by self-alignment technique using ion-implantation of phosphorus. The ion-implanted phosphorus atoms were activated by annealing at 600℃. After aluminum metallization, poly-Si TFTs were hydrogenated by furnace-annealing at 400℃ for 20 minutes in forming gases to passivate defects at grain boundaries and at interfaces of poly-Si /gate SiO₂ films.

Results and Discussion

Figure 1 shows transfer characteristics of poly-Si TFTs with channel lengths of 18 μ m, 9 μ m and 3 μ m for substrate temperatures of room temperature, 100℃, 200℃ and 300℃. The figures clearly show that drain currents depend on substrate temperature and saturated drain current for room-temperature are highest for all channel lengths. The figures also imply that poly-Si TFT characteristics of mobility, threshold voltage and subthreshold slope depend on deposition temperature of gate SiO₂ films.

Figure 2 shows mobility dependences on gate voltage for channel lengths of 18 μ m, 9 μ m and 3 μ m as a parameter of substrtae temperature. The figures clearly indicate that mobility greatly depend on gate SiO₂ film deposition temperature. Mobility for room-temperature is 90 cm²/V·s, much higher than 50 cm²/V·s for 300℃ deposition temperature.

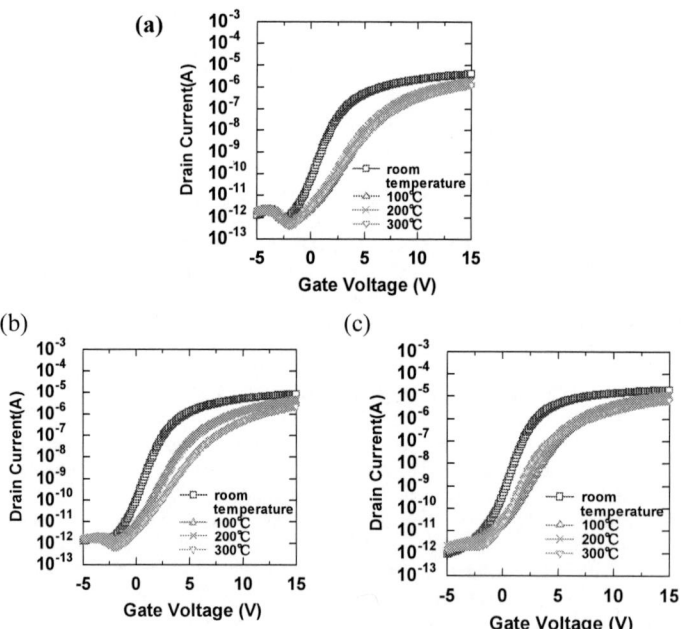

Fig.1: Transfer characteristics of poly-Si TFTs with channel lengths of 18 μ m (a), 9 μ m (b), and 3 μ m (c) as a parameter of substrate temperature.

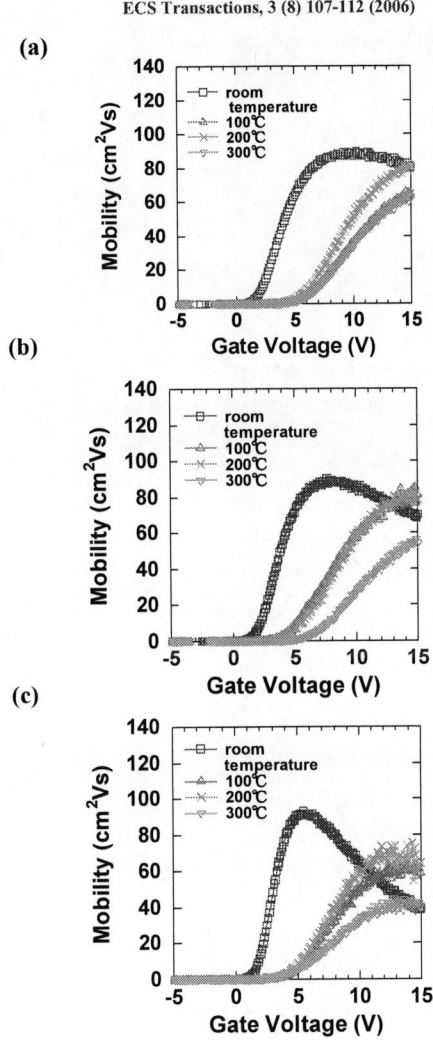

Fig.2 : Mobility changes by gate voltage for poly-Si TFTs with channel lengths of 18 μ m (a), 9 μ m (b), and 3 μ m (c) as a parameter of substrate temperature.

As shown in Fig.1, threshold voltage and subthreshold slope are also influenced by deposition temperature of gate SiO$_2$ film. Figure 3 shows dependences of threshold voltage (Vth) and subthreshold slope on channel length for various substrate temperatures. Both subthreshold voltage and subthrshold slope are greatly improved by decreasing deposition temperature down to room-temperature for all channel lengths.

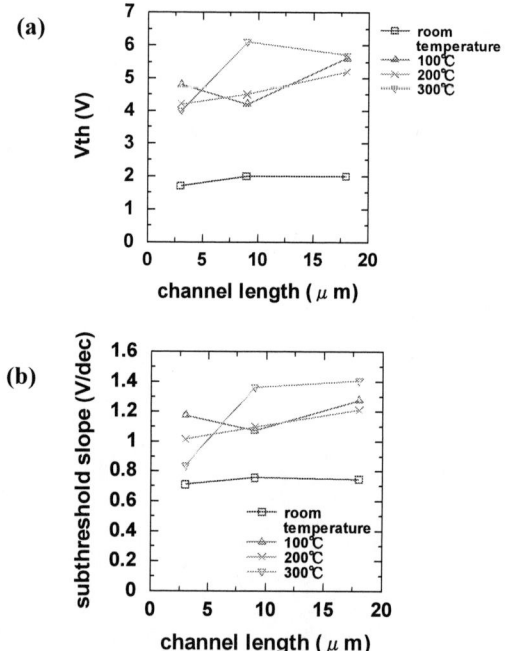

Fig.3: Dependences of threshold voltage (a) and subthreshold slope (b) on channel length in poly-Si TFTs for substrate temperature from room-temprerature to 300℃.

To investigate reliability of sputtered gate SiO_2 film poly-si TFTs, electrical DC stress was imposed on the poly-Si TFTs. The DC stress conditions were 15 V of drain voltage and fixing of gate voltage at threshold voltage. Figure 4 shows changes of transfer characteristic by DC stress time for poly-Si TFT with gate SiO_2 film deposited at room temperature. The changes by the stress are very small. Moreover, as shown in the figure, leakage drain currents at off-state in poly-Si TFT are maintained at low level. Figure 5 shows changes of mobility, threshold voltage and subthreshold slope by DC stress time as a parameter of SiO_2 film substrate temperature, compared with controlled poly-Si TFT fabricated by conventional method. The controlled poly-Si TFT shows typical degradation phenomena of mobility, threshold voltage and subthreshode slope. However, sputtered poly-Si TFTs show small degradation for all substrate temperatures. These results means that sputtered gate SiO_2 poly-Si TFTs have high immunity from hot carrier effect even for room-temperature deposition.

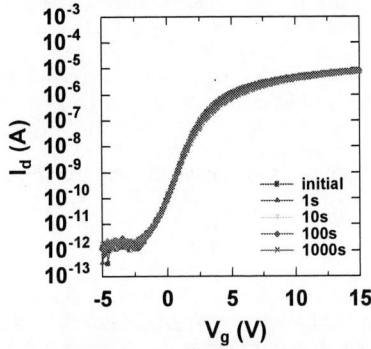

Fig.4: Changes of transfer characteristics by stress time for poly-Si TFT with room-temperature deposited gate SiO2 film poly Si TFT.

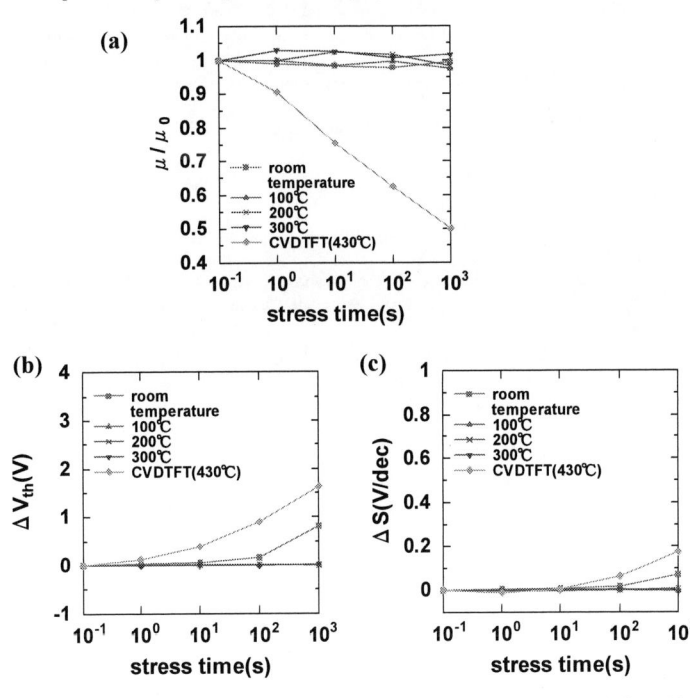

Fig. 5: Changes of mobility (a), threshold voltage (b), and subthreshold slope (c) by dc stress time for sputter-poly-Si TFTs and CVD-poly-Si TFT.

Insulator properties of sputtered gate SiO_2 films were measured by applying positive voltage on gate electrode against source electrodes in poly-Si TFT. Measurements indicated that the SiO_2 films had high breakdown field of over 7 MV/cm and leakage current as low as 2 x 10^{-8} A/cm^2. These insulator properties are almost same as high-temperature thermally grown SiO_2 film on single crystalline Si substrate, in spite of deposition at low temperature. It also indicates excellent step-coverage property of sputtered SiO_2 film that thick and high quality SiO_2 films are deposited even on walls at edges of polycrystalline Si patterns. Moreover, it was confirmed that deposition rate and optical properties of sputtered SiO_2 films are almost independent on substrate temperature differing from those by conventional CVD method.

Conclusions

Gate SiO_2 films were sputter-deposited at low substrate temperature down to room-temperature. It was first clarified that device-graded gate SiO_2 films were deposited at very low substrate temperature by sputtering. Measurements also showed that poly-Si TFT characteristics of mobility, threshold voltage and subthreshold slope were greatly improved with decreasing substrate temperature to room-temperature. Degradations of poly-Si TFTs by hot carrier effect were very small comparing with conventional CVD poly-Si TFTs. Moreover, the SiO_2 films had excellent insulator properties. Then, sputter-deposited SiO_2 films are very powerful for fabricating advanced flat panel displays at low substrate temperature.

References

1. S.Suyama, A.Okamoto, S.Shirai and T.Serikawa, Proc. Mat.Res.Soc. Symp.,**146**, 301 (1989).
2. G. K. Guist and T. W. Sigmon, IEEE Trans. Electron Devices, **47**, 207 (2000).
3. T.Serikawa et al., AM-LCD'05 (Kanazawa, Japan July 6-8,2005).
4. T.Serikawa, et al,.Jpn.J.Appl.Phys., **45** , 4358 (2006).
5. D. P. Gosain, T. Noguchi, and S. Usui, Jpn. J. Appl. Phys.**, 39**, L179 (2000).
6. T.Serikawa and F.Omata, IEEE Trans. Electron Devices, **49**, 820 (2002).
7. T. Serikawa, S. Shirai, S. Suyama and A. Okamoto, IEEE Trans. Electron Devices, **ED-36**, 929 (1989).

ECS Transactions, 3 (8) 113-118 (2006)
10.1149/1.2356343, copyright The Electrochemical Society

Novel Characterization Technique for Oxidation Processes

Yu.V. Sokolov

Fairchild Semiconductor Corp., West Jordan, Utah 84088, USA

A new application of the commercially-available Surface Charge Profiler (SCP) tool to evaluate metal ionic contaminants that can be brought in during device fabrication processes, including oxidation, is discussed. The SCP measures the surface photovoltage (SPV) signal, and associated depletion width (W_d) and the minority-carrier recombination lifetime (τ). The technique allows not only detect a risky level of the contaminants but also to distinguish between different impurities such as Fe, Cu, and Na. A behavior of these metals in the vicinity of Si-SiO$_2$ interface is discussed. A novel characterization technique should be a beneficial alternative metrology to the conventional capacitance versus voltage (CV), triangular voltage sweep (TVS), and SPV methods.

INTRODUCTION

The lifetime measurement technology is currently recognized for clean control in the device manufacturing processes. Process defects, most commonly heavy metals, cause lifetime degradation in those regions where they occur. The SCP provides some benefits (1) as compared with other light-induced SPV techniques, which are usually used for Fe contaminant, as well as with CV / TVS technique, which is used for detecting Na contaminant. First, the SCP tool detects contaminants within about 1 μm of the surface, in contrast to other SPV techniques, which measure the substrate bulk and are relatively insensitive to surface-charge dynamics. Second, the SCP is non-destructive tool, and does not require an MOS structure as in CV / TVS methods. In addition, the computed outputs are believed to represent the properties of the surface that are not modified in the course of the measurement itself, due to application of the low intensity light.

In prior study (2,3) using SCP technique, we showed, how metal contaminants affect the surface property of bare silicon. The distinctive variations of W_d and τ at the intentionally contaminated surface make known the behavior and the redistribution of the electrically active metal impurities such as Fe, Cu, Ni, and Na in subsurface. The research, in turn, allowed us to introduce a new approach to monitor metal ions in the real manufacturing condition. Moreover, it was shown that different metals can be distinguishable by using a re-association kinetics approach. The lifetime degradation during time delay after the heating was used to study relaxation kinetics of the impurity in p-type silicon. In order to positively identify the metal, however, elaborate and time consuming time relaxation experiments are required. This type of testing, though reliable, cannot feasibly be implemented into a production environment. Lately, the SCP was shown to be successfully used as metrology to qualify oxidation processing (4).

113

This work resumes a new approach to detect typical electrically active metal impurities, which happen to occur in silicon dioxide. As for particular Na contaminant, SCP is used as an alternative tool to the conventional CV method.

EXPERIMENTAL

The SCP tool supplied by QCSolutions Inc. performs process-monitoring functions without touching the wafer surface, or affecting the condition of the surface in any other way. The tool enables the measurement of the near surface doping type, the width of the surface charge region, the doping concentration/resistivity, and the minority-carrier recombination lifetime. The SCP also produces a wafer map for each of these parameters.

The SCP generates SPV signal that is associated with positive charges, which exist on the surface of bare p-type silicon, or which always present at the Si-SiO_2 interface. The measurement is performed by flashing a light onto the wafer, and evaluating the real and imaginary components of the returning ac-SPV signal to derive values for two critical parameters: W_d and τ. The tool uses the resulting W_d value to compute a value for the dopant concentration, N_{sc}, and the space charge, Q_{sc}, in the substrate beneath the oxide.

The positive charges occurring in the oxide are balanced by the negative space charges in the depleted subsurface. Any foreign, electrically active atoms change the magnitude of the SPV signal, and accordingly, change the value of depletion width W_d and the lifetime.

To study the behavior of metals in the vicinity of Si-SiO_2, thermal oxide was grown up to 96 nm thick on silicon wafers doped with boron at about $1E15/cm^3$, and then the oxide was intentionally contaminated with Fe, Cu, and Na anions. The contaminants were brought in via two methods: by implantation at different calibrated doses, and by wetting the test wafers in the metal standard solutions. Metals were implanted into SiO_2/Si to a depth of about 10 nm from the SiO_2 surface. Na^+, Fe^+, and Cu^+ were introduced into the oxide by immersing test wafers into NaOH (50%) solution and into standard Fe and Cu solutions (iron / copper in dilute nitric acid, $1000\mu g/mL$). After keeping oxidized wafers in the solution for a certain time, wafers were dried in the oven at $150^{\circ}C$. This bake should also drive metal ions into oxide at a depth that is proportional to baking time. The wafers, which were oxidized in Na- and Fe-contaminated furnace, were also investigated.

It was theorized that sodium ions can be passivated by UV illumination. UV light (the dominant line $\lambda = 254$ nm), which is one of the functional operations in the SCP tool, was applied to the oxidized wafers. The values of W_d and τ were measured before and after oxidation, before and after contamination, as well as before and after UV treatment.

RESULTS AND DISCUSSION

Test wafers with doping concentration, N_{sc}, equal to $1E15$ cm^{-3} or less has been found to be most appropriate. A heavier doped substrate causes W_d to be too shallow and the derived space charge values to become questionable. W_d and τ values were measured in the substrate before and after thermal oxidation. The combination of τ, and the percent change of W_d after thermal oxidation serve as critical parameters for oxide qualification.

The quality of SiO_2 is characterized by the interfacial and total oxide charges. They invert the lightly doped silicon subsurface inducing a depletion region beneath the oxide. The depth of the surface charge region will depend on the dopant concentration in the substrate, the amount of charge at the interface, and any charges in the oxide. First, we need to know the expected post-oxidation depletion width, W_{d-post}, for ideal, non-contaminated $Si-SiO_2$ by building the calibration curve $W_{d\ post}$ versus initial doping concentration in silicon, N_{sc}, which is, in its turn, derived from the pre-oxidation depletion width, W_{d-pre}. This calibration curve allows us to predict, with 10 percent confidence, the expected post-oxidation depletion width, W_{d-post}, versus initial N_{sc}. For the perfectly grown and uncontaminated oxide layers, the typical value of τ is 30 to 40 μsec, and the W_d value is $0.8 \pm 0.1 \mu m$ for an N_{sc} of about $1E15\ cm^{-3}$. Any departure of these parameters was considered evidence of oxide imperfection and/or a contamination occurrence.

Contaminants introduced during oxidation contribute to the interface charge state and/or induce image charge to the substrate, thereby changing W_d. Monitoring the percent change of the depletion width, $\Delta W_d/W_{d-pre}$, enables us to exclude variation in initial doping concentration as a component of the total variation in the monitor. Wafer-to-wafer variation in $\Delta W_d/W_{d-pre}$ is attributed solely to variation in interface and/or oxide charges. Another benefit of monitoring the percent change in depletion width as opposed to the post-oxidation W_d alone, is that it is conveniently insensitive to the within-wafer non-uniformity of W_d that is common in oxidized wafers.

Minority-carrier recombination lifetime is another critical parameter that significantly affects the presence of a contaminant. An abnormally low τ indicates the presence of extra recombination centers in the vicinity of the interface. Therefore, degradation of the lifetime value, coupled with a reduced W_d, as compared with pre-oxidation performance, is a signature of electrically active metal contaminants.

The extent of the lifetime value degradation and W_d narrowing should depend, not only on the impurity concentration, but also on the location of the foreign ions in the oxide film. If metal ions occupy the oxide surface as in the implanted case, it changes the capacitance of the oxide and redistributes the interface charges that influence W_d and τ in one way. If metal ions introduced into the oxide occupy the interface that is more common in practice, this causes W_d and τ to change in another way. Due to the variety of possibilities for contaminant location, and their differing effects on W_d and τ, final values for τ and W_d are not used to quantitatively deduce a value for contaminant concentration. However, they *can* be used to monitor process conditions and detect relative abnormalities.

Fe, Cu, and Na contaminants in SiO2 without UV treatment

The lifetime, depletion width, and space charge in silicon undergo significant changes after the ions of Na^+ or Fe^+, or Cu^+ occur in the SiO_2. The lifetime and W_d, which are measured in silicon beneath the oxide, decrease with regard to the impurity concentration.

The prior analysis of systematic change of W_d and τ, when metal present in the oxide, allowed to study the behavior / distribution of typical metal contaminants during oxidation and after annealing of oxidized wafers (3). Iron has a strong interaction with the thermal SiO_2 layer that is a trap for iron. Iron impurity has a strong tendency to segregate into silicon dioxide or, at least, into the $Si\text{-}SiO_2$ interface while the oxide is growing. About 75 % of iron is known to remain in the oxide layer even after high-temperature annealing at $1150^\circ C$ (5). Cu atoms have a tendency to diffuse almost totally into the p-type silicon bulk (3,5), while expected to pile up at the n-type Si/SiO_2 interface during annealing (6).

Sodium ions concentrate nearby the $Si\text{-}SiO_2$ interface. This was confirmed by both spectroscopic analysis and SCP measurements. It is interesting to note that the extra amount of Na^+ implanted into SiO_2/Si, which then were processed through oxidation, has been revealed neither by the SCP nor by the CV / TVS measurements. This means that Na^+ mobile ions diffuse out of space charge region during oxidation.

Annealing at temperatures as low as about $200^\circ C$ enables the increase of W_d and τ and the decrease of the space charge. This is explained by redistribution of metal ions in the oxide and possible neutralization of metal charges due to their reaction with oxide at elevated temperature. The SCP parameter's maps verify such re-distribution.

Fe, Cu, and Na contaminants in SiO2 after UV treatment

There is a significant change in τ and W_d after Na- and Fe-contaminated wafers undergo UV treatment. The lifetime increases up to the initial, "uncontaminated" level, and W_d becomes shallower. The closer sodium contaminants to the $Si\text{-}SiO_2$ interface, the stronger the effect. Neither parameter changes noticeably during several weeks after UV treatment. Lifetime has a positive correlation with depletion width after UV. It is noteworthy that the depletion width allocating across the wafer constitutes the specific pattern in form of regular figures of "diamond" shape. These figures gradually vanish from W_d map with time. The repeated UV treatment causes Wd to depress.

Our results mainly support the findings of other authors (7) who suggested that Na^+ ions are neutralized by electrons induced by UV light. The UV light excites electrons in silicon, which then are injected into SiO_2, and some of them may be trapped by Na^+. The effect is remarkably stable: the heat treatment at $300^\circ C$ / 60 minutes does not eliminate the passivation effect. Their results imply that the sodium atoms, neutralized during UV treatment, do not release electrons even throughout the bias-temperature test.

Copper contamination, in contrast to Na and Fe, does not show the lifetime improvement after UV illumination. This confirms the early conclusion that Na and Fe remain in the interface or above, whereas Cu diffuses out of oxide elsewhere below surface charge region.

Iron ions present in the oxide are affected by UV light in the same manner as sodium ions. That is why Fe^+ contaminants, unfortunately, can not be directly differentiated from Na^+. Furthermore, the oxidized and non-contaminated on purpose wafers also result in increase of τ after UV treatment. This property is peculiar to both wet and dry oxides. The thinner oxide is, the more essential this effect. For wet oxide 100 nm thick, when UV absorption becomes considerable, about 50% of non-contaminated test wafers show improvement of τ after UV. This may be explained by different oxide quality (imperfections, interfacial charges, metal impurities, etc.) that varies from one oxidation run to another. This ambiguity makes oxide qualification via UV treatment problematic.

Nevertheless, an attempt to monitor contaminants in the test wafers in-line, using UV illumination, shows the meaningful result (Fig.1). Wafers with high τ values, which are above 25 μsec, are not a concern of contamination; they are of good quality. Wafers with initial low τ may be of two sorts: those which recover after UV treatment (wafers 2,4,6), and which do not (wafers 1, 5). The former part may be a suspect for metal contamination; the latter is speculated as merely an indication of poor substrate or oxide quality (oxygen precipitates, oxide imperfections, etc.). The UV application used for qualification of oxidation process thus deserves further study.

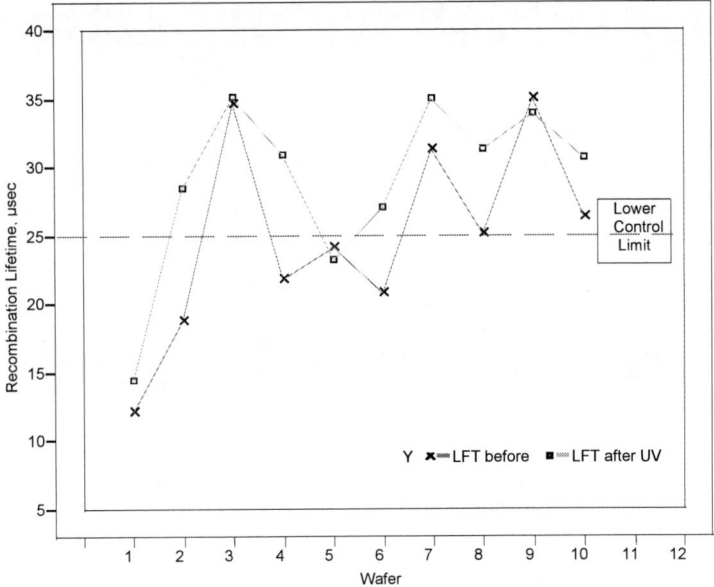

Figure1. The recombination lifetime of oxidized wafers measured before (crosses) and subsequently after (squares) UV treatment

Monitoring metals in the production line

The SCP technique was applied for the first time to qualify the pre-process furnace condition (3) as well as to detect metal contaminations introduced during oxidation process (4). It was also used to measure oxidation process improvement.

To qualify oxidation furnaces, the test wafers were pre-measured for τ and W_d, and then placed into a furnace at $700^\circ C$ for 30 seconds. Only N_2 flows into the furnace. After being removed from the furnace, the wafers were cooled and re-tested. The metal, most commonly Fe, if present in a furnace, is visualized in form of spot(s) on the SCP map. Contamination spots are of different size and amount depending on severity of the contamination. The best qualification parameter turns out to be not the lifetime mean itself but the lifetime standard deviation. In the case of localized imperfections, where τ degradation is significant but the overwhelming part of the wafer surface has still a small value of τ, an average mean of τ does not change strongly. However, the τ standard deviation is changed expressly that can be used as a qualitative qualification parameter. The ROST technique at SCP (3) is used for this kind of testing.

The SCP was also used to qualify thermo furnaces performing oxidation test (4). The pre-testing wafers were placed in a furnace for regular oxidation. After being oxidized, the values of τ and $\Delta W_d/W_d$ are measured and analyzed for indication of contaminants. Data with $\tau < 25$ µsec and $\Delta W_d/W_d < 0$ are suspected for contaminated wafers.

The new characterization technique saves an enormous amount of furnace qualification time. This non-destructive method enables detection of metal impurities in both bare silicon and oxidized test wafers. The SCP monitoring does not require MOS preparation, and it can be used as an alternative to conventional CV / TVS methods detecting mobile ions contaminants. The SCP provides even greater W_d /N_{sc} /Q_{sc} measurement accuracy that the C-V approach. The SCP testing can be used to improve all processes that are sensitive to the properties of the $Si-SiO_2$ interface.

ACKNOWLEDGMENTS

The author is grateful to Herb Robertson and Kathy Anderson from Fairchild Semiconductor Corp. for their encouragement and active contribution to this work, and also to Edward Tsidilkovski from QCSolutions, Inc for many useful discussions.

REFERENCES

1. E. Kamieniecki, PV 99-16, The Electrochemical Society Proceedings Series, (1999).
2. Yu.V. Sokolov, Solid State Technology, September 2004.
3. Yu.V. Sokolov, International Semiconductor, online exclusive issue, August 2005.
4. Yu.V. Sokolov, and K. Anderson, Micro, March 2006.
5. M. Hourai et al., Jap.J.Appl.Phys.,**28**, 2413 (1989).
6. A.L.Rotondaro et al., J.Electrochem.Soc., **143**, 3014 (1996)
7. M. Itsumi at al., J.Electrochem.Soc., **143**, 3359 (1996)

ECS Transactions, 3 (8) 119-124 (2006)
10.1149/1.2356344, copyright The Electrochemical Society

Deposition of Highly Crystallized Poly-Si Thin Films on Polymer Substrates Using Pulsed-Plasma CVD under Near-Atmospheric Pressure

M. Matsumoto[a], M. Suemitsu[a], T. Yara[b], S. Nakajima[b], T. Uehara[b], Y. Toyoshima[c], and S. Itou[d]

[a] Center for Interdisciplinary Research, Tohoku University,
Aramaki aza-Aoba, Aoba-ku, Sendai 980-8578, Japan
[b] Sekisui Chemicals Co. Ltd,
2-3-17 Toranomon, Minato-ku, Tokyo 105-8450, Japan
[c] Energy Technology Research Institute, AIST,
1-1-1 Umezono, Tsukuba 305-5568, Japan
[d] Institute for Materials Research, Tohoku University,
2-1-1 Katahira, Aoba-ku, Sendai 980-8577, Japan

Growth of highly crystalline poly-Si films on polyethylene terephthalate (PET) substrates at 150 °C has been achieved with practical growth rates by using pulsed-plasma CVD under near-atmospheric pressure. The precursor is SiH_4 diluted in H_2, and no inert gases such as He were used. A short-pulse based power system has been employed to maintain a stable discharge in the near-atmospheric pressures. By using this technique we have succeeded in growth of poly-Si thin film on the glass substrate at 180 °C with virtually no incubation layers. Good crystallinity of the poly-Si thin films were observed by Raman scattering spectroscopy and cross-sectional transmission electron microscopy.

Introduction

Si thin film formation on noncrystalline substrates, typically on glass, is widely used in macroelectronic devices such as solar cells and thin film transistors (TFTs) in flat-panel displays. If technology is established to grow crystalline Si on polymer substrates, it may exhibit an extraordinary impact on macroelectronic devices. The light weight and the flexibility of the plastic films will greatly enhance the applications of macroelectronic devices into our daily life. In the near future, plastic substrates, by greatly reducing the weight of the display, will alleviate the problem of the panel breakage (1).

A crucial challenge, however, is the fact that conventional growth techniques, such as low-pressure chemical vapor deposition (CVD), normally require temperatures that are much higher than the melting points of most polymer materials (>300°C). Several works have been reported on low temperature growth of Si on polymer substrates. McCormick *et al.* report formation of amorphous Si (a-Si) TFTs on polyethylene terephthalate (PET) substrates at 125 °C using dc reactive magnetron sputtering (2). Theiss *et al.* report formation of polycrystalline Si (poly-Si) TFTs on polymer materials by using postdeposition laser crystallization of a-Si, demonstrating the device properties superior to those of a-Si TFTs (1,3). For practical device fabrication processes, however, plasma enhanced chemical vapor deposition (PECVD) technique may be much preferred over sputtering or laser processing because PECVD in parallel plate configuration is suitable for large area device fabrication and thus commonly employed for a-Si based solar cells and TFTs. Fabricating device-compatible a-Si on polymer substrates directly by PECVD

119

is thus a matter of great interest for many researchers (4,5).

It is well known that, under H_2 dilution conditions, PECVD can also provide the poly-Si growth, with a couple of problematic features. One of them is low growth rate. The other is the formation of interfacial amorphous layer (called incubation layer) on heterogeneous (glass, typically) substrates, which can be partly conquered by further increasing the H_2 dilution, with a further sacrifice of the low growth rate (6).

We have succeeded in incubation-free high rate growth of poly-Si thin film on glass substrates by use of pulsed discharge-based PECVD operated at near atmospheric pressure (7). A key feature in our growth technique is simultaneous achievement of quite high H_2 dilution, that favors the nucleation process and thus results in incubation-free growth, while keeping the high growth rate by high density supply of growth precursors, compared to conventional low-pressure PECVD, by operating the discharge at near atmospheric pressures. In addition to the general expectation of reducing the production costs by utilizing the atmospheric processes, our pulse-based discharge operation needs no further dilution by inert gases (typically He), that are often employed in atmospheric discharge operations. In this paper, we report a high-rate growth of poly-Si on polymer substrates using our pulsed-plasma CVD under near atmospheric pressure.

Experiments

Figure 1 shows the schematic diagram of the apparatus. The process pressure (300, 500, and 700 Torr) was controlled by a balance between the hydrogen flow rate (1500 ml/min) and the evacuation rate. The discharge plasma was generated by applying a pulsed electric bias (typically 9.2kV, 30 kHz, pulse width 5 µs) onto the hot electrodes placed 1-mm apart from the substrate attached to the grounded electrode. Si films are grown on 0.2-mm-thick polyethylene-terephthalate (PET) flexible substrates. After pure water rinsing for ten minutes, the PET substrate was introduced into the chamber, and its surface impurities were etched off using a H_2 plasma treatment for five minutes. The Si growth was then started by mixing SiH_4 flow (0.0-5.0 ml/min) into the H_2 flow. The substrate temperature was varied for RT-200 °C. The grown film was characterized by using Raman scattering spectroscopy and cross-sectional transmission electron microscopy.

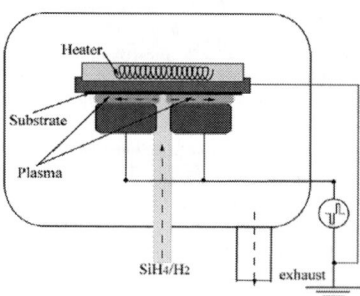

Fig1. A schematic diagram of the apparatus. The discharge plasma is generated by applying a pulsed electric bias on the hot electrode pair, located opposite to the substrate on the grounded electrode. The H_2/SiH_4 mixture gas flows through the gap between the hot electrodes and the substrate.

Results and discussion

Raman scattering spectroscopy

Figure 2 shows a Raman scattering spectrum from a film grown at 150 °C and 500 Torr. The Raman spectrum is deconvoluted into three Gaussian components: the crystalline component at 520 cm^{-1}, the nanocrystalline (<10 nm) component at 510 cm^{-1}, and the amorphous component at 480 cm^{-1} (8,9). The area-based ratio of the crystalline components at 520 and 510 cm^{-1} to the total band intensity was obtained for a measure, actually a lower limit, of the crystal volume fraction (10). A value as high as 0.83 is obtained for this film, indicating that a major part of the film is crystallized on this PET substrate.

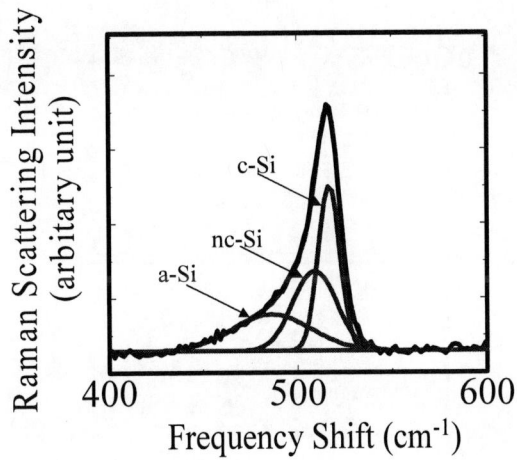

Fig.2 Raman scattering spectrum of the Si film grown on a PET substrate at 150 °C and 500 Torr.

In Figure 3, the crystalline fraction of the films obtained on PET substrates is displayed. Figure 3(a) shows the silane-flow-ratio R=SiH$_4$/(H$_2$+SiH$_4$) dependence of the crystalline fraction grown at 150 °C and 500 Torr. At lower flow ratio, no film growth was observed (denoted as "no growth" in the Figure). Highly crystallized films are obtained in the medium region ("poly-Si"). At flow ratio of 0.1 %, a maximum crystallinity is obtained. The crystalline fraction degenerates with increasing the flow ratio and results in complete "amorphous" at flow ratio of above 0.5 %.

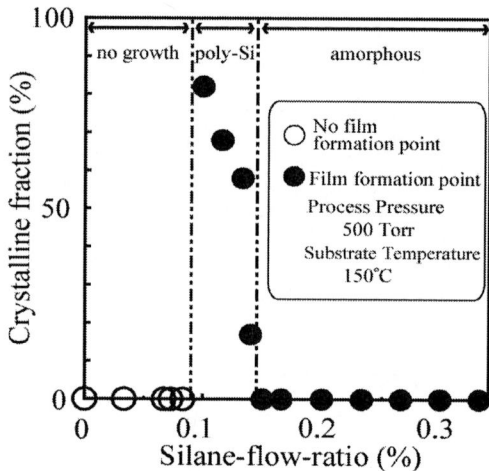

Fig3(a). Silane-flow-ratio $R=SiH_4/(H_2+SiH_4)$ dependence of the crystalline fraction for growth at 150 °C and 500 Torr.

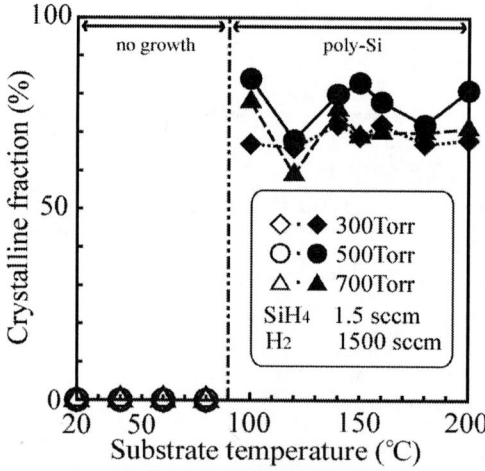

Fig3(b). substrate-temperature dependence of the crystalline fraction. The silane flow ratio R was fixed at 0.1 %

Figure 3(b) shows the temperature dependence of crystalline fraction with varying the total pressure at a fixed flow ratio of 0.1 %. Above 100 °C "poly-Si" films are obtained with no significant temperature dependen   anwhile, no film growth was observed for temperatures below 100 °C. It is known that hydrogen atoms generated in H_2 plasma can etch Si with negative temperature dependence (etch rate is higher at lower temperature) (11). No film growth below 100 °C is explained by this negative temperature dependence of etch rate, assuming a competition of growth by silane-related radicals to etching by hydrogen atoms in our reaction conditions. No film growth at lower flow rate observed in Fig. 3(a) can also be understood accordingly.

Cross-Sectional Transmission Electron Microscopy

Observations by cross-sectional transmission electron microscopy (X-TEM) were performed to confirm the crystalline nature of the poly-Si films. Figure 4 shows the typical X-TEM images of the film grown at 150 °C, 500 Torr and silane-flow-ratio of R=0.1 %.

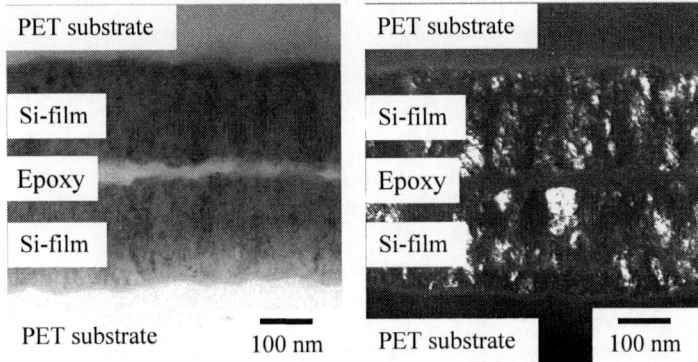

Fig.4 X-TEM bright (left) and dark (right) field images of the film grown at 150 °C. Pieces of sample films are stuck together at top surface side by epoxy cement due to observation convenience. The bright parts seen in dark field image are crystallites aligned to the <111> directions.

From these images, the thickness of the film is estimated to be about 200 nm, which corresponds to a growth rate of 0.75 nm/s. This growth rate is generally higher than those obtained in conventional PECVD.

A distinct gray zone with a thickness of about 10 nm, especially in the dark field image, is clearly observed at the interface between the poly-Si film and the PET substrate. This zone is a silica coating layer, which is often employed in these kinds of polymer films for gas barrier. The crystallites in the dark field image are almost uniformly distributed in the entire Si film, indicating that initiation of the crystalline growth is almost immediate just on the polymer substrate (actually on silica coating, to be precise), as we have already achieved on glass substrates. This is one of the major advantages in our growth technique since it is generally noticed that incubation layer exists in conventional PECVD-grown Si films in its initial part before poly-Si starts to grow.

The existence or absence of incubation layer is related to some specific features of PECVD. According to Kondo *et al.* the inc⋅⋅⋅⋅⋅⋅⋅ layer appears because the positive ions, summoned by negative surface charges, obstruct the crystal growth (6). After the incubation layer is formed to some thickness, establishing enough conductivity for the negative charges accumulated in the growing film surface to release, a crystalline film is now able to grow. Based on this scheme, reduction of positive ions impinging onto the growing surface would reduce the incubation layer thickness. In our plasma operation at atmospheric pressure the plasma is in the viscous-flow regime, whose numerous collisions may contribute to suppress the positive ion impingement onto the surface. The high dilution with hydrogen can also boost up the nucleation process. Pulsed operation of discharge may have some additional benefits on positive ion behavior since there are no steady-state plasma potentials that are likely to push positive ions away to electrodes and sidewalls (and also to growing surface) because the turn-off period is much longer than discharge period in our setup.

In summary, by using a pulsed-discharge based PECVD operated at near atmospheric pressures, we obtained highly crystalline polycrystalline Si films on polymer substrates (PET films). The highly crystallized features are shown by Raman scattering spectroscopy and cross-sectional transmission electron microscopy for the film grown at 150 °C.

References

1. S. D. Theiss, P. G. Carey, P.M. Smith, P. Wickboldt and T. W. Sigmon, *IEEE International Electron Devices Meeting,* **98**, 257 (1998).
2. C. S. McCormick, C. E. Weber, J. R. Abelsona and S. M. Gates, *Appl. Phys. Lett.,* **70**, 13 (1997)
3. N. D. Young, G. Harkin, R. M. Bunn, D. J. McCulloch, R. W. Wilks, and A. G. Knapp, *IEEE Electron Device Letters,* **18**, 19 (1997)
4. C.-S. Yang, L. L. Smith, C. B. Arthur, and G. N. Parsonsa, *J. Vac. Sci. Technol.,* **18**, 683 (2000)
5. H. Gleskova and S. Wagner, *IEEE Electron Device Letters,* **20**, 473 (1999)
6. M. Kondo, Y. Toyoshima, and A. Matsuda, *J. Appl. Phys.,* **80**, 6061 (1996)
7. H. Kitabatake, M. Suemitsu, H. Kitahata, S. Nakajima, T. Uehara and Y. Toyoshima, *Jpn. J. Appl. Phys.,* **44**, L683 (2005).
8. Z. Iqbal and S. Veprec, *J. Phys. C.,* **15**,377 (1982).
9. Hua Xia, Y. L. He, L. C. Wang, W. Zhang, X. N. Liu, X. K. Zhang, and D. Feng, *J. Appl. Phys.,* **78**, 6705 (1995)
10. L. Houben, M. Luysberg, P. Hapke, R. Carius and F. Finger, *Philos. Mag.,* **77**, 1447 (1998)
11. Y.Toyoshima, K.Arai, and A.Matsuda, *J. Non-cryst. Solids,* **114**, 819 (1989).

CHAPTER 4

POLY-Si TFTs FROM LASER CRYSTALLIZATION PROCESSES

Nanosecond Monitoring of Lateral Crystallization
Dynamics induced by ELA

Y. Takami[a], T. Warabisako[b], M. Matsumura

Advanced LCD Technologies Development Center Co., Ltd. 292 Yoshida-cho, Totsuka-ku, Yokohama, 244-0817, Japan

In this paper we report the direct observation of the lateral crystallizing process induced by single-pulse excimer laser annealing (ELA) for the first time. In order to intentionally modulate a beam profile on amorphous-silicon thin films, a mask projection ELA system was used. Incorporating an optical microscope, high-speed gated image intensifier, and streak camera imaging system, the novel monitoring system successfully revealed the dynamic lateral motion of a liquid-solid Si interface with micrometer spatial resolution and nanosecond time domain.

Furthermore, it was confirmed that the obtained streak images were well consistent with beam profiles simulated to induce lateral growth crystallization, and a lateral solidifying velocity of 13 m/sec was directly measured when 4-μm-long grain had grown by using phase-modulated excimer laser annealing (PMELA).

Introduction

Currently, excimer laser annealing (ELA) technologies have generally been used to fabricate low-temperature poly-silicon (LTPS) thin film transistors (TFTs) for use in advanced flat panel display applications. In the conventional ELA method[1], a spatially homogenized, uniform laser beam is used to irradiate an amorphous-silicon thin film deposited on a glass substrate. Given this condition, the excimer-laser-initiated melt-solidifying process originates from the bottom unmolten a-Si seed layer, then proceeds vertically upward due to the large heat flow downward to the substrate. During this process, a lot of Si grains simultaneously start to grow from the unmolten layer to the top surface. Since collision with neighboring grains interrupts their lateral growth, the vertically grown grains have a relatively small columnar shape, measuring about 0.1 micrometer in diameter. Thus, the fabricated LTPS-TFTs have many grain boundaries within the active channel region, which seriously degrade the device performance. To improve TFT performance, crystalline-Si film with large grain sizes and controlled locations is required. To accomplish this objective, laser induced lateral growth mode[2] is a key technical issue because it can easily yield grains over 1 micrometer long. Several attempts have been made to achieve single-crystalline-Si TFTs by using sequential lateral solidification (SLS), diode pumped solid-state (DPSS) CW laser crystallization (CLC), or phase-modulated excimer laser annealing (PMELA).[3, 4, 5, 6]

Meanwhile, since the latter half of the '70s, there has been continuous interest in investigating laser-annealing processes of semiconductors. At that time, the main theme was recrystallization and impurity redistribution control of ion-implanted crystal silicon substrates. To clarify the laser-induced rapid melt-resolidifying process, measurements

---

Present addresses: [a] Shimadzu Corporation, 380-1, Horiyamashita, Hadano-city, Kanagawa, 259-1304, Japan; [b] National Institute of Advanced Industrial Science and Technology, Center2, 1-1-1 Umezono,   a, Ibaraki 305-8568, Japan

of transient electrical conductance[7] and time-resolved reflectivity[8] were frequently performed. For example, Lowndes[9] reported that the vertical motion of the KrF laser-induced molten layers on crystal silicon was clearly perceived, and solid-liquid interface velocities of 8-14 m/s were measured.

Thompson and his co-workers intensively investigated the vertical melt-regrowth process of silicon thin film deposited on thermally isolated substrates to study the thermodynamics of the liquid-solid interface far from equilibrium.[10, 11, 12]

Though these measurement methods provide powerful, relatively simple means for studying the vertical propagation process, they are not applicable to characterize the above-mentioned micro-scale lateral crystallizing process because there is no positional information in these methods.

In recent years, to induce lateral crystallization on glass substrates effectively, Gupta[13] and Yeh[14, 15] separately simulated and proposed beam profiles based on heat flow calculations. Though a lot of experimental results indicate that the lateral melt-solid interface propagation process has occurred in amorphous-silicon thin films, conclusive experimental evidence in support of these simulations has not yet been presented. It is still important and necessary to investigate the correlation between the beam intensity profile and the lateral motion of solid-liquid interfaces or the lateral crystallization process.

In an earlier work on a related subject, snapshots of a Nd:YLF laser-induced solidification process were taken by Lee[16]. However, they did not have a way to control the beam profiles on the film, nor did they intend to measure the continuous motion of solid-liquid interface motion directly.

In this report, we have intentionally modulated the beam profile with micro-scale spatial resolution by using a high-resolution projection lens. Additionally, we have not only taken snapshots of an ELA-induced melt-solid interface of the film, but also directly visualized the continuous motion of solid-liquid interfaces in Si thin films using a streak camera imaging system, which was newly installed for detecting nanosecond transient phenomena.

Monitoring System and Imaging Principle

Figure 1 illustrates a nanosecond solid-liquid interface monitoring system installed in the ELA system. A KrF excimer laser (wavelength: 248 nm, pulse-width: 20 nsec) was used as a film-melting source. The raw beam of the excimer laser was cut into segments by crossed cylindrical array lenses. These segments were overplayed and uniformly homogenized on a mask. Then the mask pattern was imaged onto a film on glass substrate via a 5x demagnifying projection lens to trigger melt-regrowth of the Si. Numerical aperture of the projection lens was 0.13, and σ-value of the illumination optics alternated from 0.5 to 1. The excimer laser energy fluence on the film was controlled by a variable attenuator system, which is not shown in the figure.

Xenon-flash lamp (pulse width: 2 μsec) illuminates the a-Si film at the same time the excimer laser irradiates the film. The transmitted lamp light through the film on glass substrate was magnified by an objective microscope (x100, NA=0.8), and imaged on the photocathode of the streak camera or CCD camera with a high-speed gated image intensifier (ICCD camera). The photocathode converts the incident light to a number of electrons proportional to the light intensity, and these electrons were accelerated, intensified, and imaged on the phosphor screen. In case of the streak camera, the image on the photocathode is limited by a slit in the sweep direction. Then the slit image is converted to electron flow by the photocathode and accelerated towards the screen. By applying the synchronized signal to the deflecting electrodes equipped between the photocathode and the screen, the streak camera can take one-dimensional continuous motion in the slit image, the contrast of which reflects the solid-liquid interface. The ICCD camera obtained two-dimensional snapshots of the resolidification process. A trigger delay between the excimer laser, the streak camera, and the ICCD camera was set by a function generator to take images at the appropriate time. The exposure time of the ICCD camera was set at 10 nsec, and the measuring time of the streak camera was set at 500 nsec in this experiment. Images were acquired by a computer that was connected to the streak and ICCD cameras.

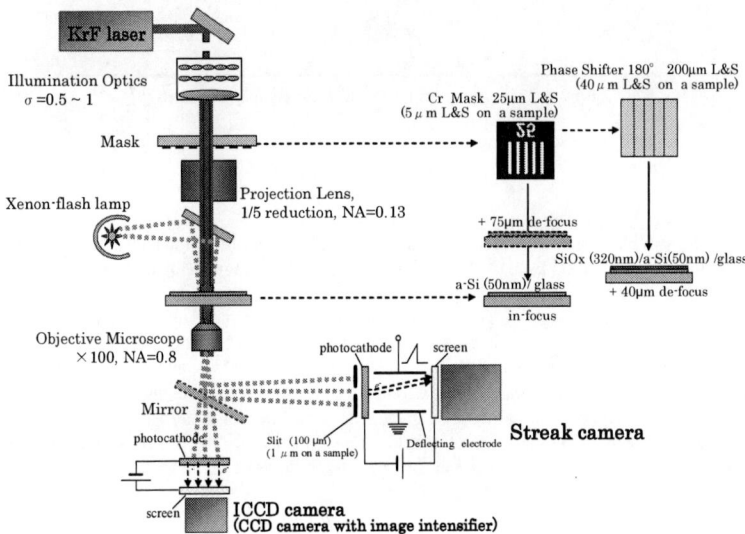

Figure 1. Schematic view of ELA system and nanosecond monitoring system.

The refractive index (n) and extinction coefficient (k) of a-Si (n=4.2, k=0.46), liquid Si (n=3.8, k=5.2), and poly-Si (n=3.8, k=0.03 at λ = 633 nm) were previously reported by G.E. Jellison, Jr. et al.[15] It is well known that as the value of "k" increases, the transmissivity of the film decreases, in general. Thus, we were able to distinguish the area of solid-Si, liquid-Si, and Poly-Si in the film image, as shown in Figure 2. It is worthy to note that an image becomes dark even though only the region near the Si surface is molten because of the extremely high k value of liquid Si.

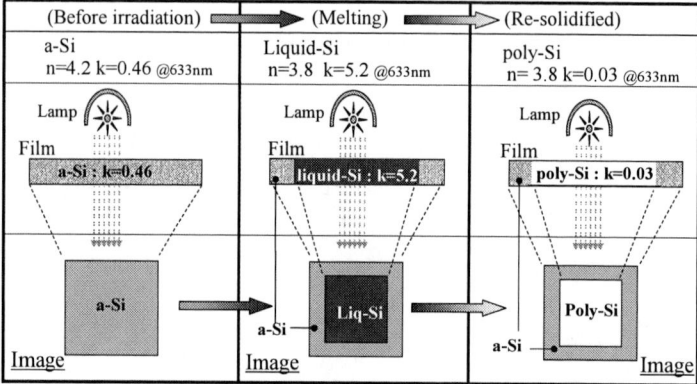

Figure 2. Imaging principle of the measuring system.

Results and Discussion

Observation of lateral motion of solid-liquid interface

At the start, we tried to obtain images of the melting and resolidifying phenomena by delaying the acquisition time of the ICCD camera. In this experiment, we used a Cr mask with 25-μm lines and spaces (L&S) (5-μm L&S on a sample) patterned with five openings because the light intensity distribution on the sample is easily understandable based on geometrical optics. Samples were 50-nm-thick a-Si film deposited on glass and were set either at just focused or at 75-μm defocused position to change the beam profile. The σ-value of the illumination optics was set at unity in this experiment.

Calculated one-dimensional laser beam profiles at focused and defocused positions are shown in Figure 3. There are five rectangular-shaped peaks for the in-focus image on one hand, but only four peaks with slopes for the 75-μm defocused image, on the other. The four sloping peaks have the potential to cause lateral crystallization.

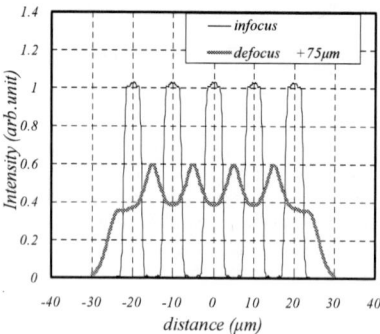

Figure 3. Calculated beam profiles (λ =248nm, NA=0.13, σ =1, x5 demagnification).

The obtained snapshot images are shown in Figure 4. To confirm that the objective microscope was set at the focus position, there are pre-patterned 2-μm L&S marks in the upper right position in (a) and (e). Then the sample was moved 75 μm upwards for defocusing (f), (g), and (h). The (b) and (f) images were taken 45 nsec after the onset of melting, whereas (c) and (g) were at 65 nsec, and (d) and (h) at a few seconds after ELA. Dark areas in (b), (c), (f), and (g) indicate that the Si film was molten, and bright areas in (d) and (h) indicate that the Si was crystallized. Grey areas in (a) – (h) indicate a-Si since no excimer laser beam was illuminated there. These eight images show clearly the sequence of the melting and resolidifying process. By 75-μm defocusing upward of the a-Si film, four bright stripes ((f) ~ (h)) were revealed on the Si film instead of the five bright stripes. Further, the edge of the melting zone of (b) was clear, while that of (f) was somewhat vague, and both sides of the peak position in the beam profile has begun to crystallize. This suggests that lateral motion of the solid-liquid interface occurred.

From these results it was confirmed that the installed ELA optics performed sufficiently well to modulate a beam profile in micro-scale spatial resolution, as the calculation predicted. Also, it was confirmed that the melting, resolidifying, and crystallized areas were distinguishable in the obtained images.

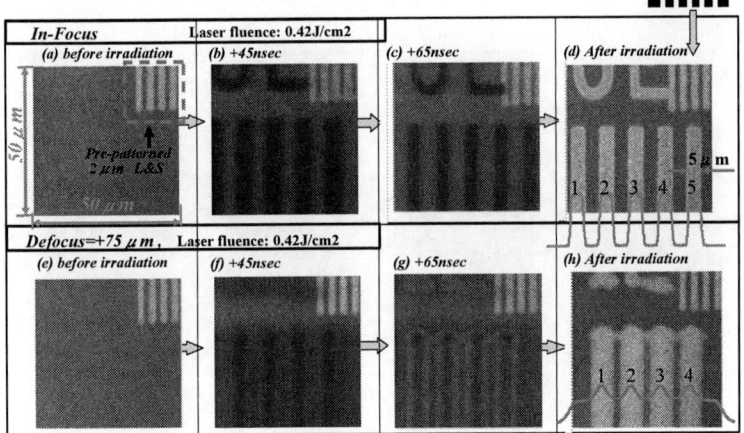

Figure 4: Snapshot images obtained by ICCD camera. Acquisition time was set at 10 nsec.

The images in Figure 4 suggest that lateral crystallization occurred in the film. However, to be exact, the time-series aligned images are obtained with a different process induced by another shot of ELA. This is because the ICCD camera cannot take images continuously at intervals of several tens of nanoseconds. In order to directly measure continuous motion of the solid-liquid interface induced by a single shot of ELA, we installed the streak camera imaging system. The image on the photocathode was limited to 100 μm in height, which is equ   um on a sample. While a synchronized signal was applied to the sweeping electrodes, the image on the photocathode was recorded on the phosphor screen. In other words, the streak image is the vertically-stacked image of the time-resolved one-dimensional image.

The obtained space-time streak photographs at the focused position (left) and at the 75-μm defocused position (right) are shown in Figure 5. The vertical axis represents a time coordinate, and the horizontal axis represents a spatial coordinate. The position with the long melting time corresponds to the position with the high beam intensity. It can be seen that the vertically flipped solid-liquid interface shapes of the streak images resembled the shapes of the micro-scale laser beam profiles shown in Figure 3.

These results are now inferred below. When a-Si is set at the focus position, the laser beam profile is anticipated to be rectangular, with five flat shapes at the top. In this case, the heat-melted a-Si film was almost uniformly transferred to the underlayer, which was glass. In contrast, at the 75-μm defocused position, the laser beam profile is modulated to a phase-reversed four-peak pattern, with valleys between the four peaks. In that case, the position of the melting point traversed laterally, as the space-time streak photographs showed lateral motion of the solid-liquid interface, which was 3-μm-long in 82 nsec, equaling an interface motion of about 36 m/sec.

Scanning electron microscopy (SEM) images showing surface morphology for the film irradiated at the 75-μm defocused position are in Figure 6. Laterally grown grains appeared 2 μm apart from the peak points and valleys. The lateral motion of the solid-liquid interface was so fast that the lateral growth length was up to 0.4 μm long.

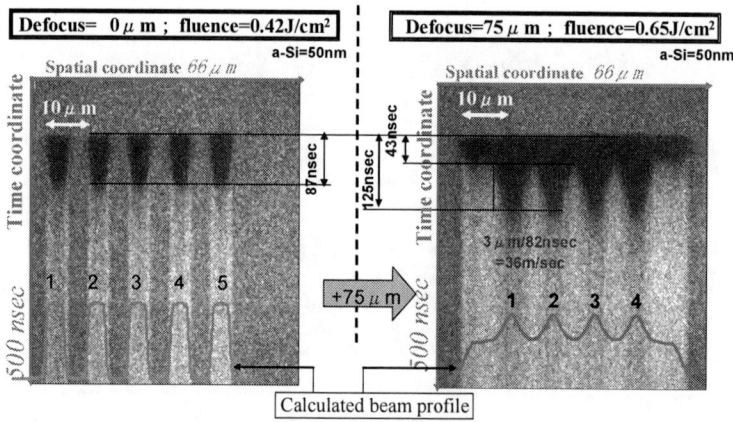

Figure 5. Obtained space-time streak photographs; left: In-focus 420 mJ/cm^2, right: Defocus 650 mJ/cm^2

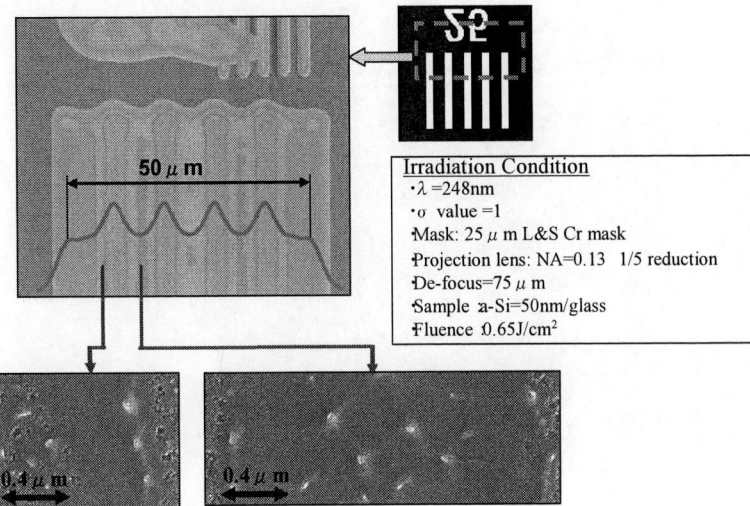

Figure 6. SEM images of surface morphology (after Secco-etching) for sample irradiated at 75-μm defocused position.

Observation of lateral crystallization

Using a phase shifter as an ELA mask is advantageous in terms of energy efficiency, since there is no fundamental loss at the mask. Moreover, micro-scale beam profiles on a sample can be adequately controlled by designing the diffracted light from a phase shifter and utilizing diffraction-limited reduction optics.[7]

In this experiment, we used a basic one-dimensional 180° phase shifter with an L&S of 200 μm (40 μm on a sample); the σ-value of the illumination optics was set at 0.5.

A sample with a stacked structure consisting of a 320-nm-thick SiOx capping layer and a 50-nm a-Si layer deposited on glass was set at a defocused position 60 μm upward from the just focus point to effectively induce lateral crystallization.

Figures 7 shows the obtained streak image (a), calculated beam profile (b), and SEM image (c), respectively. The dark area in (a) indicates molten Si with low transmittance. As expected, the position where the melting time is long corresponds to the position where the light intensity is high. The melt-resolidifying border is consistent with the calculated beam profile with micro-scale spatial resolution. The top Si surface was crystallized 54 nsec after laser irradiation. The nuclei seem to be formed a little earlier and continue to grow until 320 nsec after the irradiation. Laterally-running traces in the Secco-etched sample indicate that 4-μm-long Si grains were grown.

From these results, it is clear that lateral motion of the solid-liquid interface represents lateral crystallization, and thus the lateral crystallization velocity was calculated to be about 13 m/sec.

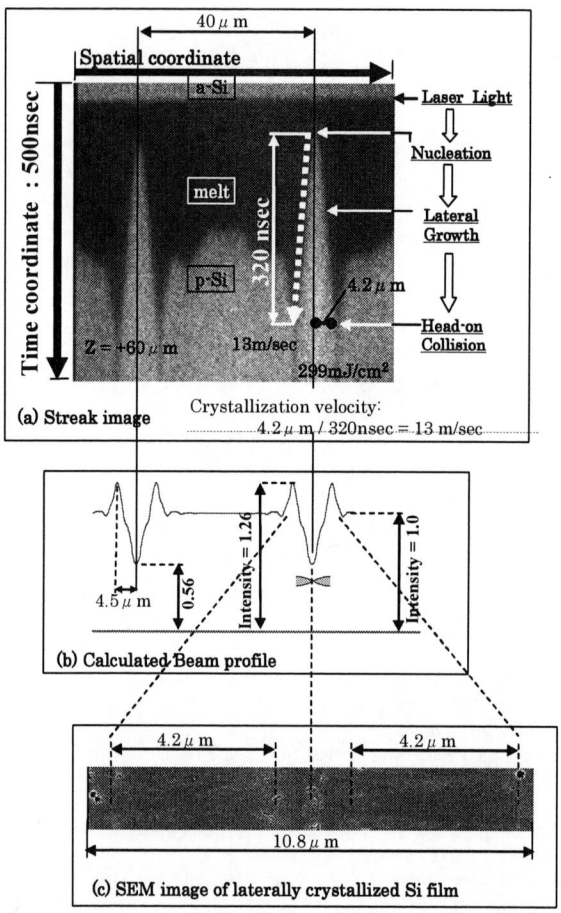

Figure 7 (a) Space-time streak photograph (b) Calculated beam intensity profile (c) SEM image of laterally crystallized Si film (after Secco-etching).

Conclusion

Incorporating flash lamp illumination and a micro-scale spatial resolution streak camera imaging system, lateral motion of a solid-liquid interface has been directly visualized.

It was confirmed that the obtained space-time streak photographs were well consistent with simulated beam profiles for inducing lateral growth crystallization, and we found a lateral solidifying velocity of 13 m/sec through direct measurement when a 4-μm-long grain was grown using phase-modulated excimer laser annealing (PMELA).

By simultaneously measuring the lateral crystallization process with real spatial resolution and in the nanosecond time domain, we believe we can apply our findings to discuss fundamental technologies not only including ELA optics and film structures for achieving single crystalline-TFTS, but also thermodynamics and material science.

Acknowledgments

We would like to express our thanks to Mr. M. Hiramatsu for his technical assistance and for the preparation of the deposited samples. Our gratitude also goes to Dr. H. Ogawa and Mr. N. Akita for the design and preparation of the precisely fabricated masks. Author Y.T. thanks Messrs. T. Taguchi, T. Ohno, M. Hiratsuka, S. Ishida, and T. Nishimura in the Semiconductor Equipment Division of Shimadzu Corporation for their kind supportand also wishes to acknowledge the valuable technical help from Mr. K. Yamada of Shimadzu Device Corporation.

This study was supported in part by the New Energy and Industrial Technology Development Organization (NEDO) and the Ministry of Economy, Trade, and Industry (METI) of Japan.

References

1. T. Sameshima, S. Usui, and M. Sekiya, *IEEE Electron Device Lett.*, **7**, 276 (1986).
2. D. H. Choi, E. Sadayuki, O. Sugiura, and M. Matsumura, *Jpn. J. Appl. Phys.*, **33**, 70 (1994).
3. C. H. Oh and M. Matsumura, *Jpn. J. Appl. Phys.*, **37** 5474 (1998).
4. T. Katou, Y. Taniguchi, M. Hiramatsu, K. Azuma, and M. Matsumura, Proceedings of International Display Workshop, 1219 (2005).
5. A. Hara, K. Yoshino, F. Takeuchi, and N. Sasaki, *Jpn. J. Appl. Phys.*, **42**, 23 (2003).
6. J. S. Im, M. A. Crowder, R. S. Sponsil, J. P. Leonard, H. J. Kim, J. H. Yoon, V. V. Gupta, and H. S. Cho, *Phys. Status. Solidi*, *A*, **166**, 603 (1998).
7. G. Galvin, M. O. Thompson, J. W. Mayer, R. B. Hammond, N. Paulter, and P. S. Peercy, *Phys. Rev. Lett.* **48**, 33 (1982).
8. D. H. Auston, C. M. Surko, T. N. C. Venkatesan, R. E. Slusher, and J. A. Golovchenko, *Appl. Phys. Lett.*, **33**, 437 (1978).
9. D. H. Lowndes, G. E. Jellison, Jr., S. J. Pennycook, S. P. Withrow, and D. N. Mashburn, *Appl. Phys. Lett.* **48**, 1389 (1986).
10. M. O. Thompson, G, J. Glavin, J. W. Mayer, and P. S. Peercy, *Phys. Rev. Lett.* **52,** 2360, (1984).
11. S. R. Stiffler, M. O. Thompson, and P. S. Peercy, *Phys. Rev. Lett.* **60**, 2519, (1988).
12. S. R. Stiffler, M. O. Thompson, and P. S. Peercy, *Phys. Rev. B.* **43**, 9851, (1991).
13. V. V. Gupta, H. J. Song, and J. S. Im, *Appl. Phys. Lett*, **71**, 99 (1997).
14. W. Yeh and M. Matsumura, *Jpn. J. Appl. Phys*, **40**, 492 (2001).
15. W. Yeh and M. Matsumura, *Jpn. J. Appl. Phys*, **41**, 1909 (2002).
16. M. Lee, S. Moon, C. P. Grigoropoulos, *J. Cryst. Growth.*, **8**, 226 (2001).
17. G. E. Jellison, Jr. and D. H. Lowndes, *Appl. Phys. Lett*, **47**, 718 (1985).

18. Y. Taniguchi, M. Matsumura, M. Jyumonji, H. Ogawa, and M. Hiramatsu, *J. Electrochem. Soc.*, **G67**, 153 (2006).

ECS Transactions, 3 (8) 137-142 (2006)
10.1149/1.2356346, copyright The Electrochemical Society

The Geometry Effect of A Counter-Doped Lateral Body Terminal on Poly-Si TFTs

Sang-Myeon Han, In-Hwan Ji, Joong-Hyun Park, Sung-Hwan Choi, Juhn-Suk Yoo[a],
Byoung-Duk Choi[b], Ki-Yong Lee[b] and Min-Koo Han

School of Electrical Engineering, Seoul National University, Seoul, Korea
Electrical Engineering and Computer Science, University of Michigan, U. S. A.[a]
Display R&D Center, Samsung SDI, Gyeonggi-do, Korea[b]

We have designed and fabricated poly-Si TFTs employing
various counter-doped lateral body terminals (LBT) which
suppress a kink current and improve the reliability under the
various electrical stresses. Our experimental results show that the
TFT with a LBT structure connected near the drain region was
more effective to suppress the kink current and to reduce the
leakage current. The suppression of the kink effect employing LBT
structure can be attributed to the decrease of E-field at the drain
junction due to the potential redistribution in the channel of the
TFT by the counter doped LBT and the collecting holes, which
have been generated by impact ionization at the drain.

Introduction

Low temperature poly-Si TFTs on the glass substrates employing excimer laser
annealing (ELA) is promising for the high-resolution AMLCD and AMOLED backplane
as well as pixel element due to their high current driving capability [1][2]. Although the
electrical characteristics such as field effect mobility and threshold voltage of LTPS have
been improved considerably, the kink current and device reliability due to the electrical
bias may be a critical problems for designing advanced pixel circuits and peripheral
integrated circuits [3]. It is well known that the kink current is caused by parasitic bipolar
transistor action in the floating body where the holes generated in the drain junction are
accumulated [4]. Various structures such as lightly doped drain (LDD) and gate
overlapped LDD (GOLDD), dual gate and asymmetric dual gate structure may reduce the
kink effect [5-8], but it is rather difficult to apply LDD and GOLDD structures to the
short channel LTPS TFTs without misalignment. We have already reported that the kink
current in poly-Si TFT can be reduced by the grounded 'fourth' terminal called lateral
body terminal (LBT) which collects the opposite polarity carriers [9]. The poly-Si TFT
with LBT exhibited suppressed kink current, reduced leakage current and improved
electrical stability [9]. However, characteristics of LBT TFTs according to the geometry
of LBT have not been studied. The optimized LBT design can be required to lead better
electrical characteristics of short channel LBT TFTs.

The purpose of our work is to design various LBTs and to report the characteristics of
the poly-Si TFT with the counter doped LBT which suppresses kink effect in poly-Si
TFT. The TFT with a LBT structure near drain region was more effective to suppress the
kink current and to reduce the leakage current. It should be noted that LDD was not
employed to suppress troublesome kink effect. The suppression of the kink effect
employing LBT structure can be attributed to the hole collection by LBT structure and
decrease of E-field at the drain by P-I-N structure in the LBT TFTs.

ECS Transactions, 3 (8) 137-142 (2006)

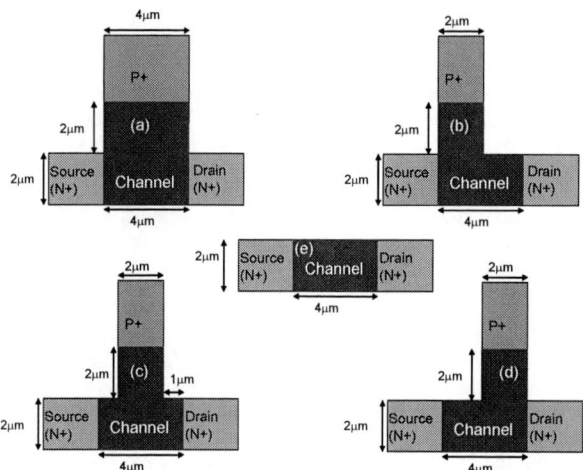

Fig. 1 The schematic diagrams of the fabricated poly-Si TFTs. (W/L=2/4 μm)
(a) LBT with a width of 4 μm (Type 1), (b) 2 μm width LBT near the source (Type 2), (c)
2 μm width LBT at the center (Type 3), (d) 2 μm width LBT near the drain (Type 4), (e)
conventional poly-Si TFT

Experiment

We have fabricated n-channel poly-Si TFT with p-type LBT on the glass substrate
employing standard LTPS process. The 70nm thick a-Si film was deposited and
crystallized by excimer laser annealing. Source, drain and P+ LBT doping was performed
by ion shower followed by thermal activation. We designed the counter-doped LBT TFT
with various dimensions as shown in Fig. 1. The width and length of the LBT TFT's
channel was 2μm and 4μm. The P type terminal was self-aligned with the gate electrode
and connected to the side of the channel of the poly-Si TFT. The 2μm length poly-Si
channel is extended to the P+ LBT so that the channel can be connected to the P+ LBT
electrically. Type 1 TFT, which have 4μm width LBT, is shown in Fig. 1(a). The 2μm
width LBT is connected to the channel near the source in Type 2 TFT as shown in Fig. 1
(b). Type 3 is the TFT where the LBT is connected at the center of the channel as shown
in Fig. 1 (c) and Type 4 have the LBT connected to the channel near the drain of the TFT
as shown in Fig. 1 (d). Fig. 1 (e) exhibits the conventional poly-Si TFT without LBT
structure. The electrical characteristics such as transfer curve and output curve were
measured to observe the effect of the LBT structure. The ISE simulation program was
used to study the 3-dimensional potential distribution and E-field of the poly-Si TFTs
during the operation.

Results and Discussion

The transfer characteristics of the TFTs with various LBT geometries are shown in
Fig. 2. The drain was set to 5V, the source and the LBT were set to 0V and the gate was

138

varied from -10V to 20V. The mobility and threshold voltage of the fabricated poly-Si TFT are summarized in the Table.1. The slightly increased mobility of Type 1 TFT (LBT with whole channel length) compared with other structures can be attributed to the increase of effective channel width which is extended to the P+ LBT structure. The LBT TFT shows low leakage current compared with the conventional structure TFT at negative gate bias and the increase slope of leakage current is also considerably small as more negative gate bias is applied. The decreased leakage current at negative gate bias of LBT structure TFTs can be attributed to the hole collecting function of LBT structure [9].

In the conventional poly-Si TFTs, a large number of electrons are emitted by the vertical field in the drain depletion the hole accumulation in the channel increases the body potential and lowers the junction barrier at the source region so that a large amount of holes may be collected by the source [10, 11]. The leakage current is the sum of the electron current by field-emission at the drain region and the hole current by P-N forward bias at the source region. In the proposed LBT TFT, the holes are collected by the LBT and a junction barrier is formed between the body and source region, so that the hole current at the source region is reduced. In this transfer curve graph, it should be noted that the effective decrease of leakage current is feasible employing proposed LBT structure without any additional LDD masks or processes.

Table. 1 The mobility and threshold voltage of the fabricated poly-Si TFTs

	Type 1	Type 2	Type 3	Type 4	Conventional
Field effect mobility	185.3 cm^2/Vs	150 cm^2/Vs	160 cm^2/Vs	154.5 cm^2/Vs	144.6 cm^2/Vs
V_{TH}	0.4V	0.4V	0.4V	0.4V	0.7V

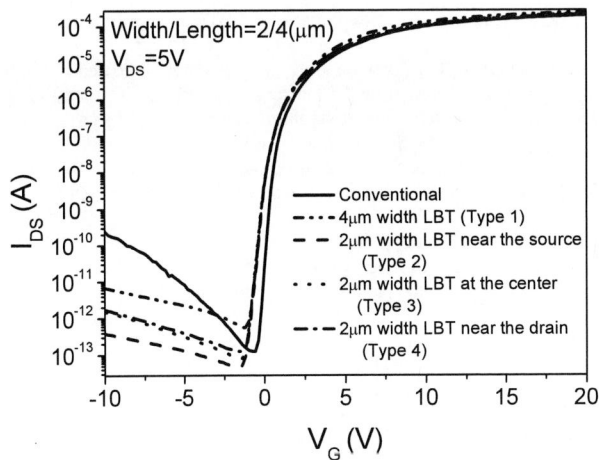

Fig. 2 Transfer characteristics of the TFTs with LBT and conventional TFT

Fig. 3 shows the output characteristics of the fabricated TFTs. The gate was set to 5V, the source and the LBT were set to 0V and the drain was varied from 0V to 13V. The kink effect is very severe in conventional structure poly-Si TFT due to its inherent floating body structure and accumulated holes in the source region [4]. However, the kink current at high drain bias was suppressed in the LBT structure TFTs (Type1~4). Among various LBT structures, the ability suppression of kink current was almost identical, but the type 4 (LBT near the drain) is most effective to suppress the kink current since lateral E-field redistribution can be maximized and hole collection is performed by P+ LBT simultaneously [9].

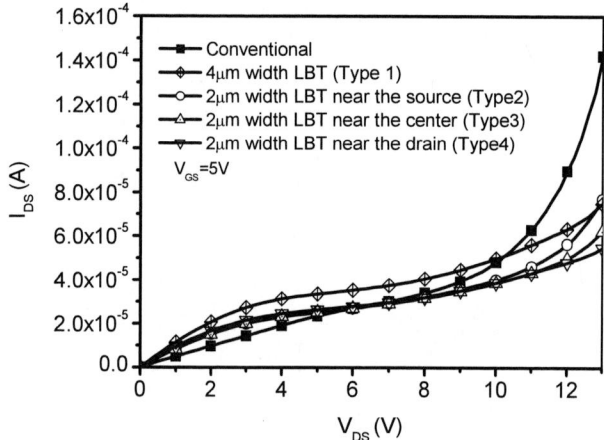

Fig. 3 Output characteristics of the TFTs with LBT and conventional TFT

To verify the effect of LBT on reduction of the electrical field near the drain junction, 3-dimensional device simulation was performed. The active layer and gate pattern was designed to be identical to those of the the fabricated poly-Si TFTs. The poly-Si model in the ISE 3-dimensional simulation program was used for simulation. Fig. 4 shows the potential distribution in poly-Si active layer of the poly-Si TFTs with the various geometries. In the conventional TFT, there is only typical N-I-N structure, that is, source-body-drain, so that most of field can be applied to the drain junction. However, P+ LBT is located in the side of channel in the proposed LBT TFTs, so that P-I-N structure is also formed in the channel to bring the effect of an E-field distribution to the LBT/channel junction. As shown in Fig. 4 the potential distribution near drain region of Type 4 spreads and extends to the direction of P+ LBT and the potential near the drain region is also higher than other LBT structures while the potential distribution of Type 2 LBT poly-Si TFT (LBT near source) is almost identical to that of conventional structure poly-Si TFT. The E-field peak at drain junction of Type 4 calculated employing ISE simulator was lower than that of Type 2 and conventional TFT by 10%. And the E-field peak of Type 4 spreads along the X-axis of drain junction. This means the electrical field can be decreased near the drain region, thus the troublesome impact ionization causing hole generation can be suppressed in the Type 4. The simulation results correspond to the kink current suppression of the Type 4 LBT poly-Si TFT in the Fig. 3 and low hole current at LBT node of the Type 4 LBT poly-Si TFT in the Fig. 5, which will be explained later.

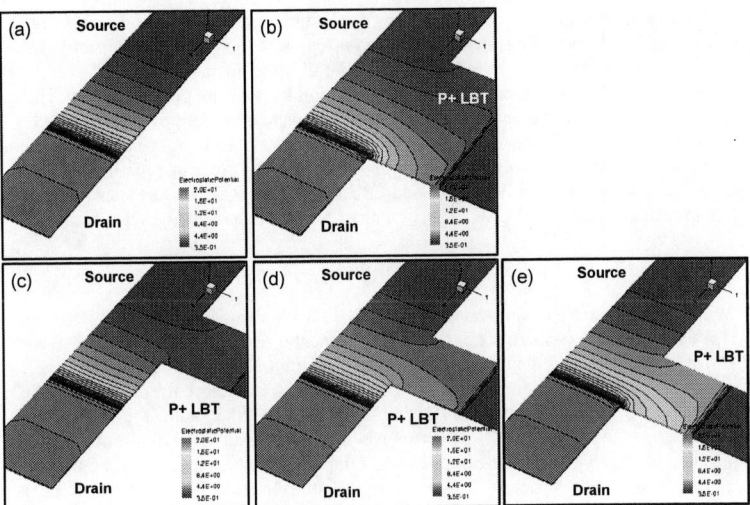

Fig. 4 Simulated 3 dimensional potential distributions in poly-Si layers of the poly-Si TFTs. The 5V was applied to the gate. The drain was set to 20V and the source and the LBT.

(a) conventional poly-Si TFT (b) LBT with a width of 4 μm, which is connected through the n-channel (Type 1), (c) LBT with a width of 2 μm near the source (Type 2), (d) LBT with a width of 2 μm at the center (Type 3), (e) LBT with a width of 2 μm near the drain (Type 4)

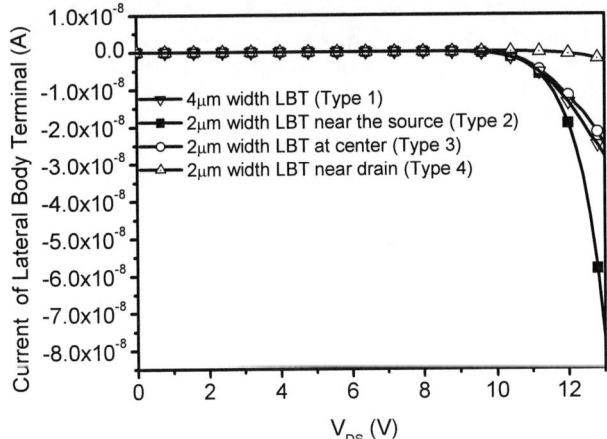

Fig. 5 Current of p-type lateral body terminal during output curve measurement

The current measured at the LBT during the output characteristics measurement is shown in Fig. 5. The type 2 has the largest hole current in the LBT and the Type 4 has the smallest hole current, which is induced by hole collection in the channel. As Type 4 (LBT near drain) has a reduced E-field near the drain region as plotted in Fig. 4, the impact ionization can be suppressed, so that less holes were generated in the channel. The type 4 poly-Si TFT (LBT near the drain) have the lowest hole current in the proposed LBT structures, which is corresponding to the smaller kink effect in Fig. 3. This result implies the kink current can be suppressed effectively by LBT structure near the drain region just employing typical CMOS poly-Si process without any additional processes.

Conclusion

We have designed and fabricated poly-Si TFTs with various lateral body terminals (LBTs) which suppress the kink effect. Our experimental results show that the TFT with a LBT structure connected to the channel near the drain region was most effective to suppress the troublesome kink current and to reduce the leakage current without LDD structure. In the proposed poly-Si TFT with LBT located near drain, the hole current measured at the LBT was small, which means the less impact ionization occurs at the drain junction, so that the kink effect can be suppressed. The suppression of the kink effect employing LBT structure can be attributed to collecting holes generated in the drain junction by LBT structure and the decreasing of E-field at the drain junction by LBT which was verified by 3-dimensional numerical simulation. In poly-Si TFTs where LDD structure is not adopted typically, the proposed LBT structure located near drain junction can be useful to suppress kink current without any additional process.

Acknowledgments

This research was supported by a grant (ASD1) from Information Display R&D Center, one of the 21st Century Frontier R&D Program funded by the Ministry of Commerce, Industry and Energy of Korean government.

References

1. N. Kubo, N. Kusumoto, T. Inushima, S. Yamazaki, *IEEE Trans. On Electron Device*, **41**(10), 1876 (1994).
2. J. H. Jeon, M.C. Lee, K. C. Park, and M, K. Han, *IEEE Electron Device Letters*, **22** (9), 429 (2001).
3. A. G. Lewis, D. D. Lee, R. H. Bruce, *IEEE Journal of solid state circuits*, **27** (12), 1833 (1992).
4. M. Hack, *et al, IEEE IEDM Tech. Digest*, 252 (1988).
5. S. Seki, *et al., IEEE Electron Device Lett.*, **8**, 434 (1987).
6. K. Nakazawa, *et al., Tech. Digest of SID '90*, 311 (1990).
7. C-W Lin, *et al., IEEE Electron Device Lett.*, **23** (3), 133 (2002).
8. Min-Cheol Lee and Min-Koo Han, *IEEE Electron Device Lett.*, **25** (1), 25 (2004).
9. J. S. Yoo, C. H. Kim, M. C. Lee, M. K. Han, H. J. Kim, *IEEE IEDM Tech. Digest*, 217 (2000).
10. C. M. Park, J. S. Yoo and M. K. Han, *SSDM Ext. Abs.*, 350 (1997).
11. J. G. Fossum, A. Ortiz-Conde, H. Shichijo, *S*. K. Banejee, *IEEE Trans. Electron. Devices*, **32** (9), 1878 (1985).

High-Performance Low Temperature Poly-Silicon Thin Film Transistors Fabricated by Excimer Laser Irradiation with Bottom-Gate Scheme

Chun-Chien Tsai[a], Hsu-Hsin Chen[a], Bo-Ting Chen[a], Yao-Jen Lee[b], and Huang-Chung Cheng[a]

[a] Department of Electronics Engineering and Institute of Electronics, National Chiao Tung University, Hsinchu, Taiwan
Address: Dept. of Electronics Engineering, National Chiao Tung University, Hsinchu, Taiwan, ROC.
[b] National Nano Device Laboratories, Hsinchu, Taiwan, ROC.
Phone: +886-3-5712121-54218; Fax: +886-3-5738343
E-mail: cctsai.ee92g@nctu.edu.tw

> In this work, high-performance low-temperature poly-Silicon bottom-gate TFTs with single grain boundary perpendicular to the current flow in the channel regions have been demonstrated. A lateral grain growth of 0.75 μm in length could be artificially grown via the super lateral growth phenomenon using excimer laser irradiation with the plateau structure. Consequently, bottom-gate TFTs made by this method exhibit higher field-effect-mobility, lower leakage current, steeper subthreshold slope, larger on/off current ratios and improved device uniformity owing to this location-manipulated lateral grains. In addition, it shows better reliability because of the smooth interface between gate dielectric and poly-Si channel films.

Introduction

Low-temperature polycrystalline silicon (LTPS) thin film transistors (TFTs) fabricated by excimer laser crystallization (ELC) have been extensively studied for active matrix liquid crystal displays (AMLCDs) and active matrix organic light emitting displays (AMOLEDs) owing to their high driving-current capability.[1,2] In the early stage of the development of LTPS TFTs, bottom-gate (BG) TFT structure was attractive because the excimer laser annealing was thought as an additional process step to the a-Si TFTs. However, bottom-gate TFTs suffered from worse electrical performance than top-gate (TG) TFTs because of the smaller grain size and poor grain quality resulting from the bottom-gate metal acting as a heat sink during excimer laser crystallization.[3,4] Although field-effect-mobility of 200 cm^2/Vs for TG-TFTs has been attained by ELC, it is difficult to make the laser energy density hit the super lateral growth regime everywhere due to the fluctuation of pulse-to-pulse energy and amorphous silicon (a-Si) layer thickness.[5-7] Furthermore, in the applications of system-on-panel (SOP), high-performance LTPS TFTs are still needed to integrate memory and controller with driver circuits on a single substrate. Thus, there is a great interest in improving the performance of LTPS TFTs by laser crystallization approaches, including sequential lateral solidification by laser beam scanning within several micrometers step by step,[8] μ-Czochralski (grain filters) method,[9] capping the reflective or anti-reflective layer,[10] dual beam ELA,[11] double–pulsed laser annealing,[12] CLC method using the diode-pumped solid state continuous wave laser,[13] and selectively enlarging laser crystallization

(SELAX).[14] However, most of them need complex fabrication process or not compatible with the existing excimer laser annealing systems.

In this work, a lateral grain growth method with the same conventional fabrication process of bottom gate TFT is proposed to produce high-mobility poly-Si TFTs. Only single grain boundary perpendicular to the current flow can be artificially grown in the channel regions via the amorphous silicon plateau structure with excimer laser irradiation. It leads to the enhancement of device performance and the improvement of device uniformity.

Experimental Procedures

Figure 1 illustrates the key processes for the fabrication of LTPS TFTs crystallized with bottom gate structure. At first, a 1000 Å-thick in-situ doping poly silicon layer was deposited by LPCVD at 550°C on oxidized silicon wafer with oxide thickness of 1μm. After defining the gate region, a 500 or 1000 Å-thick tetraethyl orthosilicate (TEOS) gate oxide layer was deposited following by 1000 Å-thick a-Si layer deposition by LPCVD at 550°C. Laser crystallization (ELC) was performed by KrF excimer laser (λ=248 nm). During the laser irradiation, the samples were located on a substrate in a vacuum chamber pumped down to 10^{-3} Torr and substrate was maintained at room temperature. The number of laser shots per area was 20 (i.e. 95% overlapping) and laser energy density was varied. A self-aligned phosphorous ion implantation with dose of $5 \times 10^{15} cm^{-2}$ was carried out to form source and drain regions. Then, the device active region was etched by TCP-RIE. Next, a 3000 Å-thick TEOS passivation oxide was deposited and the implanted dopants were activated by thermal annealing at 600°C. Finally, contact holes opening and metallization were carried out to complete the fabrication of TFTs with bottom gate structure. Any troublesome hydrogenation plasma treatment was not performed during the device fabrication process. For the sake of comparison, the conventional Excimer-Laser-Crystallized LTPS top-gate TFTs with a channel thickness of 1000Å were also fabricated.

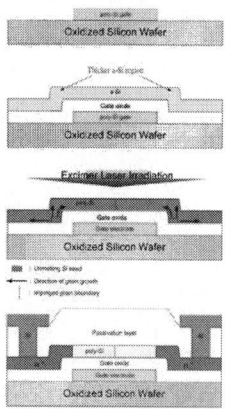

Figure 1. The key processes for fabricating small-dimension LTPS TFTs with bottom-gate structure.

Results and Discussion

Figure 2 displays the SEM photograph of excimer laser crystallized poly-Si with bottom-gate structure after Secco etching and the length of bottom-gate is 1.5 μm. The location of a-Si spacer is indicated by white dash lines. It can be observed that the lateral grains with 0.75 μm in length formed in the channel region and outside the channel, while small and fine grains are located at the corner of the bottom-gate. In the short channel length device using this crystallization, only single grain boundary perpendicular to the current folw in the channel regions indicated by white dash lines is formed. It has been reported that lateral thermal gradient could arise as a result of the heat generated at moving solid-melting interfaces.[7, 15] As a proper laser energy density is performed on the amorphous silicon thin film containing different thickness, the thin silicon region is completely melted while the thick region is partially melted, and the lateral grain growth starts from the un-melted solid Si toward the completely melted thin region. In this experiment, as excimer laser irradiation is performed on the amorphous silicon thin film with bottom-gate structure, the laser energy densities can cause complete melting 1000 Å-thick silicon thin film in the channel region but partial melting the thicker a-Si film at the edge of the bottom-gate corner . Therefore, the lateral grain growth started from the un-melted silicon solid seed at the base of the bottom-gate corner, and extended toward the completely melted region until the solid-melt interface from opposite direction impinges. If the bottom-gate plateau were arranged in a proper distance, lateral grain growth will be manufactured without any spontaneous nucleation. Thus, the grain boundaries in the channel region can be controlled and reduced. Since the number of spontaneous small grain and grain boundary is reduced, the uniformity of TFTs performance can be improved with artificially lateral grains.

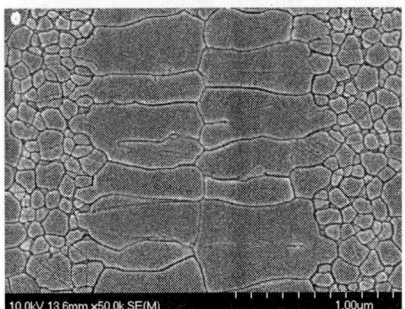

Figure 2. Scanning Electron Microscope micrograph of excimer laser crystallized poly-Si film with bottom-gate structure after Secco etch.

Typical transfer characteristics and output characteristics of LTPS BG-TFTs with lateral silicon grains and conventional TG ones for $W = L = 1.5$ μm are shown in Figure 3a and Figure 3b, respectively. Owing to the uniformly large transverse grains grown in the device channel region, BG-TFTs with lateral grains exhibit better electrical characteristics than the TG-TFTs. BG-TFT with field effect mobility of 220 cm^2/Vs can be achieved using this crystallization method while the mobility of the conventional TG-TFT counterpart is about 103 cm^2/Vs. It is generally believed that the grain boundary acts as a strong trapping center which degrades the performance of TFTs resulting from grain boundary potential barrier height.[16] The high field effect mobility is attributed to that the reduced grain boundary in the channel regions. Table 1 lists the average values of several important electrical characteristics of the two different TFT structures with the standard deviations in brackets. The improved uniformity of bottom-gate devices is attributed to this location-controlled lateral silicon grains.

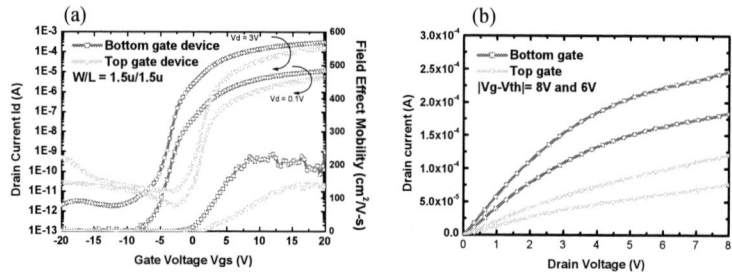

Figure 3. *I-V* curves of LTPS BG-TFT with lateral silicon grain and conventional TG structure. (a) Transfer characteristics. (b) Output characteristics.

Besides the improvement of device performance, the BG-TFTs with lateral silicon grains exhibit better reliability than TG-TFTs. Figure 4a shows the gate-breakdown voltage of two different structures TFTs with 500 Å-thick TEOS gate oxide. Bottom-gate TFTs exhibit higher breakdown voltage than top-gate ones, because the smooth interface of bottom-gate TFT, which is shown in Figure 4b. The cross-sectional TEM image reveals that a flat interface morphology between the gate dielectric and poly-Si channel films in the bottom-gate TFTs, while a rough surface roughness caused by the grain boundary in the top-gate ones shown in Figure 4c. The protruded grain boundaries due to the freezing of capillary waves excited in the silicon melt during the crystallization profoundly affect the reliability and gate dielectric integration of poly-Si TFTs.[17,18] The improved breakdown characteristic implies that the proposed BG-TFT structure is more suitable for device-size scaled-down application.

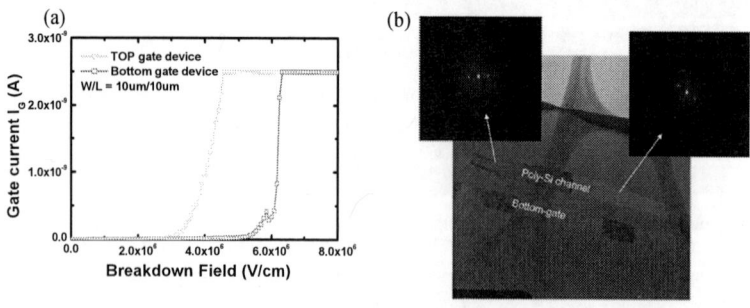

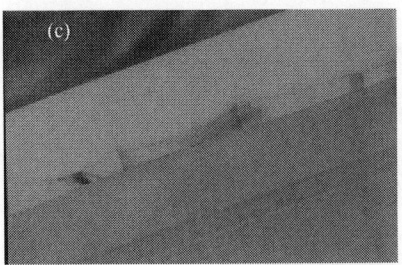

Figure 4. (a) Experimental gate-breakdown voltage of two different TFTs structures with 500 Å-thick TEOS gate oxide. FIB-prepared cross-sectional TEM images of laser-crystallized poly-Si thin films. (b) Bottom-gate devices. (c) Conventional top-gate devices.

Table I. Measured electrical characteristics of LTPS TFTs crystallized with a-Si spacer and conventional structures. The threshold voltage was defined as the gate voltage required to achieve a normalized drain current of $I_{ds} = (W/L) \times 10^{-8}$ A at $V_{ds} = 0.1$ V. The field effect mobility and subthreshold swing were extracted at $V_{ds} = 0.1$ V, and the I_{on}/I_{off} current ratio was defined at $V_{ds} = 5$ V.

TFT Structure	Threshold Voltage (V)	Mobility (cm^2/V-s)	Subthreshold Swing (V/dec)	On/off current ratio (10^7)
Proposed Bottom-Gate	-1.10	220	1.26	7.2
Conventional Top-Gate	3.33	105	1.27	1.3

Conclusions

A new crystallization technology for producing lateral silicon gains has been developed by excimer laser irradiation with bottom gate structure. Consequently, in addition to the high-performance n-channel LTPS TFTs with field-effect-mobility reaching 220 cm^2/Vs in 1.5 μm design rule, excellent uniformity of device performance was also demonstrated owing to the artificially-controlled lateral grain growth. Moreover, the experimental results revealed higher breakdown voltage and better reliability due to the smooth interface between gate dielectric and poly-Si channel films. LTPS BG-TFTs with lateral silicon grains were therefore promising for future system-on-panel applications.

Acknowledgments

The authors thank the National Science Council of Taiwan for financial support of this research under Contract no. NSC 94-2218-E-009-027 and NSC 94-2218-E-009-028, and the Nano Facility Center (NFC) of National Chiao Tung University and National Nano Device Laboratories (NDL) for providing process equipment.

References

1. H. J. Kim, D. Kim, J. H. Lee, I. G. Kim, G. S. Moon, J. H. Huh, J. W. Hwang, S. Y. Joo, K. W. Kim, and J. H. Souk, in *SID Symposium Digest 30*, p. 184-187 (1999).
2. Y. I. Park, T. J. Ahn, S. K. Kim, J. Y. Park, J. S. Yoo, C. Y. Kim, and C. D. Kim, in *SID Symposium Digest 34*, p. 384 (2001).
3. Y. Mishima, K. Yoshino, M. Takei, and N. Sasaki, *IEEE Trans. Electron Devices*, **48**, 1087 (2001).
4. I. Wei Wu, Solid State Phenomena, **37~38**, 553 (1994).
5. A. Hara, F. Takeuchi, and N. Sasaki, *J. Appl. Phys.*, **91**, 708 (2002).
6. J. S. Im, H. J. Kim, and M. O. Thompson, *Appl. Phys. Lett.*, **63**, 1969 (1993).
7. J. S. Im and H. J. Kim, *Appl. Phys. Lett.*, **64**, 2303 (1994).
8. M. A. Crowder, P. G. Carey, P. M. Smith, R. S. Sposili, H. S. Cho, and J. S. Im, *IEEE Electron Device Lett.*, **19**, 306 (1998).
9. Paul Ch. van der Wilt, B. D. van Dijk, G. J. Bertens, R. Ishihara, and C. I. M. Beenakker, *Appl. Phys. Lett.*, **79**, 1819 (2001).
10. L. Mariucci, R. Carluccio, A. Pecora, V. Foglietti, G. Fortunato, P. Legagneux, D. Pribat, D. Della Sala, and J. Stoemenos, *Thin Solid Films*, **337**, 137 (1999).
11. R. Ishihara, A. Burtsev, and Paul F. A. Alkemade, *Jpn. J. Appl. Phys. Part 1*, **39**,. 3873 (2000).
12. S. Sakuragi, T. Kudo, K.Yamazaki and T. Asano, in the proceeding of the 12th International Display Workshops, p.965-968 (2005)
13. A. Hara, M. Takei, F. Takeuchi, K. Suga, K. Yoshino, M. Chida, T. Kakehi, Y. Ebiko, Y. Sano and N. Sasaki, *Jpn. J. Appl. Phys. Part 1*, **43**, 1269 (2004).
14. M. Tai, M. Hatano, S. Yamaguchi, T. Noda, S. K. Park, T. Shiba, and M. Ohkura, *IEEE Trans. Electron Devices*, **51**, 934 (2004).
15. M .H .Lee, S. J. Moon, M. Hatano, K. Suzuki, and C. P. Grigoropoulos, *J. Appl. Phys.*, **88**, 4994 (2000).
16. R. Ishihara, in *AMLCD Tech Digest 8*, p. 259-260 (2001).
17. D. K. Fork, G. B. Anderson, J. B. Boyce, R. I. Johnson, and P. Mei, *Appl. Phys. Lett.*, **68**, 2138 (1996).
18. T. Fujimura, A. Takami, A. Ishida, S. Kawamura, and T. Nishibe, in *AMLCD Tech Dig 8*, p. 175-178 (2001).

ECS Transactions, 3 (8) 149-156 (2006)
10.1149/1.2356348, copyright The Electrochemical Society

Excimer Laser Crystallization of Amorphous Silicon Film with Artificially Designed Spatial Intensity Profile Beam

Eok Su Kim[a,b], Ki-Bum Kim[a], Myung-Kwan Ryu[b], Hyuk Soon Kwon[b], Cheon-Hong Kim[b], Gon Son[b], and Jung Yeal Lee[b]

[a] School of Materials Science and Engineering, Seoul National University, Seoul, 151-742, Korea
[b] Research Center, BOE HYDIS TECHNOLOGY Co. Ltd., Ichon-si, Gyeonggi-do, 467-702, Korea

A new method to form polycrystalline silicon film with large grains at the controlled location by using a single irradiation of excimer laser beam is proposed for polycrystalline silicon thin film transistor. The excimer laser beam is modified to have a spatial intensity profile – periodic spatial variation of intensity maxima (I_{Max}) and minima (I_{Min}) – by the specially designed mask composed of the opaque and the transparent patterns where the opaque pattern size is less than the optical resolution of projection lens. Based on the melt depth of amorphous silicon at the location of I_{Min}, one can obtain three different regimes, namely, the partial melting, the near-complete melting, and the complete melting. Polycrystalline silicon grains grow vertically and laterally from the seeds at the location of I_{Min} to the complete molten region of I_{Max}. The evolution of polycrystalline silicon microstructure is investigated and elucidated in each regime.

Introduction

Polycrystalline silicon (poly-Si) thin film transistor (TFT) is an attractive device for active matrix liquid crystal display (AMLCD) and active matrix organic light emitting diode (AMOLED). The carrier mobility of poly-Si TFT, which is much higher than that of amorphous silicon (a-Si) TFT (~ 1 cm²/Vs), facilitates the integration of the pixel TFTs and the external driver ICs on the substrate. In particular, the higher stability of poly-Si TFT against the threshold voltage shift as well as the higher carrier mobility makes it possible to supply the stable and high current to the organic layer in AMOLED. However, in order to improve the quality of system-on-glass displays and to integrate the memory and the CPU on the substrate(1) in AMLCD and AMOLED, not only the superior electrical characteristics but also the device-to-device uniformity of poly-Si TFTs is strongly required.

Excimer laser crystallization (ELC) of a-Si film has been proposed(2,3) to fabricate poly-Si film at a low temperature compatible with the glass or flexible substrate. In ELC methods, the enlargement of poly-Si grains is necessary to improve the carrier mobility by reducing the number of grain boundaries within TFT channel, but it often resulted in degradation of the device-to-device uniformity due to the difference of the number and the location of grain boundaries within each TFT channel.(4,5) For this reason, several ELC methods have been proposed not only to enlarge poly-Si grains but also to control the location of grains.(6-21) Most of these techniques utilize the lateral growth or the super lateral growth of poly-Si grains from the seeds. Depending on how to get the seeds,

149

ELC methods can be classified as follows: (a) the modulation of the spatial intensity profile of excimer laser beam by the mask and/or optics,(6-10) (b) the modulation of a-Si film thickness by the substrate structures,(11-14) (c) the modulation of heat conduction by the heterogeneous layers on or under a-Si film,(15-17) (d) the grain filtering by the neck-shape patterns,(18,19) (e) the modulation of melting temperature between a-Si and poly-Si by the amorphization(20) or crystallization.(9,21)

In this study, we propose a new ELC method to enlarge poly-Si grains and to control their location at the same time by a single irradiation of excimer laser beam. Figure 1 shows a schematic diagram of a proposed ELC method which is based on the modulation of the spatial intensity of excimer laser beam by using the designed mask. Here, the mask is composed of the opaque and transparent line patterns, which are arranged periodically. The opaque pattern size is less than the optical resolution of projection lens, whereas the transparent pattern size is larger than that. By using these patterns, the spatial intensity profile of excimer laser beam is modulated to have the minimum intensity (I_{Min}), which is not zero, under the opaque patterns and the maximum intensity (I_{Max}) under the transparent patterns. It is obvious that one can control the relative intensity of I_{Max} and I_{Min} by adjusting the relative pattern sizes and, therefore, control the relative melt depth of a-Si film. Our intension is to grow grains laterally from the seed at the location of I_{Min} to the complete molten region of I_{Max} with the same crystallographic orientation.

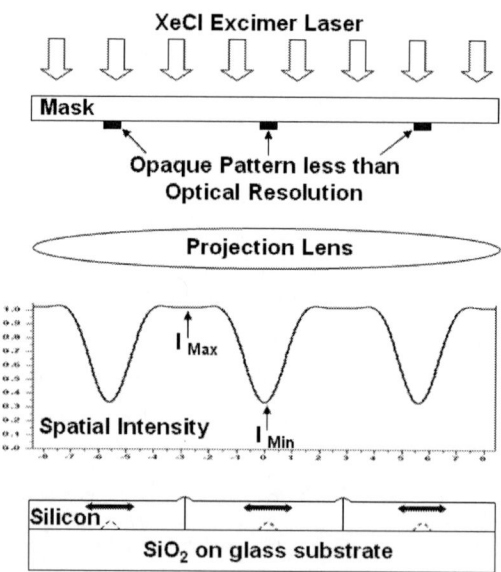

Figure 1. The schematic diagram of a new ELC method

Experimental

A 300 nm-thick silicon oxide (SiO_2) for buffer layer is deposited on the glass substrate by plasma enhanced chemical vapor deposition (PECVD) using silane (SiH_4) and nitrous oxide gas (N_2O) at 400℃ under 2 Torr. Then, a 50 nm-thick hydrogenated a-Si (a-Si:H) film is deposited successively by PECVD using silane (SiH_4) and hydrogen gas (H_2) at 400℃ under 3.5 Torr. The as-deposited a-Si:H film is dehydrogenated by furnace annealing at 500℃ for 2 hours in order to prevent the rapid evolution of hydrogen during the laser beam irradiation. The dehydrogenated a-Si film is irradiated by XeCl excimer laser beam from 0.30 J/cm² to 1.30 J/cm² of the energy density with the various opaque and transparent pattern sizes at room temperature. The wavelength (λ) is 308 nm and the original pulse duration time is 30 ns of full width half maximum (FWHM), but is extended to 240 ns of FWHM with the apparatus composed of mirrors and beam splitters in order to promote the lateral growth of poly-Si grains.

Results and Discussion

Excimer laser beam is screened by the mask composed of the periodic opaque and transparent line patterns. Excimer laser beam is diffracted by these mask patterns, but the projection lens with numerical aperture (NA) of 0.12 can not recollect the diffracted beams to reconstruct the image of mask patterns on a-Si film since the opaque pattern size is less than the optical resolution of projection lens. The optical resolution of the system is determined as 1.3 μm using the Rayleigh criterion(22) as R=0.5λ/NA. Therefore, a considerable intensity appears under the opaque pattern with the minimum intensity (I_{Min}) whereas the maximum intensity (I_{Max}) appears under the transparent pattern.

The spatial intensity profiles with the various opaque and transparent pattern sizes are calculated by utilizing the Fourier series analysis and the scalar diffraction theory.(23,24) With the decrease of opaque pattern size, I_{Min} becomes larger and approaches to I_{Max} gradually, but I_{Max} is almost unchanged and uniform regardless of the various transparent pattern sizes. The quantitative relation between I_{Max} and I_{Min} with the pattern size is expressed conveniently as the modulation function defined by $M=(I_{Max}-I_{Min})/(I_{Max}+I_{Min})$. Figure 2 shows the modulation values with the various opaque pattern sizes at fixed transparent pattern size of 5.0 μm.

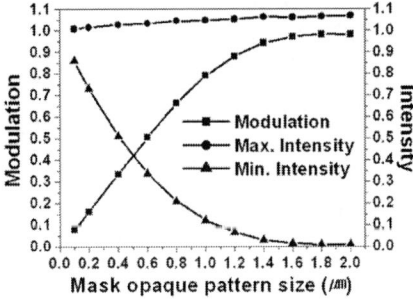

Figure 2. The modulation values with the various opaque pattern sizes

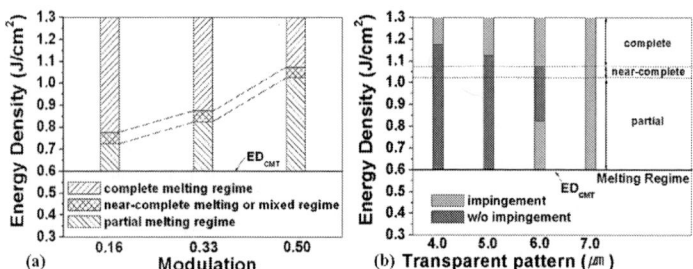

(a) Modulation   (b) Transparent pattern (μm)

Figure 3. The diagram of process window with the energy density
(a) three regimes for the various modulation values
(b) lateral growth for the various transparent pattern sizes

Regardless of the specific modulation value, a-Si at the location of I_{Max} is always melted completely over the threshold energy density (ED_{CMT}) of 0.60 J/cm². However, based on the melt depth at the location of I_{Min}, one can classify three different regimes, namely, the partial melting regime, the near-complete melting regime, and the complete melting regime. Here, the near-complete melting regime is defined such that a discrete few islands of Si, which are not melted completely, remain and the super lateral growth proceeds from those islands.(25) Figure 3(a) shows the process window of the aforementioned three regimes with the energy density for the various modulation values. With the increase of the modulation value, the partial melting regime enlarges since the intensity difference between I_{Max} and I_{Min} becomes larger. The mechanism of grain growth in each regime is shown with the schematic cross-sectional view in figure 4. Figure 5 shows the top-view SEM images of microstructure after secco-etching in each regime at the modulation value of 0.50 corresponding to the transparent / opaque pattern size of 5.0 μm / 0.6 μm

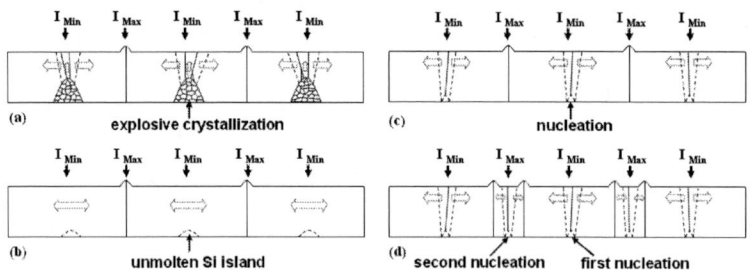

Figure 4. The mechanism of grain growth in each regime
(a) the partial melting regime
(b) the near-complete melting regime
(c) the complete melting regime
(d) the complete melting regime

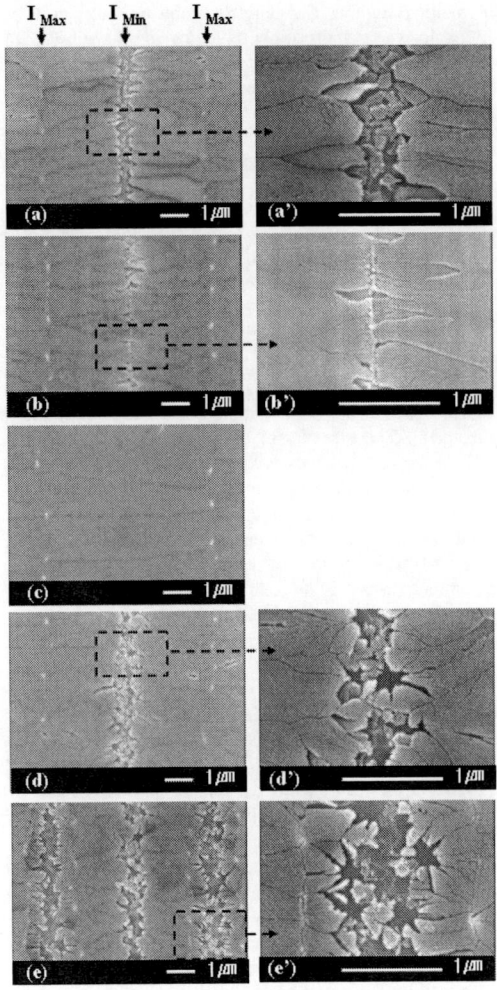

Figure 5. Top-view SEM images of microstructure after secco-etching in each regime
(a) at 0.90 J/cm² in the partial melting regime
(b) at 1.00 J/cm² in the partial melting regime
(c) at 1.05 J/cm² in the near-complete melting regime
(d) at 1.10 J/cm² in the complete melting regime
(e) at 1.20 J/cm² in the complete melting regime

In the partial melting regime, the poly-Si grains initially grow vertically at the location of I_{Min} by explosive crystallization(26,27) during the subsequent cooling. Then, the grains grow laterally to the complete molten region of I_{Max}. During the course of lateral growth, if nucleation starts to form at the location of I_{Max} by the supercooling of liquid Si, the lateral growth fronts are impinged by these nucleated grains at the location of I_{Max}. If not, the lateral growth fronts collide with each other at the location of I_{Max}. The extent of lateral growth depends on the energy density and the distance between the seeds which, again, is determined by the distance between the mask patterns. Therefore, the location and the size of grains can be controlled by the mask pattern configuration. Figure 3(b) shows the process window where the lateral growth fronts are impinged by the nucleated grains at the location of I_{Max} at the modulation value of 0.50 for the various transparent pattern sizes. The impingement does not occur for the transparent pattern size of 4.0 μm and 5.0 μm. However, the impingement occurs at low energy density for 6.0 μm pattern size and it occurs at every energy density for 7.0 μm pattern size. These results imply that the thermal budget is not sufficient to prevent the nucleation from liquid Si at the location of I_{Max} since the grains grow laterally with a relatively long distance. The longer is the distance between the mask patterns, the narrower is the process window without the impingement. The mechanism of grain growth is shown schematically in figure 4(a). Figure 5(a) shows the microstructure composed of small grains by the vertical growth at the location of I_{Min} and the large grains by the lateral growth at the location of I_{Max} at the energy density of 0.90 J/cm². Figure 5(a') zooms in on the small grains at the location of I_{Min}. With the increase of the energy density, the size of grains by the vertical growth at the location of I_{Min} becomes larger, but the region becomes narrower since the melt depth of a-Si is much deeper. Figure 5(b) and 5(b') show the ultimate microstructure in the partial melting regime at the energy density of 1.00 J/cm².

In the near-complete melting regime, a few Si islands remain without complete melt at the location of I_{Min} and serve as a single crystalline seed during the subsequent cooling. The grains grow bi-laterally from the surviving seed to the complete molten region of I_{Max}. This type of grain growth mode is called as the super lateral growth.[25] Like in the partial melting regime, the location and the size of grains can be controlled by the mask pattern configuration. As shown in figure 3(b), the bi-lateral growth fronts are impinged by the nucleated grains at the location of I_{Max} for the transparent pattern size of 7.0 μm. The mechanism of grain growth is shown schematically in figure 4(b). Figure 5(c) shows the array of large single grains formed by the super lateral growth at the energy density of 1.05 J/cm². However, it is difficult to have the uniform surviving seeds in the whole area since the near-complete melting is the transitional regime of going from the partial melting to the complete melting and corresponds to a singular phenomenon. There would be the mixed regime of partial melting, near-complete melting and complete melting rather than the near-complete melting regime.

In the complete melting regime, a-Si at the location of I_{Min} as well as I_{Max} is also melted completely. Nucleation occurs first at the location of I_{Min} by the supercooling of liquid Si since the melt temperature is the lowest at this location. The grains grow laterally from the nucleated grains to the complete molten region of I_{Max} until the growth fronts collide with each other in the center, which is shown schematically in figure 4(c). Figure 5(d) shows the microstructure composed of the nucleated grains at the location of I_{Min} and the large grains at the location of I_{Max} which grow laterally from the nucleated grains at the energy density of 1.10 J/cm². Figure 5(d') zooms in on the nucleated grains at the location of I_{Min}. With the further increase of the energy density, however, it takes a longer time to reach the nucleation condition by the supercooling of liquid Si. As it cools down for a longer time, the temperature profile tends to be flat rapidly by the heat transfer in the melt, so the temperature difference between the location of I_{Min} and I_{Max} becomes smaller when nucleation occurs at the location of I_{Min}. While the nucleation occurs first at the location of I_{Min} and the grains grow laterally from the first nucleated grains, the second nucleation occurs around the location of I_{Max} and impinges the lateral growth fronts from the first nucleated grains finally. With the increase of the transparent pattern size, as shown in figure 3(b), the lateral growth fronts are impinged by the second nucleated grains at the lower energy density due to the long distance and the rapid heat transfer. The mechanism of grain growth is shown schematically in figure 4(d). Figure 5(e) shows the microstructure composed of the first nucleated grains at the location of I_{Min}, the large grains which grow laterally and the second nucleated grains around the location of I_{Max} at the energy density of 1.20 J/cm². Figure 5(e') zooms in on the second nucleated grains at the location of I_{Max}.

Conclusion

We present a new ELC method to enlarge poly-Si grains and to control their location by a single irradiation of excimer laser beam without the additional processes. The spatial intensity profile of excimer laser beam is designed artificially to have I_{Max} and I_{Min} periodically by the mask opaque pattern with the size less than the optical resolution of projection lens. By these spatial intensity profile beam, the evolution of microstructures with the energy density is investigated and classified into three regimes and the growth mechanism is elucidated in each regime. By controlling the relative intensity of I_{Max} and I_{Min}, we can control the relative melt depth of a-Si film. Therefore, the grains grow vertically and laterally from the seed at the location of I_{Min} to the complete molten region of I_{Max} with the same crystallographic orientation. This method can be expanded to grow poly-Si grains at selective positions and, thereby, constitute the array of poly-Si grains of the square or hexagon shape by the various mask pattern configurations.

Acknowledgments

This work was supported from the National Program for Tera-level Nano Devices which is one of the 21st Century Frontier R&D Programs funded by the Ministry of Science and Technology of Korea.

References

1. B. Lee, Y. Hirayama, Y. Kubota, S. Imai, A. Imaya, M. Katayama, K. Kato, A. Ishikawa, T. Ikeda, Y. Kurokawa, T. Ozaki, K. Mutaguch, and S. Yamazaki, *ISSCC 2003*, paper 9.4, 164 (2003).
2. T. Sameshima, S. Usui, and M. Sekiya, *IEEE Electron Device Lett.*, **7**, 276 (1986).
3. K. Sera, F. Okumura, H. Uchida, S. Itoh, S. Kaneko, and K. Hotta, *IEEE Trans. Electron Devices*, **36**, 2868 (1989).
4. N. Yamauchi, J. -J. J. Hajjar, and R. Reif, *IEEE Trans. Electron Devices*, **38**, 55 (1991).
5. S. D. Brotherton, D. J. McCulloch, J. P. Gowers, J. R. Ayres, and M. J. Trainor, *J. Appl. Phys.*, **82**, 4086 (1997).
6. R. S. Sposili and J. S. Im, *Appl. Ph.. ¹ .··* , **69**, 2864 (1996).

7. J. S. Im, R. S. Sposili, and M. A. Crowder, *Appl. Phys. Lett.*, **70**, 3434 (1997).
8. C. -H. Oh, M. Ozawa, and M. Matsumura, *Jpn. J. Appl. Phys.*, **37**, L492 (1998).
9. L. Mariucci, A. Pecora, R. Carluccio, and G. Fortunato, *Thin Solid Films*, **383**, 39 (2001).
10. J. -Y. Park, C. I. Im, T. Hofmann, and D. S. Knowles, *SID 05 Digest*, P-60, 507 (2005).
11. K. -C. Park, S. H. Jung, W. -J. Nam, and M.-K. Han, *Mater. Res. Soc. Symp.*, J.1.5.1, 617 (2000).
12. C. -W. Lin, L. -J. Cheng, Y. -L. Lu, Y. -S. Lee, and H.-C. Cheng, *IEEE Electron Device Lett.*, **22**, 269 (2001).
13. A. Pecora, L. Mariucci, S. Piperno, G. Fortunato, *Thin Solid Films*, **427**, 314 (2001).
14. T. -F. Chen, C. -F. Yeh, C. -Y. Liu, and J. -C. Lou, *IEEE Electron Device Lett.*, **25**, 396 (2004).
15. H. J. Kim and J. S. Im, *Appl. Phys. Lett.*, **68**, 1513 (1996).
16. C. -H. Kim, I. -H. Song, W. -J. Nam, and M. -K. Han, *IEEE Electron Device Lett.*, **23**, 315 (2003).
17. D. -H. Choi, E. Sadayuki, O. Sugiura, and M. Matsumura, *Jpn. J. Appl. Phys.*, **33**, 70 (1994).
18. H. J. Song and J. S. Im, *Appl. Phys. Lett.*, **68**, 3166 (1996).
19. P. Ch. van der Wilt, B. D. van Dijk, G. J. Bertens, and R. Ishihara, *Appl. Phys. Lett.*, **79**, 1819 (2001).
20. H. Kumomi, *Appl. Phys. Lett.*, **83**, 434 (2003).
21. J. -H. Jeon, M. -C. Lee, K. -C. Park, and M. -K. Han, *Jpn. J. Appl. Phys.*, **39**, 2012 (2000).
22. Lord Rayleigh, *Philos. Mag.(5)*, **8**, 403 (1879).
23. B. M. Watraslewicz, *Optica Acta.*, **12**, 167 (1965).
24. B. Richards, E. Wolf, *Proc. Phys Soc.*, A253, 358 (1959).
25. J. S. Im, H. J. Kim, and M. O. Thompson, *Appl. Phys. Lett.*, **63**, 1969 (1993).
26. M. O. Tompson, G. J. Galvin, J. W. Mayer, P. S. Peercy, J. M. Poate, D. C. Jacobson, A. G. Cullis, and N. G. Chew, *Phys. Rev. Lett.*, **52**, 2360 (1984).
27. W. Sinke and F. W. Saris, *Phys. Rev. Lett.*, **53**, 2121 (1984).

ECS Transactions, 3 (8) 157-165 (2006)
10.1149/1.2356349, copyright The Electrochemical Society

Poly-Si TFT Technology: Advances in Material, Process and Device Technology

A. T. Voutsas

LCD Process Technology Lab, Sharp Labs of America, Camas, Washington 98607, USA

> In this paper we present advances in poly-Si TFT technology
> driven by the demands of the next SOP generations. Improved Si
> crystallization technology enables significant reductions of the
> variability in TFT characteristics, due to improved defect control.
> Advanced gate insulator technology allows thickness scaling of the
> gate dielectric film (concomitant to device scaling) without loss of
> performance or reliability. New plasma deposition technology
> developed and optimized by our group provides for such
> improvements due to virtual elimination of inadvertent plasma
> damage in the dielectric film. Research on novel device
> architectures enables the integration of such process improvements
> with device structures that complement and customize device
> performance according to function. SLA's research and
> development will be the main thrust behind the next generation of
> novel display products and applications.

Introduction

Polysilicon thin-film-transistors (TFT) are key building blocks for active-matrix-driven flat panel displays (FPDs). Many studies have demonstrated the ability of poly-Si based transistors to support a variety of functions beyond pixel switching, which has been the traditional role of TFTs in FPD applications (1-2). Poly-Si material enables the design of smaller TFTs that offer higher current and faster switching characteristics. As a result, pixel-driving circuits can be monolithically integrated on the display substrate (3). Such integration not only reduces the amount of external interconnections to the panel, but also improves the form-factor of the resulting display. The improved performance of poly-Si TFTs is expected to further yield increasing levels of component integration that will enable the fabrication of unique display systems (4).

To achieve high performance poly-Si TFTs, concomitant improvements at various levels are required. New elemental process technology is needed for the formation of high quality critical layers, such as the device active and the gate-insulator layers. In addition, improvements in the device architecture are vital in two aspects: (a) enable the fabrication of sub-μm channel dimensions with technology compatible with LCD manufacturing and (b) provide an additional means to compensate for deficiencies and variation in key layer, material properties.

In terms of active layer, high quality poly-Si microstructure is needed to increase device performance. The crystallization process is a very critical step of the TFT fabrication process, as it needs to satisfy conflicting requirements on material quality and cost and, at the same time, comply with the thermal-budget constraints imposed by the display substrate. Over the past 10 years, laser-based crystallization has been intensely studied and developed for poly-Si TFTs (5-7). Despite of its shortcomings, laser crystallization is the only technology with the ability to produce very high crystal-quality poly-Si films. This potential has been finally realized, in recent years, by the advent of a variety of laser-based lateral-crystallization technologies that have been shown to yield

157

extremely high-performance transistors with good uniformity (8-10). However, as the channel dimensions continue to shrink, issues of uniformity emerge even for optimally crystallized Si-films.

In the area of gate insulator, TEOS-based dielectric has been the technology of choice in display manufacturing. As in the case of VLSI, device scaling in poly-Si TFTs necessitates reductions in the gate insulator thickness. With the GI thickness gradually decreasing to 50nm and beyond, TEOS technology seems incapable of meeting the challenge, in terms of maintaining high quality for increasingly thinner GI layer.

Although improvements in elemental process technology can substantially elevate device performance, further complications exist, specific to LCD operations. For example, device scaling into the sub-μm domain is more challenging due to limitations in the resolution of lithography equipment that are capable of handling large substrates. Additionally, restrictions in the control of junction depth and doping accuracy are more prevalent in TFT technology due to limitations in process technology for doping and activation, imposed by equipment and substrate constraints. As a result, novel TFT device architectures are also needed to overcome such challenges.

Our group at Sharp Labs of America has been working on next generation poly-Si TFT materials and device technology. In this work we present recent developments in key elemental process technology and TFT device architecture, consistent with the issues of performance and scaling for SOP applications.

System-on-Panel Requirements

Figure 1 shows the anticipated evolution of poly-Si technology development and its impact on the degree of on-panel integration (adopted from ref. (2) and other published information). From the point of view of device performance, keywords are <u>high speed</u> and <u>low power consumption</u>. From the point of view of system, keywords are <u>high resolution</u> and <u>added value/functionality</u>.

Year	2004	2006	2008	2010
SOP Generation (arbitrary)	1st	2nd	3rd	4th
Display Resolution	300ppi	400ppi		
Power Supply	12V	3-5V	1.5-3V	1.5V
TFT Mobility	200-300cm^2/Vs	300-500cm^2/Vs	500cm^2/Vs	
Design Rule	3μm	1.5μm	0.8μm	0.5μm
Logic Frequency	~3MHz	10MHz	20-50MHz	50-100MHz
Key Process *Crystallization* *Gate Insulator* *Patterning*	CGSi	Advanced CGSi Thin GI Fine Patterning	Texture-control CGSi Ultra-thin GI Sub-μm Patterning	
Monolithic Integration	Digital Driver	Timing Generator Photo Sensor Amplifier	Display controller Image processor LN Amplifier	ULC Driver RF Capability Advanced MPU

Figure 1: Roadmap of poly-Si TFT technology

To achieve high speed, significant advances are needed in carrier mobility and fine lithography. For low power consumption, low and centered (between nMOS and pMOS device types) threshold voltage is needed with exceptionally tight distribution. Such characteristics will enable the realization of ultra-high resolution displays with on-board processing capability and small form-factor for novel products and applications.

Figure 2 illustrates the evolution of device design rules, in response to system requirements presented in figure 1. Current mass production technologies can apply to 1st SOP generation and be possibly extended to 2nd SOP generation. However, the development of more advanced panels/systems (i.e. 3rd generation and beyond) requires the development and introduction of new technologies in several key areas of process and device.

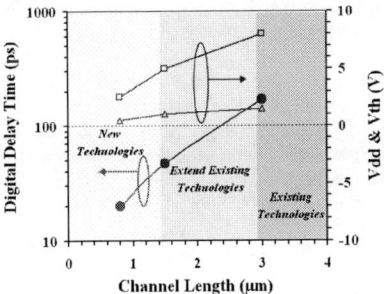

Figure 2: Technical specifications corresponding to SOP requirements. New technologies are needed beyond the 1.5µm design rule node.

Figure 3 shows a cross-section of a poly-Si TFT device and illustrates the specific requirements for key device layers to meet the technological demands presented in figure 2. Development of new elemental process technologies is needed to achieve high quality active and gate insulator layers in response to the itemized, specific requirements.

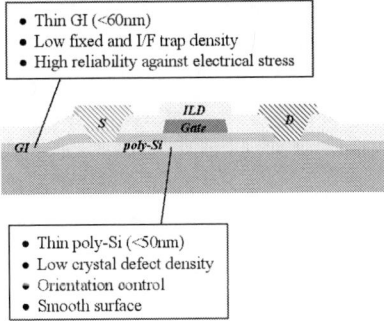

Figure 3: poly-Si TFT cross-section showing key device layers and associated, specific requirements.

Novel process Technologies

Silicon Crystallization

SLA has been developing novel, laser-based crystallization technology, dubbed "advanced-CGSi" (adv-CGSi). adv-CGSi enables significant advances in the microstructural characteristics of poly-Si material, such as long lateral growth length (LGL) and decreased defect density. The process relies on an advanced heating scheme that enables longer solidification time for the laser-irradiated (and molten) Si film. A representative microstructure of such poly-Si film is shown in figure 4(left). Using such poly-Si material for the active layer, TFT devices were fabricated and characterized.

Figure 4(right) summarizes the results, in terms of TFT mobility and threshold voltage. As shown, wide variation in TFT characteristics is observed. The reasons for this variation were carefully analyzed and ascribed to two main causes: (1) occasional incorporation of defective regions within the device active layer (i.e. device ❸) and (2) when the device channel region is formed on single grains, variation in the film crystallographic orientation (i.e. devices ❶ and ❷). To improve the absolute value and the uniformity of TFT characteristics, both causes have to be addressed. Clearly, out of the two, the former cause has a more profound impact on TFT performance and uniformity.

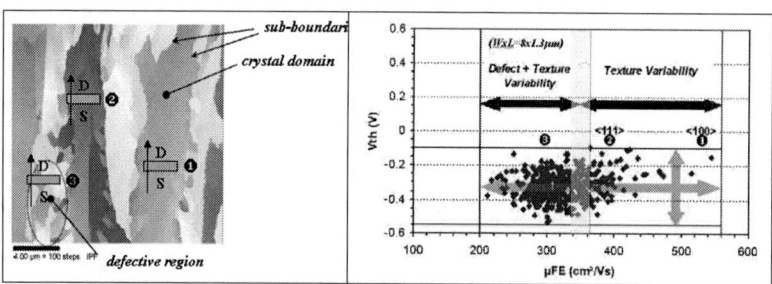

Figure 4: (left) Advanced poly-Si film microstructure. (right) Threshold voltage versus mobility for poly-Si TFTs fabricated by adv-CGSi process at SLA.

Variation in carrier mobility with Si crystallographic orientation is well-known and is attributed to the anisotropy in the effective electron/hole mass (11). This effect can explain electron mobility variation of up to ~150cm²/Vs between devices that happen to fall on pure <111> or <100> regions. Further decrease in mobility is attributed to the inclusion of material defects, within the active region, in addition to crystallographic variation. It follows that the impact of material defects is more severe on performance. Therefore, to improve mobility, reduction (and ultimately elimination) of material defects is the first priority, before attacking texture variation.

Threshold voltage is also affected by crystallographic orientation, although this effect is much weaker compared to that on mobility. The effect of texture is linked to variations in the density of surface states. "Bulk" defects in the poly-Si active layer are also contributing to variations in Vth. It again appears that the latter effect (bulk defects) is stronger. In that sense, measures that reduce bulk states should be first sought to combat threshold voltage variation.

Starting from these premises, SLA has further advanced adv-CGSi process by applying a localized crystallization scheme that enables better control of the microstructure. As a result, substantial improvement on the uniformity of TFT characteristics has been achieved, as shown in fig. 5.

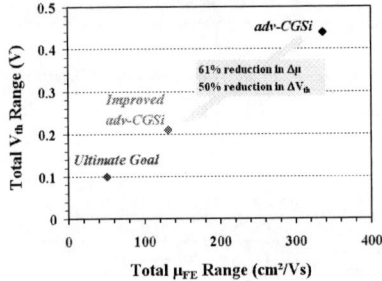

Figure 5: Improvements in variability control of TFT characteristics by adv-CGSi process family.

The achieved improvement is the result of the consistently reduced defect density within the crystal region that comprises the active region. Additional improvements, towards the set goal, are expected by employing partial (and ultimately total) texture control technologies. Such efforts are currently ongoing at SLA (for example see fig. 6).

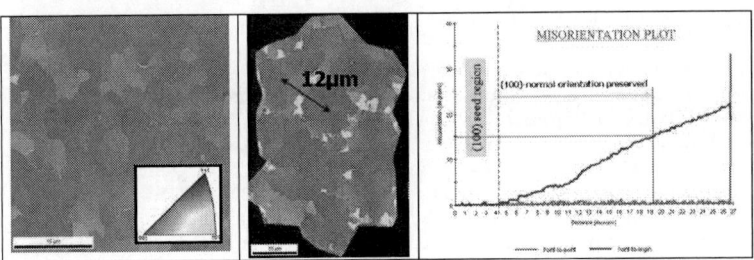

Figure 6: EBSP image of adv-CGSi crystallized Si island showing development of preferred crystallographic orientation. Left panel: seeded layer, demonstrating high yield of (100) normally-oriented grains (see inset Kikuchi triangle for color coding). Center Panel: same film after laser irradiation to achieve lateral growth (note the preservation of normal orientation to a typical distance of 12µm). Right Panel: detailed analysis of crystal misorientation, as a function of distance from seeded region.

Gate Insulator Formation

Gate insulator technology is a critical area for next generation poly-Si TFTs. This is driven by the stringent requirements in GI thickness (physical or electrical equivalent) and film quality – i.e. fixed and interface trap density. GI thickness reduction is necessitated by device scaling. An intermediate target of 50-70nm, followed by even more drastic thickness reduction to 20-30nm is shown in roadmaps of display device technology. As GI thickness decreases, issues of step coverage become increasingly more severe. With current GI technology (based on TEOS-SiO2) a trade off exists between film quality and step coverage. Film deposition temperature, in this case, is the key parameter. While TEOS films deposited at low temperature demonstrate excellent step coverage, they lack sufficient quality. On the flip side, high deposition temperature improves film quality, but degrades step coverage. Figure 7 demonstrates this trade-off.

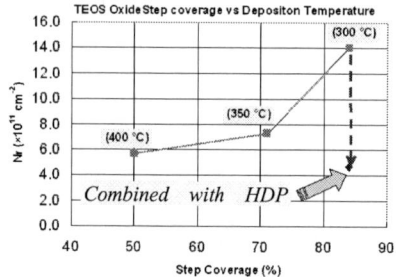

Figure 7: Trade off between TEOS-SiO2 step coverage and film quality (in terms of fixed-trap density) as a function of deposition temperature. New technology developed at SLA aims at resolving this trade-off in support of next generation TFT device requirements.

SLA has been developing High-Density-Plasma based gate-insulator technology in support of next generation TFT requirements (12-13). Based on the development and process optimization at SLA, superior quality dielectric films have been demonstrated, rivaling in quality that of thermal oxide. The excellent material characteristics are attributed to favorable plasma conditions that enable low plasma damage, especially around the interface region. This is, in part, explained by the ion energy distribution within the plasma configuration employed by SLA (see figure 8). Under optimal conditions we can obtain high concentration of desirable radicals, with narrow electron/ion energy distribution. The lack of distribution "tail" into the high energy region is particularly beneficial for keeping the plasma potential low and essentially eliminating plasma damage due to ion acceleration towards the deposited film.

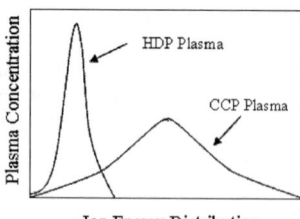

Figure 8: Active species concentration vs. ion energy distribution comparison between conventional (CCP) plasma and high-densi   P) plasma.

The proprietary HDP technology developed by SLA can be applied by itself, or in combination with conventional TEOS technology. When combined with TEOS-SiO2, HDP process can substantially improve the combined GI characteristics. Returning to figure 7, we observe a significant reduction in fixed-states when TEOS-SiO2 is optimally combined with HDP SiO2. Under these conditions, excellent step coverage AND film quality can be realized. Recent data indicate that such combination has further beneficial implications, as it is even more effective than conventional annealing steps in terms of achieving uniform film characteristics. In that sense, increasing the uniformity in flat-band voltage provides a direct means for improving the uniformity of the TFT threshold voltage.

Device Technology & Architecture

In addition to developing elemental process technologies, SLA has been studying modification of the basic device architecture to enable additional degrees of freedom and strategies for performance optimization. The thrust of this activity is function-driven architecture; in other words TFT architecture is tailored according to the device function. This can be better understood by placing different device architectures on speed versus power consumption map (figure 9).

Using a combination of different architectures on the panel we can resolve the usual trade-off between power consumption and speed. Moreover, SLA embodiments of double-gate and bottom-gate architectures are uniquely tied to advances in crystallization and gate insulator process technologies. At the very high end of the speed range, novel device concepts (dubbed "V-TFT") are being developed to enable deep-sub-μm design rules with existing (LCD) lithographic capabilities. The mixing and co-integration of different device architectures is aiming at:

1. Threshold voltage control and centering (without need of precise channel doping steps).
2. Control of hot-carrier (kink) effects (without additional drain engineering steps).
3. Integration of low-voltage and high-voltage TFTs without additional lithographic steps.
4. Realization of very high-speed TFTs for the ultimate integration of basic processing capabilities, on-panel.

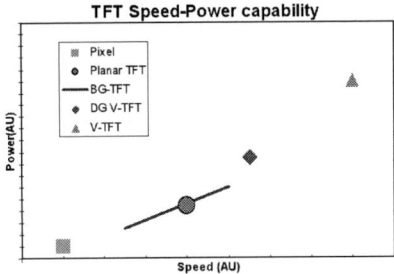

Figure 9: Power-vs.-Speed capability for various TFT architectures.

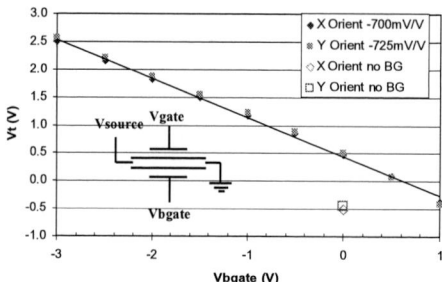

Figure 10: nMOS TFT threshold voltage adjustment as a function of back-gate voltage.

Items #1 and #2 can be achieved by simple modifications of the basic, planar TFT architecture. For example, some sort of double-gate structure can be used. SLA has been researching this technology and developing proprietary architectures. Examples of Vth swings by double gate structures fabricated at SLA can be seen in figure 10. Items #3 & #4 are being addressed by novel process technologies, among which, devices featuring a unique vertical-channel architecture. A notable example is the recent fabrication of a world-first 0.25µm-channel TFT by an i-line lithography tool-set.

Conclusions

SLA is developing novel process and device technologies in support to the requirements set forth for the several next SOP generations. New crystallization and gate insulator technologies are aiming at improving the quality of the respective device layers. Novel device architectures are also being studied to complement material quality enhancement measures and provide additional controls for system optimization. Based on the overall improvement plan, SLA technology is aiming at providing solutions for on-glass system integration and low power consumption. At the same time our technologies will achieve high quality, new function and lower processing cost. SLA's goals are both short-term and long-term. Short term goals are focusing in supporting the implementation of new technology into pilot manufacturing. Long term goals are the identification and research of the types of display-related technologies that Sharp ought to pursue to procure the next generation of "one-of-a-kind" display products.

Acknowledgments

The author would like to thank all the individuals who contributed to this work. Specifically, Dr. B. Crowder and Dr. R. Sposili on advanced crystallization technology, Dr. P. Joshi on advanced gate insulator technology and Dr. P. Schuele, Dr. H. Kisdarjono and Dr. T. Afentakis on device technology. The author is also indebted to many colleagues from Sharp Corporation for their guidance and support. Among them, Dr. T. Muramatsu, Dr. H. Komiya, Dr. Y. Takafuji, Mr. M. Maekawa and Mr. M. Moriguchi.

References

1. H. Sakamoto, N. Makita, M. Hijikigawa, M. Osame, Y. Tanada and S. Yamazaki, in Proceedings of SID 00 Digest, **XXXI**, 1190 (2000).
2. T. Matsuo and T. Muramatsu, in Proceedings of SID 04 Digest, **XXXV**, 865 (2004).
3. T. Nishibe, Mat. Res. Soc. Symp. Proc., **685E**, D6.1.1 (2001).
4. M. Brownlow, G. Cairns, C. Dachs, Y. Kubota, H. Washio and H. Yamashita, in Proceedings of SPIE, **4295**, 85 (2001).
5. S.D. Brotherton, D.J. McCulloch, J. B. Clegg and J.P. Gowers, *IEEE Trans. Electron Dev.*, **40**, 407 (1993).
6. A.M. Marmorstein, A.T. Voutsas and R. Solanki, *Solid State Electronics*, **43**, 305 (1999).
7. A.T. Voutsas, *Appl. Surf. Sci.*, **208-209C**, 250 (2003).
8. J.S. Im, R.S. Sposili and M.A. Crowder, *Appl. Phys. Lett.*, **70**, 3434 (1997).
9. J.S. Im, M.A. Crowder, R.S. Sposili, J.P. Leonard, H.J. Kim, J.H. Yoon, V.V. Gupta, H.J. Song and H.S. Cho, *Phys. Stat. Sol. A*, **166**, 603 (1998).
10. S.D. Brotherton, M.A. Crowder, A.B. Limanov, B. Turk and J.S. Im, in Proceedings of Asia Display/IDW '01, 387 (2001).
11. T. Sato, Y. Takeishi, H. Hara and Y. Okamoto, *Phys. Rev. B*, **4**, 1950 (1971).
12. P. C. Joshi, S. Droes, J. Flores, A.T. Voutsas, and J.W. Hartzell, in Proceedings of the Electrochemical Society, Chemical Vapor Deposition XVI and EUROCVD 14, **1**, 638 (2003).
13. P. C. Joshi, Y. Ono, A.T. Voutsas, and J.W. Hartzell, *Electrochem. Solid-State Lett.*, **7**, G62 (2004).

ECS Transactions, 3 (8) 167-172 (2006)
10.1149/1.2356350, copyright The Electrochemical Society

Preferred <100> Surface and In-plane Orientations in Self-assembled Poly-Si by Multiple Excimer Laser Irradiation

Ming He, Ryoichi Ishihara, Wim Metselaar, Kees Beenakker

Delft University of Technology, Delft Institute of Microelectronics and Submicron Technology (DIMES),
Laboratory of Electronic Components, Technology and Materials (ECTM)
Feldmannweg 17, 2628CT Delft, the Netherlands
Email: mhe@dimes.tudelft.nl

Clear preference of <100>- orientation in self-assembled poly-Si is observed for the first time not only in surface, but also in-plane orientations. This textured poly-Si can be used for TFT active channel, expecting a high performance with an excellent uniformity; or can be used as a seed layer for orientation controlling.

Introduction

With μ–Czochraski (grain filter) process, two-dimensional location-controlled single grains can be prepared [1]. Thin film transistors (TFTs) fabricated inside the above single grains have shown excellent performances. a field-effect mobility of 510 cm^2/Vs, subthreshold swing of 0.33 V/dec and off-current of 0.04 pA have been reported [2]. It has been found that crystallographic orientation of the grains is random [3]. It is well known that electronic properties of MOSFETs ([4], [5]) have a pronounced dependence on the surface and in-plane crystal orientations with respect to the direction of the current flow due to the anisotropy of the effective mass. The performance of single grain TFTs (*SG-Si* TFTs) can be improved further if crystallographic orientation of location-controlled single grains can be controlled in some preferred crystal orientation. Furthermore, by this means, the uniformity of TFTs will be improved as well.

In this study, a clearly preferred <100> orientation is observed for both surface and in-plane in square-shaped grains by multiple shots excimer-laser. Laser induced periodic surface structure (LIPSS) is observed and could be the reason for self-assembled square-shaped grains with three-dimensional texture. This textured poly-Si film can be directly used as active material for poly-Si TFT channel or can be used as a seed layer, combined with μ–Czochraski process, to prepare the orientation and location-controlled grains and SG-TFTs.

Experiment details

A 30 nm thick α-Si layer is deposited on thermally oxidized Si wafer (1 μm thick SiO_2) in a conventional horizontal hot-wall LPCVD reactor using pure silane as a source gas at a pressure of 20 Pa and a temperature of 547°C. Subsequently, the α-Si layer is irradiated with an excimer laser in a vacuum chamber by the XMR 5121 laser system (XeCl laser, $\lambda = 308$ nm, FWHM=50 ns) with an energy density varying from 250 to 280 mJ/cm^2 in 5 mJ/cm^2 steps. This energy density of the laser is slightly below the super-lateral growth (SLG) region [6]. The number of shots varies from 100 to 500, with a pulse repeating frequency of 5 Hz. There is about 3 degree between the incidence light and the normal direction of the surface. No intentional heating of the substrate is applied during the laser irradiation. Within the light path, a tilted quartz attenuator is used to reduce the energy density of the laser and to partly polarize the light.

Results and Discussion

<u>Self-assembly poly-grains:</u>

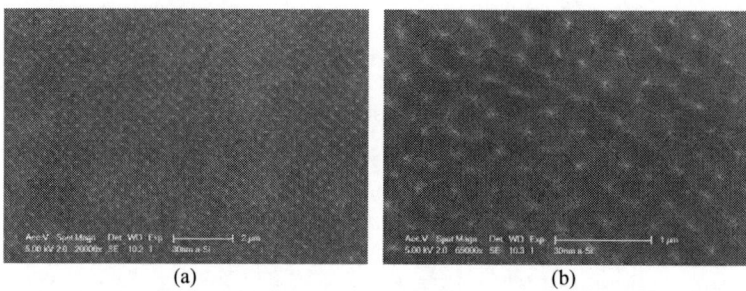

(a)   (b)

Figure 1 SEM images of self-assembly poly-Si grains, crystallized at 260 mJ/cm^2 after 500 shots: (a) periodic grain boundaries; (b) square-shaped grains.

After the excimer laser crystallization and Secco etching, morphologies of grains are investigated by SEM. Fig. 1 (a) is a SEM image showing the morphology of the poly-Si grains. Parallel, spatially periodic grain boundaries are visible and are well aligned in one direction with a uniform periodicity of about 300 nm. With a high-resolution view [Fig. 1(b)], it is clearly shown that grains are nearly square-shaped, with a grain size of 300 nm, comparatively equal to the wavelength of excimer laser (λ). It is found that the periodicity is independent of the number of shots. These characteristics indicate a very close connection to LIPSS. S. Horita et al. [7] have reported that periodic grain boundaries were created during multiple shots solid-state laser crystallization of an α-Si layer with linearly polarized light. Although excimer-laser light is known to have a minor degree of polarization, in the current experiment settings, the tilted quartz attenuator makes the light partly polarized.

The LIPSS has been confirmed by atomic force microcopy (AFM) analysis. Fig. 2 shows the AFM image of 5 μm by 5 μm area of poly-Si grains. The neighboring grains form a hillock, which represents the grain boundary. The periodic grain boundaries are aligned on one line, with a uniform spacing of about 300 nm. Square-shaped grains are self-assembled, and form a hillock at the each corner of the square.

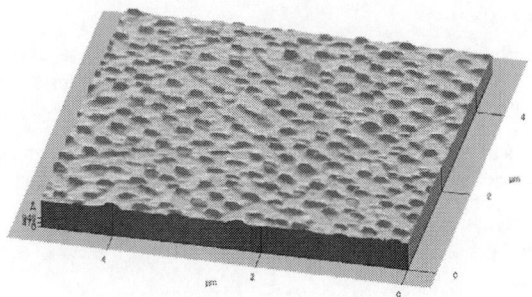

Figure 2 AFM image of the poly-Si grains, crystallized at 260 mJ/cm^2 after 500 shots.

Crystallographic orientation:

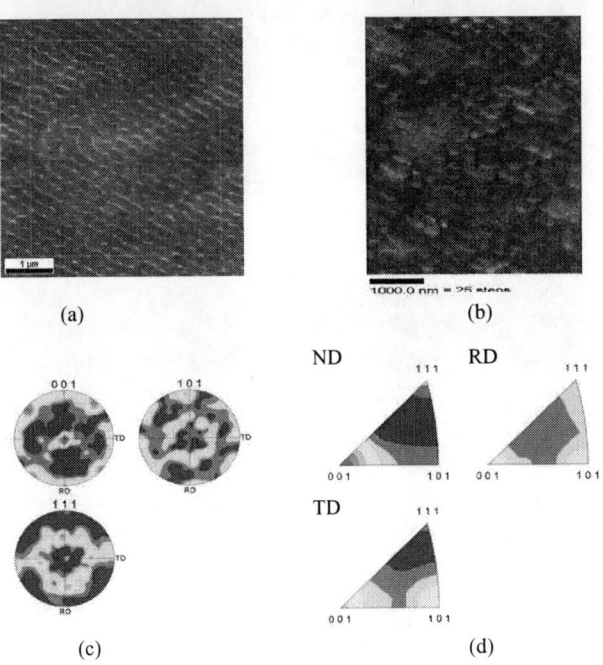

Figure 3 Texture of poly-Si grains, crystallized at of 260 mJ/cm^2 after 500 shots:(a) SEM image of poly-Si, mapping area is shown inside the rectangle; (b) crystal direction map of the analyzed area, red color indicates <100> crystal orientation in normal direction (ND); (c) pole figure of mapped area and (d) inverse pole figure in ND, rolling direction (RD) and transverse direction (Tl

The texture of the film is analyzed by electron backscatter diffraction (EBSD) with an automatic mapping method. Figure 3 shows the pole figures and inverse pole figure of poly-Si grains in a region as indicated in Fig.3 (a). Fig. 3(a) shows that the square-shaped grains are well-aligned. The clear preference of <100> texture is visible not only in the surface orientation [Fig. 3(d)] but also in the in-plane orientation [Fig. 3(c)]. This preferred in-plane orientation is achieved for the first time for lateral growth of Si grains. Furthermore, the preferred in-plane <100> orientation is either parallel or perpendicular to the LIPSS direction, i.e., other 4 <100> directions are perpendicular to the 4 sides of the square shaped grains.

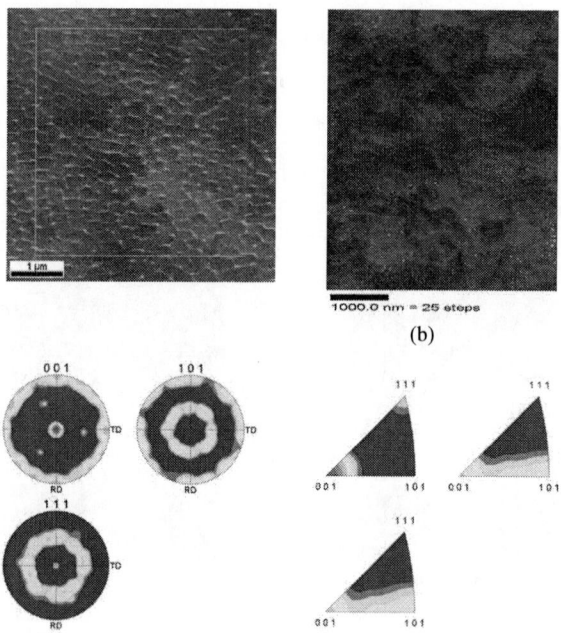

Figure 4 Texture of poly-Si grains, crystallized at 280 mJ/cm² after 500 shots: (a) the selected area for EBSD mapping; (b) crystal direction map of the analyzed area, red color show <100> surface orientation (ND); (c) pole figure of poly-Si and (d) inverse pole in ND, rolling direction (RD) and transverse direction (TD).

Figure 4 shows the texture of the poly-Si grains, crystallized at a higher energy. As shown in Fig. 4(a), the periodic grain boundaries are faceted, which means that the periodic grain boundaries, such as LIPSS in Fig 1 or in Fig. 3 (a) are not well aligned in a line. The grain boundaries of square-shaped poly-Si grains are aligned in a zigzagged line. The only difference between Fig.3 and Fig. 4 is the energy density, which indicates that the facet is easily formed at a higher energy density of laser. When the periodic grain boundaries are faceted, there is weak in-plane preferred orientation for these square-shaped grains, shown in Fig.4(c) and Fig.4 (d). However, the preferred surface orientation of <100> is still maintained. It should be noted that the poly-Si grains with square shape have the <100> preferred orientation; otherwise the preferred orientation changes to <111> orientation, as indicated in Fig. 4(b).

Mechanism for preferred orientation

In liquid phase crystallization, the <100>-texture has been explained by anisotropy of the melting temperature [8]. It is argued that <100>–orientated grains have a higher melting temperature and can coexist in molten Si liquid, as a result of which the new solidified poly-Si grains have <100>-preferred orientation. However, the difference of the melting temperature is negligibly small. For a 0.5 µm thick film, it is about 10^{-2} K. At the melting temperature, the heat conductivity of solid or molten Si is high enough to annihilate the small difference. In this study (Fig.3 and Fig. 4), the <100>-texture can be obtained within a relatively wide energy density window. Furthermore the liquid-solid coexistence is reported to be stable only when the Si contacts with oxide on both sides [9]. Another explanation for the preferred orientation is the anisotropy of surface free energy [9], which is argued for the strong <111>-texture obtained by multiple shots ELC [11]. However, the above anisotropy of the melting temperature is caused by and calculated from the anisotropy of the surface free energy [10]. Thus by the same reason, it could be possible to obtain both <111> and <100>-texture after the multiple shots ELC. In this study, strongly preferred <100> in-plane orientations are successfully obtained as well as the <100> preferred surface orientation in square-shaped poly-grains by multiple shots excimer-laser. LIPSS is observed and could be the reason for self-assembled square-shaped grains with three-dimensional texture.

During the LIPSS formation, the alternate directions of melting and solidification occur on a shot-to-shot basis [12]. Thus melting-solidification cycle requires in-plane 4-fold symmetrical lateral growth direction, the bipolarities of the directions perpendicular and parallel to the LIPSS. It is suggested here that the growth rate, for a given undercooling, is the fastest for <100> direction than other directions. With hundreds of melting-solidification cycles, the preferred in-plane <100> orientation is selected.

This textured poly-Si film can be used as active materials for TFT channels. This textured poly-Si film could be used as seeding layer, combined with µ–Czochraski (grain filter) process, to prepare orientation and location-controlled single grains for TFT fabrication.

Conclusions

In summary, strong <100>-texture in self-assembled square-shaped poly-Si grains are observed in both surface and in-plane orientation. This is achieved with for 30 nm α-Si precursor by multiple shots ELC. The strong texture co-exists with the LIPSS and can be obtained in a relatively wide energy density window. It is speculated that the <100>

orientation has fastest growth rate. During the LIPSS formation, in-plane directions of solidification, perpendicular and parallel to the LIPSS, require in-plane 4-fold symmetry and the <100>-orientation is selected and other orientations are occluded. This strong textured film can be used for TFT fabrication or seeds to prepare the orientation controlled grains.

Acknowledgments

The authors would like to thank Ellen Neihof, Yvonne Andel, and John Slabbekoorn for technical help for ELC. The authors would give the great thanks to Dr.Yasushi Hiroshima, Macro Krogt for the EBSD and AFM analysis, respectively. This work is part of the research program of the "Stichting voor Fundamenteel Onderzoek der Materie (FOM)", Project No. 97TF05, which is financially supported by the "Nederlandse Organisatie voor Wetenschappelijk Onderzoek (NWO)".

References

1. P.C.van der Wilt, B.D. van Dijk, G.J.Bertens, R. Ishihara, and C.I.M. Beenakker: Appl. Phys. Lett. 79 1819 (2001).
2. R. Vikas, R. Ishihara, Y. Hiroshima, D. Abe, S. Inoue, T. Shimoda, J.W. Metselaar and C. I. M. Beenakker, IEEE Proc. of IEDM'05, p. 37-3 (2005).
3. D. Danciu, R. Ishihara, F. Tichelaar, Y. Hiroshima, S. Inoue, T. Shimada, J. W. Metselaar, C. I. M. Beenakker, Digest of tech. papers of AMLCD'04, p: 145 (2004)
4. T. Sato, Y. Takeishi, H. Hara and Y. Okamoto, Phys. Rev. B 4 1950 (1971).
5. T. Sato, Y. Takeishi and H. Hara, Jpn. J. Appl. Phys. 8 588 (1969).
6. J. S. Im and H. J. Kim, Appl. Phys. Lett. 64 2303 (1994).
7. H. Kaki and S. Horita, J. Appl. Phys. 97 014904 (2005).
8. D. P. Gosain, A. Machida, T. Fujino, Y. Hitsuda, K. Nakano and J. Sato, Jpn. J. Appl. Phys. 42 L135 (2003).
9. H. I. Smith, C.V. Thompson, M. W. Geis, R. A. Lemons and M. A. Bosch, J. Electr. Soc. 130 2050 (1983).
10. R. J. Jacodin, J. Electrochem. Soc. 110 524 (1963).
11. H. Kuriyama, T. Nohda, S. Ishida, T. Kuwahara, S. Noguchi, S. Kiyama, S. Tsuda and S. Nakano, Jpn. J. Appl. Phys. 32 6190 (1993).
12. M. He, R. Ishihara, W. Metselaar and Kees Beenakker, <100>-textured self-assembled square-shaped poly-Si grains by multiple shots excimer laser crystallization, submitted to J. Appl. Phys.

ECS Transactions, 3 (8) 173-178 (2006)
10.1149/1.2356351, copyright The Electrochemical Society

Crystallization of Double-Layered Silicon Thin Films by Solid Green Laser Annealing and Its Application to Low Temperature poly-Si Thin Film Transistors

Y. Sugawara[a], Y. Uraoka[a], H. Yano[a], T. Hatayama[a], T. Fuyuki[a], and A. Mimura[b]

[a] Nara Institute of Science and Technology 8916-5, Takayama, Ikoma, Nara 630-0192, Japan
[b] National Institute of Advanced Industrial Science and Technology Tsukuba Central 5 1-1-1, Higashi, Tsukuba, Ibaraki 305-8565, Japan

> In this study, we propose a novel laser-crystallization technique by employing a double-layered Si thin films substrate for solid green laser anneal (GLA) crystallization, named GLADLAX (Green Laser anneal Double Layer X'tallization). Simultaneous crystallization of both the upper and the lower a-Si films of the substrate in a single laser scan was successfully achieved, with the upper a-Si becoming poly-Si with fine crystallinity, and the lower one, μc-Si. In addition, it was found that approximately 30% of the laser energy was reduced to obtain nearly the same grain size of the upper poly-Si formed through the GLADLAX technique, compared with that of the conventional GLA crystallized single layer poly-Si. Furthermore, we fabricated thin-film transistors using the upper Si film of GLADLAX poly-Si as their active channels, and found they had excellent switching performance, with their mobility exceeding $350 cm^2/Vs$.

Introduction

Solid Green Laser Annealing System

Laser-anneal-crystallized polycrystalline Si (poly-Si) thin films have attracted much attention due to its applicability to thin film transistors (TFTs) for flexible displays or future displays, such as electric papers or system displays. Recently, a solid green laser annealing (GLA) system employing the 2nd harmonics of Nd:YAG laser ($\lambda = 532nm$) has been developed for mass production use, featuring its advantages over excimer laser annealing systems, such as reduction of the maintenance cost and laser power stability (1). However, the absorption coefficient of poly-Si for the green laser is relatively smaller than for the excimer laser (2), so that approximately 80% of its laser energy penetrates through poly-Si films of 50nm-thick, resulting in a rather greater loss of laser energy in the case of a successive green laser irradiation to a-Si thin film substrates.

The Concept of GLADLAX

In this study, we propose a novel crystallization technique by employing a double-layered Si thin films substrate for green laser annealing crystallization, aiming to utilize the penetrating green laser light through the poly-Si film. It has two a-Si layers with an interlayer SiO_2 between them, so that, once the upper a-Si crystallizes to form poly-Si, the a-Si layer absorbs the penetrating light through the poly-Si and then crystallizes. We

173

named this laser-crystallization technique GLADLAX (Green Laser anneal Double Layer X'tallization) . The crystallization process of GLADLAX is shown in Fig.1.

In this paper, we investigated and discussed the crystallinity of both the upper and the lower Si films of the poly-Si substrates formed through the GLADLAX technique (GLADLAX poly-Si) depending on the crystallization laser energy. Furthermore, we fabricated the TFTs using the upper Si films of GLADLAX poly-Si, and their electrical characteristics are reported.

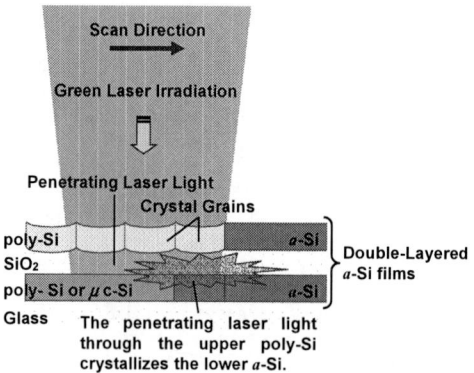

Figure 1. Schematic image explaining the crystallization process of the GLADLAX technique.

GLA Crystallization of Double-Layered a-Si Thin Films

<u>Sample Preparation</u>

For GLA crystallization, two kinds of a-Si substrates were prepared, one was the double-layered one, and the other was the conventional single layer one for the reference. Fig.2. shows a schematic view of the two substrates. A plasma-enhanced chemical vapor deposition (PECVD) system was utilized for the deposition of SiN, SiO_2 and a-Si films as shown in the figure. The 50nm-thick SiN and 100nm-thick SiO_2 films on the quartz were serving as the buffer layers. The two 50nm-thick a-Si films were separated by a 50nm-thick interlayer SiO_2 film in the substrate of double-layered a-Si films. Next, these a-Si substrates were dehydrogenated by annealing at 490 degrees in vacuum ambient. The samples were then crystallized by using the solid green laser annealing system developed by ULVAC, Inc. The pulsed laser irradiation was performed at a frequency of 40kHz. The laser irradiation pitch was $2\mu m$. The size of laser spot was 105mm times $40\mu m$. The energy density of laser was varied from 449 to 496 mJ/cm² for the double-layered a-Si films, and from 567 to 638 mJ/cm² for the single layer one.

<u>Evaluation Method</u>

The grain size and the surface morphology of poly-Si formed through the GLA crystallization were examined by scanning electron microscopy (SEM) and atomic force microscopy (AFM) respectively. The crystallinity of the lower Si films of crystallized double-layered a-Si substrates were evaluated by SEM and Raman spectroscopy after both the upper poly-Si films and the interlayer SiO_2 films were wet-etched by KOH and HF solution, respectively. Before SEM observation, the poly-Si surface was treated by Secco-etching solution.

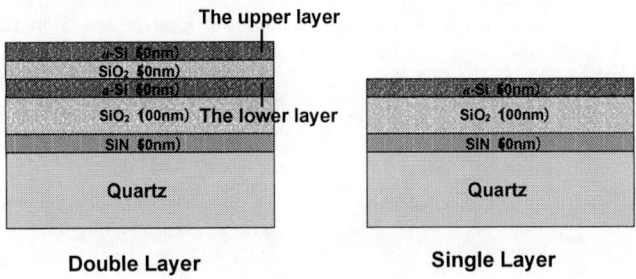

Figure 2. Schematic cross sectional view of the substrates of double-layered a-Si films (Left) and conventional single layer a-Si film (Right).

Crystallinity of GLADLAX poly-Si

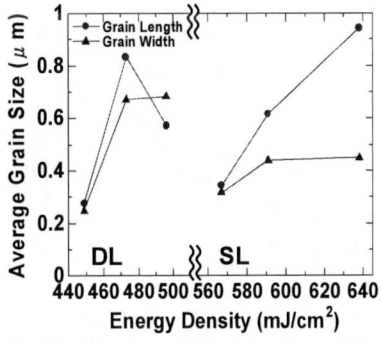

Figure 3. Relationship between average grain size and laser energy density.

Grain Size Fig.3 shows the dependence of the average grain size of the upper Si film of GLADLAX poly-Si (in region DL) and the GLA crystallized single layer poly-Si (in region SL) on the crystallization laser energy density. Here, the grain length denotes the size of grains measured in parallel with the laser scanning direction, while the grain width denotes the size of grains measured perpendicularly to it. A drastic enlargement of grain size of the upper poly-Si is observed at a lower energy density as compared with the crystallization of single layer a-Si, the grain size of which enlarged mainly toward the laser scanning direction. From this figure, it is found that approximately 30% energy is reduced to obtain nearly the same grain size of the upper Si film of GLADLAX poly-Si compared with that of the GLA crystallized single layer poly-Si.

The Lower Layer Fig.4 shows Raman spectra of the lower Si film of GLADLAX poly-Si. Peaks are observed over the energy density of 473mJ/cm^2, indicating the crystallization of the lower a-Si over that energy condition. Also, SEM observation reveals that the lower a-Si crystallizes to form μc-Si. These results demonstrate that the simultaneous crystallization of both the upper and lower films of double-layered a-Si substrate is successfully achieved. Furthermore, it is noted that the enlargement of the upper poly-Si occurs at the energy condition of the crystallization of the lower a-Si, implying that the crystallization of the lower a-Si plays a special role in the grain-enlargement of the upper Si film of. GLADLAX poly-Si

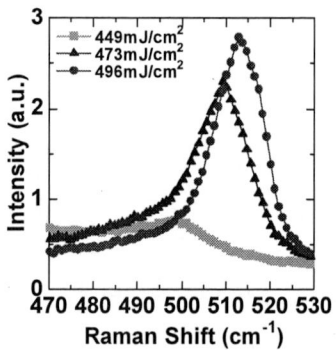

Figure 4. Raman spectra of the lower Si films of GLADLAX poly-Si substrates.

Discussion It is supposed that when the penetrating laser light through the upper poly-Si film has enough energy to crystallize the lower a-Si to form μc-Si, a heat generation occurs at the crystallizing region, which decreases the thermal gradient in the vicinity of the melting part of the upper a-Si, causing the extension of its melting time and thus, its grain-enlargement. In short, the lower a-Si layer is acting as a thermal reservoir when it crystallizes. In addition, it is also supposed that the lower a-Si and/or μc-Si layer act as a reflector of the penetrating laser light, so that the reflected laser light also contributes to the annealing of the upper a-Si film, which will results in the reduction of the energy of the incident laser light required to melt the upper a-Si film.

GLADLAX poly-Si TFT

TFT Fabrication

TFTs using the upper Si films of GLADLAX poly-Si were fabricated. Also fabricated were the conventional GLA single layer poly-Si TFTs for the comparison. These TFT were fabricated through a conventional self-align process the maximum process temperature of which was 500 degree. Fabrication process flow is as follows. After the poly-Si film was patterned to form islands, a 100nm-thick SiO_2 as the gate insulator was deposited by PECVD using a mixture of Tetraethoxysilane (TEOS) and O_2 as a source gas. A 250nm-thick Mo film was deposited and patterned to form gate electrodes. Phosphorous was implanted to form source and drain regions. After a 400nm-thick SiO_2 was deposited, the samples were furnace-annealed by 6 hours at 500℃ for impurity activation. After contact holes were opened, the source/drain electrodes (Ti/Al) were formed. Finally, a post metallization anneal was performed for 1 hour at 450 degree in a forming gas atmosphere.

The transfer characteristics of TFTs with both the channel length and the channel width of 5μm were evaluated by a semiconductor parameter analyzer.

Electrical Characteristics

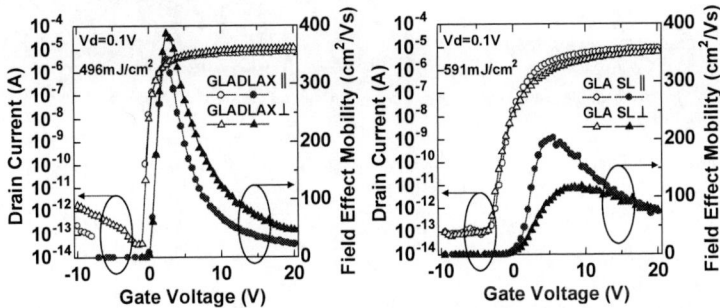

Figure 5. Transfer characteristics of TFTs (W/L=5/5[μm]) using the upper Si film of GLADLAX poly-Si (Left), and the GLA crystallized single layer poly-Si (Right).

Fig. 5 shows typical transfer characteristics of TFTs with their active channels made of the upper Si film of GLADLAX poly-Si (GLADLAX-TFTs) and the GLA crystallized single layer poly-Si (GLA SL-TFTs). The crystallization laser energy density is indicated in each figure. Here, GLA SL $\parallel$ and GLADLAX$\parallel$ denote the TFTs with their channels direction in parallel with the laser scan direction, while GLA SL$\perp$ and GLADLAX$\perp$ denote the TFTs with their channels in perpendicular to it. The electrical characteristics of GLA SL-TFTs have clear dependence on their channel direction, with the mobility of GLA SL$\perp$ considerably lower than that of GLA SL$\parallel$. This is explained by the fact that the channel of GLA SL$\perp$ crosses more grain boundaries, which impede the carrier-transport, than GLA SL $\parallel$ (3, 4). On the other hand, an excellent switching performance and its independence on their channel direction are remarkable for GLADLAX-TFTs, with their mobility over 350cm^2/Vs, strongly reflecting the isotropic morphology and fine crystallinity of the upper Si film of GLADLAX poly-Si.

Application of GLADLAX poly-Si

In this section, a few possible applications of GLADLAX poly-Si is briefly considered. Fig.6 shows images of application of poly-Si formed through GLADLAX technique. The GLADLAX poly-Si is able to be applied to devices such as double-gate TFTs by using the lower μc-Si as a lower gate material of them. The upper poly-Si can be used as the channel of the TFTs, and the fine crystallinity of the poly-Si will ensure superior performance of the TFTs with fine gates, indicating their applicability to reliable large scale integration. The lower μc-Si film can also be used as a sacrificial layer for a device transferring process, so that the integrate TFT-circuits utilizing the upper layer of GLADLAX poly-Si can be transferred on to any substrates, such as flexible plastic films, proposing a novel method for the production of flexible IC chips.

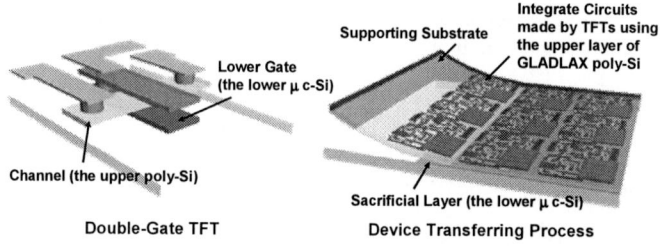

Figure 6. Applications utilizing the double-layered poly-Si films made by GLADLAX: double-gate TFT (Left), device transferring process (Right).

Conclusions

GLA crystallization of double-layered a-Si thin films using a solid green laser anneal system was performed and crystallinity of the obtained poly-Si films was investigated with respect to the crystallization laser energy. Both the upper and the lower a-Si films of the GLA double-layered a-Si substrates were successfully crystallized in a single laser scan, with crystallization laser energy reduced approximately 30% compared with that of conventional single layer a-Si films, suggesting the laser reflection and thermal reserving role of the lower Si layer in GLA crystallization of double-layered a-Si substrates. The TFTs with their active channels made of upper layer of the GLADLAX poly-Si show excellent transistor characteristics, with their mobility higher than 350cm^2/Vs, reflecting the improved crystallinity of the poly-Si. Thus, the double-layered poly-Si films, formed by this technique, GLADLAX, will be promising for the future thin-film electronics.

Acknowledgments

The authors wish to thank ULVAC, Inc. for the preparation of GLA samples.

References

1. M. Hayama, et al., *Extended Abstracts (The 52nd Spring Meeting); The Japan Society of Applied Physics and Related Societies*, No.0, 78 (2005).
2. R. F. Wood, et al., *Semiconductor and Semimetals* vol.23, *Pulsed Laser Processing of Semiconductors*, p.107, Academic Press, Inc. (1984).
3. Y. Helen, et al. *Thin Solid Film* **383**, 143 (2001).
4. T. Nishimura, et al., *Jpn. J. Appl. Phys.* **22**, suppl.22-1, 217 (1983).

ECS Transactions, 3 (8) 179-183 (2006)
10.1149/1.2356352, copyright The Electrochemical Society

Poly-Si Thin Fim Transistor with Multiple Nanowire Channels Prepared by Excimer Laser Annealing

Po-Chuan Yang[a], Chao-Yu Meng[b], Ming-Wei Tsai[a], Si-Chen Lee[a], Fellow, IEEE

[a] Graduate Institute of Electronics Engineering , National Taiwan University, Taipei, Taiwan , R.O.C.
[b] Graduate Institute of Electrical Engineering , National Taiwan University, Taipei, Taiwan , R.O.C.

Low temperature (<500°C) polysilicon thin-film transistors (LTPS TFTs) prepared by KrF excimer laser annealing with multiple nanowire channels have been fabricated successfully. Grain boundary defects of poly-Si can be reduced effectively by multichannel structure. The ten 40 nm split channels TFT passivated by oxygen and NH_3 plasma exhibits good electrical performance, i.e., high ON/OFF current ratio (>10^6), better subthreshold swing and suppressed kink effect.

Introduction

Low temperature polysilicon thin film transistors (TFTs) have drawn much attention because they can accomplish various functional devices for displays [1-3], sensors [4], and other complicated circuits [5]. It is expected that the next generation display will integrate all functional components monolithically on a substrate.

Three techniques have been applied to fabricate low temperature poly-Si film, i.e., solid-phase crystallization (SPC) [6], metal-induced lateral crystallization (MILC) [7] and excimer-laser annealing (ELA) [8]. ELA was found to be easier and more effective to use in low temperature and large area preparation of poly-Si because the crystallization occurs rapidly. Moreover, unlike MILC, the poly-Si film prepared by ELA can avoid the problem of metal contamination and thermal damage in the glass substrate.

The presence of the grain boundary defects in the poly-Si film and SiO_2/Si interface deeply affect the electrical characteristics of the poly-Si TFTs. It is reported that the multi-channel structure can effectively reduce grain boundary defects [8,9]. In addition, oxygen plasma pre-treatment [10-12] and NH_3 plasma post-treatment [13] not only terminate the dangling bonds in the laser crystallized poly-Si films, but also form stronger Si-N bonds at the SiO_2/Si interface.

Experiments

Figure 1(a) displays the schematic diagram of the fabrication processes of the poly-Si TFT. First, a 50 nm thick undoped amorphous Si (a-Si:H) and 20 nm thick n-type phosphorus-doped amorphous Si layer were deposited by plasma enhanced chemical vapor deposition (PECVD) on oxidized Si substrate. During deposition, the substrate temperature, RF power density and chamber pressure were fixed at 250°C, 0.11 W/cm^2 and 0.45 torr, respectively. The gas flow rate for undoped a-Si:H layer was 2(H_2) and 5(SiH_4) sccm, and n-type doped a-Si:H layer 0.8(H_2), 5(SiH_4) and 1.2(PH_3) sccm,

179

respectively. Second, the channel region was defined and samples were sent to furnace to expel hydrogen from the film for 10 minutes at 500 °C under nitrogen environment. After annealing by the excimer laser (KrF, duration time 28ns, 330 mJ/cm²), multiple nanowires were patterned by electron beam lithography and transferred by reactive ion etching (RIE). Then, a 100 nm thick SiO$_x$ was deposited by PECVD at 250 °C as a gate insulator following oxygen plasma treatment of the dangling bonds in the laser crystallized poly-Si films for 30 seconds. The conditions for oxygen plasma treatment were RF power of 200W, temperature of 250 °C, pressure of 67 Pa and 200 sccm of oxygen flow. The contact windows for source and drain were opened. The samples were treated by NH₃ plasma at 250 °C to passivate the grain boundary for 3 hours with pressure of 2 torr and RF power density of 0.44 W/cm². Finally, a 150 nm thick aluminum was evaporated on the devices to form ohmic contact and gate/source/drain electrodes.

The I-V characteristics of fabricated devices were measured by a Keithley 2361 semiconductor parameter analyzer. The threshold voltage is defined from the intercept on the voltage axis of the straight line in the square root of drain current ($I_D^{1/2}$) versus the gate voltage (V_{GS}) plot. The leakage current I_{OFF} is the minimum value of the drain current at $V_{ds} = 0.1$ V.

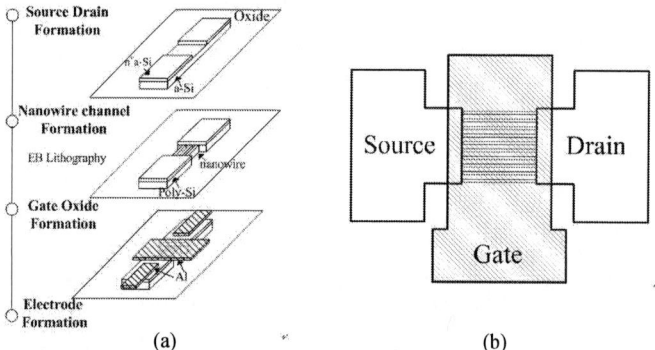

(a)   (b)

Figure 1. (a) The schematic diagram showing the fabrication processes of the multiple channel poly-Si TFT. (b) Top view of the proposed device stricture.

Results and Discussion

Figure 1(b) demonstrates the top view of the proposed device structure. In this work, devices are divided into three groups based on their channel number and width, i.e., 400 nm single channel (E1), 200 nm dual channel (E2), and ten 40 nm multi-channel (E3), all with the same effective channel width, and channel length (5µm), as listed in Table I.

TABLE I. Device dimension of E1,E2 and E3 poly-Si TFT.

Device Name	Gate length (μm)	Channel Number	Each channel width (nm)
E1	5	1	400
E2	5	2	200
E3	5	10	40

Figure 2(a) shows the optical microscope picture of the ten 40 nm multi-channel poly-Si TFT before the deposition of aluminum electrodes. The major advantage of this process lies in the saving of time and its beneficial effect. In contrast to the traditional method which usually takes four electron beam lithography steps, only one step is used in this case and a special photo mask pattern has been designed to make tolerance smaller than 1 μm. Figure 2(b) depicts the scanning electron microscope (SEM) image of the active pattern of E3 device. As the inset displays, each nanowire channel is 40 nm wide, and the distance between each neighboring channel is 1 μm.

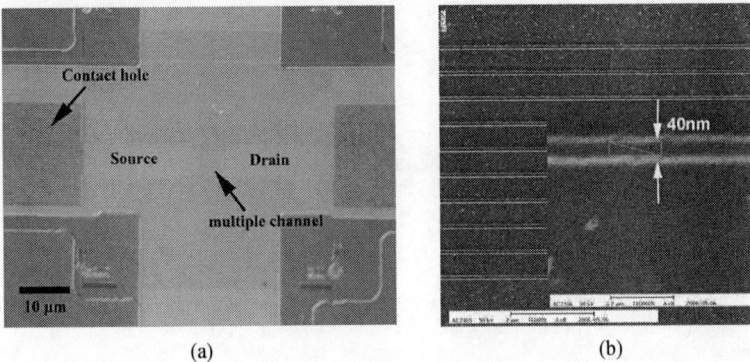

(a)   (b)

Fig. 2 (a) The optical microscope picture of the E3 device. (b) The SEM image of the active pattern .The inset displays the 40 nm nanowire channel.

In order to examine the effect of multi-channel, the transfer curves of multiple and single channel devices are compared. In Fig. 3, it is clear that the multi-channel devices exhibit higher ON/OFF current ratio ($>10^6$) and a better subthreshold swing. It is because that more channel areas were exposed to the oxygen and NH_3 plasma, and the dangling bonds in the poly-Si films and SiO_2/Si interface defects were also passivated. Hence, there are less grain boundary defects in the channels.

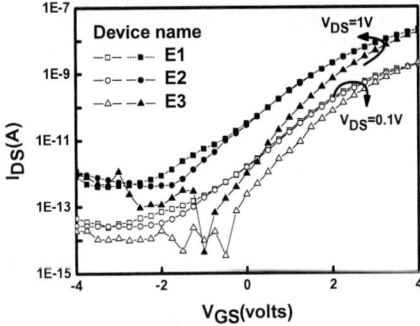

Fig. 3 The transfer curves of three poly-Si TFT devices with different channel number and width.

Fig. 4 displays the output characteristics of E1, E2 and E3 devices. E3 shows a better result mainly because the nanoscale channels are fully controlled by the gate metal. Accordingly, the kink effect is suppressed.

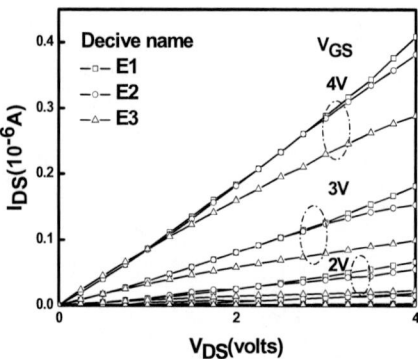

Fig. 4 Output electrical characteristics of E1,E2 and E3 devices .

Table II lists the device performance of three types of poly-Si TFTs. The mobility and threshold voltage (V_T) of the TFTs are almost the same, but ON/OFF current ratio of E3 device is the best.

TABLE II. Comparison of poly-Si TFTs devices parameters

Device Name	I_{ON}/I_{OFF}	V_{TH} (V)	Mobility (cm2/V-s)
E1	5.1×10^4	1.5	78
E2	4.9×10^4	1.6	80
E3	4.2×10^6	1.9	89

Conclusions

In conclusions, LTPS TFTs prepared by KrF excimer laser annealing with multiple nanowire channels have been fabricated successfully. After plasma treatment, TFTs with multi-channel structure and smaller channel width result in less grain boundary defects in poly-Si. The ten 40 nm split channels TFT exhibits a good electrical performance, i.e., high ON/OFF current ratio ($>10^6$), a better subthreshold swing and suppressed kink effect.

Acknowledgments

This work was supported by the National Science Council of the Republic of China under Contract No. NSC 94-2120-M-002-013.

References

1. J. W. Park, M. C. Lee, W. J. Nam, I. H. Song, and M. K. Han, *IEEE Electron Device Lett.*, **22,** 402 (2001).
2. C. C. Wu, S. D. Theiss, G. Gu, M. H. Lu, J. C. Sturm, S. Wagner and S. R. Forrest, *IEEE Electron Device Lett.*, **18**, 609 (1997).

3. J. Y. Lee, J. H. Kwon and H. K. Chung, *Organic Electronics*, **4**, 143 (2003).
4. F. Yan, P. Estrela, Y. Mo, et al. *Applied Physics Lett.*,**86**, 053901(2005)
5. T. Sameshima and S.Usui, *Journal of Applied Physics*, **74**, 6592 (1993).
6. Nbriyoshi Yamauchi, and Rafael Reif, , *Journal of Applied Physics*, **75**, 3235 (1994).
7. Seok-Woon Lee and Seung-Ki Joo, *IEEE Electron Device Lett.*, **17**, 160 (1996).
8. I. H. Song, C. H. Kim, S. H. Kang, W. J. Nam and M. K. Han, *IEDM*, pp. 561-564, (2002).
9. Yung-Chun Wu, Chun-Yen Chang, Ting-Chang Chang, Po-Tsun Liu, Chi-Shen Chen, Chun-Hao Tu, Hsiao-Wen Zan, Ya-Hsiang Tai, and Simon Min Sze. *IEDM*, (2004).
10. Y. Tsunoda, T. Sameshima, and S. Higashi, *Jpn. J. Appl. Phys.*, **39**, 1656 (2000).
11. Seiichiro HIGASHI, Daisuke ABE, Yasushi HIROSHIMA, et al. *Jpn. J. Appl. Phys.*, **41**, 3646(2002).
12. Horng Nan Chern, Chung Len Lee, and Tan Fu Lei, *IEEE Transactions on Electron Devices*, **40**, 2301(1993).
13. Huang-Chung Cheng, Fang-Shing Wang, and Chun-Yao Huang, *IEEE Transactions on Electron Devices*, **44**, 64 (1997).

ECS Transactions, 3 (8) 185-191 (2006)
10.1149/1.2356353, copyright The Electrochemical Society

Green Laser Crystallization of α-Si Films Using Preformed α-Si Lines

I. Brunets[a], J. Holleman[a], A.Y. Kovalgin[a],
A.A.I. Aarnink[a], A. Boogaard[a], P. Oesterlin[b], J. Schmitz[a]

[a] MESA+ Institute for Nanotechnology, Chair of Semiconductor Components,
University of Twente, P.O. Box 217, 7500 AE Enschede, The Netherlands.
[b] Innovavent GmbH, Bertha-von-Suttner-Str. 5, 37085 Goettingen, Germany.

In this work, amorphous silicon films with preformed α-Si lines
were crystallized using a diode pumped solid state green laser
irradiating at 532 nm. The possibility of controllable formation of
grain boundaries was investigated. The crystallization processes in
the rapidly melted silicon films were discussed. The influence of
the crystallization parameters (i.e., energy density, scan velocity,
etc.) and structure type (i.e., with and without preformed lines) on
properties of the crystallized films was studied. The laser treatment
with an energy density of 1.00 J/cm^2 at a laser pulse overlapping of
90% provided the optimal crystallization process with predefined
grain boundary location. X-ray diffraction (XRD), SEM and AFM
microscopy have been used to characterize the crystallized silicon
films.

Introduction

Low thermal budgets are required to realize 3-D integrated circuits (IC) such as stacked
memory devices (1, 2). Various integrated memory designs are investigated in (3-5). It is
concluded that for a better performance of the circuit components, an additional quality
improvement of low temperature silicon films is needed. Laser crystallization of low-
temperature deposited amorphous silicon can provide polysilicon films at low substrate
temperature, with sufficiently large grains (6-11). However, inherent to this technique is
the random position of grain boundaries, leading to large device-to-device variations.
Recently, novel techniques were reported allowing the controlled formation of grain
boundaries in a-Si laser crystallization through air-gap formation (12), two-pass laser
crystallization (13), or the introduction of buried crystallization seeds (14). Bearing
manufacturing cost and yield considerations in mind, it is at this time unclear what is the
most effective solution for controlled grain formation.
In this paper, a novel approach for the controlled location of grain boundaries is
described. The re-crystallization seeds are amorphous silicon strips fabricated with
standard methods. The preformed α-Si lines patterned prior to the deposition of α-Si
introduce an additional temperature gradient and provide controlled lateral crystallization
of the molten material. This method can easily be introduced in silicon-compatible
process flows and can allow a 3-D integration of devices with significantly improved
characteristics.
The applied laser wavelength plays an important role during crystallization process. The
glass substrates and interlayer oxide must be transparent at this wavelength to prevent
their heating which can cause cracks and other damage or deformations. At the same time
the absorption coefficient for amorphous Si film must be high enough to provide
complete melting of the irradiated film. In ---- of excimer laser crystallization (308 nm

wavelength) the penetration depth is ≈ 10 nm (9), so primarily a thin top layer of silicon film is melted and only further thermal diffusion provides complete melting of the film. For this reason we chose for an irradiation at 532 nm wavelength, giving a lower absorption coefficient (15) leading to a more homogeneous melt (16). This increases the process window for re-crystallization in terms of the laser power and pulse duration.

Experimental

Silicon films with preformed α-Si lines

To realize a controllable crystallization of the amorphous silicon films we used preformed α-Si lines patterned prior to the deposition of the amorphous films (see Figure 1). First, a 50 nm thick α-Si film was deposited by LPCVD at 550 °C on top of a 0.7-µm thick thermally grown SiO_2 layer. After patterning the film, the lines with different width (from 0.4 up to 2.0 µm) were formed. Subsequently, a 100-nm thick α-Si layer was deposited using the same technique, which resulted in an amorphous film with a periodically varied thickness.

During the following laser crystallization, this varied film thickness influenced the temperature gradient in the lateral direction perpendicular to the laser scan direction. Re-crystallization will start where the temperature is lowest, and through this mechanism, the preformed lines determine where the main grain boundaries will occur.

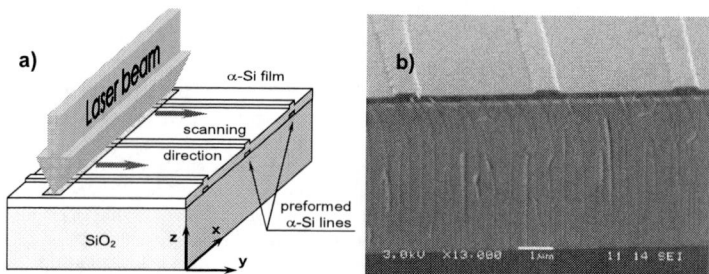

Figure 1. The laser crystallization process (a) and a cross-section SEM image of a silicon film with preformed lines (b).

Description of the laser system

In this work, the film crystallization was carried out using the laser optical system LAVA available at Innovavent GmbH. The system provided an up to 54 mm wide green laser beam (532 nm). A schematic block diagram of the optical system is shown in Figure 2. The laser system created a laser beam with a uniform top-hat profile along the x axis and a Gaussian profile along the y axis (see Figure 3). The beam's intensity profile was measured on the wafer plane using a beam profiling system equipped with a CCD camera, and a 40× microscope. The applied beam length was 5.15 mm and the width was 5.8 µm, both Full-Width Half-Maximum (FWHM) values. The average energy density in the beam was adjusted by an optical attenuator. The energy densities were calculated by dividing the total pulse energy by the FWF··· ι of the beam.

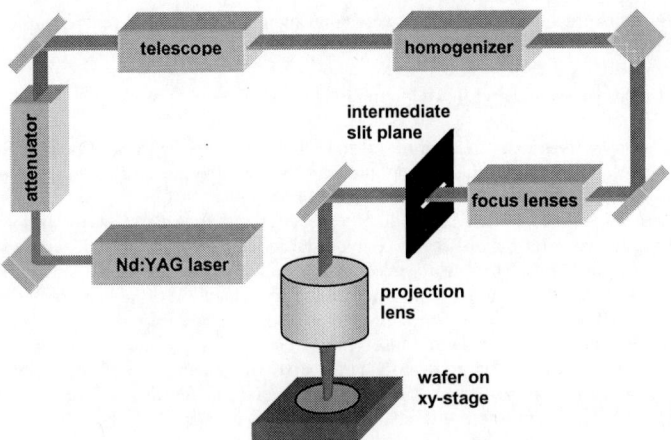

Figure 2. Block diagram of the optical system.

We used a frequency doubled Nd:YAG laser (model LDP-100MQG from Lee Laser) irradiating with an average power of 42.5 W at a repetition rate of 8.8 kHz. The pulse duration was 200 ns. The scan velocity was varied between 1 and 5 mm/second. To define and shape the line edges, an adjustable mechanical slit embedded in the intermediate slit plate was used. It was imaged on the wafer surface with a 10× demagnification.

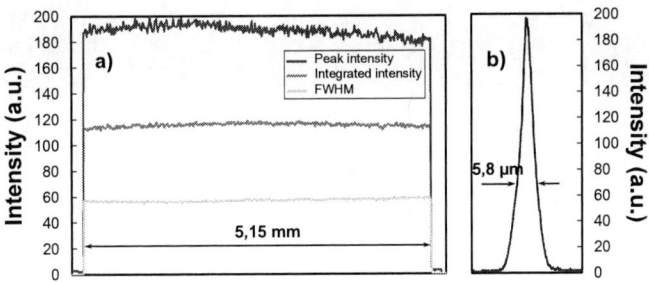

Figure 3. Laser beam intensity profiles: (a) along the x axis and (b) along the short y axis, see Figure 1 for details.

The wafer was located on a high accuracy motorized xy-stage. The projection lens was mounted on a high precision z-stage providing accurate focus adjustment. The depth of focus was ± 10 µm. The adjustment was done by making single pulse imprints on a test wafer followed by optical microscope inspection. The wafer stage, laser shutter and optical attenuator were computer controlled, so that the treatment steps with different energy densities and pulse overlaps were done automatically, at different locations on the wafer.

Laser treatment of silicon films with preformed α-Si lines

The laser treatments of uniform silicon films and films with preformed α-Si lines were investigated and compared. For uniform films, the irradiated amorphous silicon film region was rapidly melted during a short laser pulse. After the pulse, the lateral crystallization process started at the interfaces between the solid and molten regions (13). The mentioned solid region was the polycrystalline silicon layer crystallized during the previous laser pulses. There were small crystallites with different orientations, serving as the seeds at the solid-liquid interface, where the crystallization of molten regions started. It was possible to obtain good quality films in the scanning direction (i.e., along the *y* axis) by optimizing the laser pulse length, energy density, beam overlapping, etc. However, the number of grain boundaries and, in general, the film quality in the perpendicular direction (along the *x* axis) was difficult to control (see Figure 4a). Due to the uncontrollable grain size, a dramatic device-to-device spread can be expected, caused by the random number and position of the grain boundaries localized in the device area. This problem is eliminated in a finely-grained crystalline film, but unfortunately such a film has poorer electrical properties and herewith impaired device functionality.

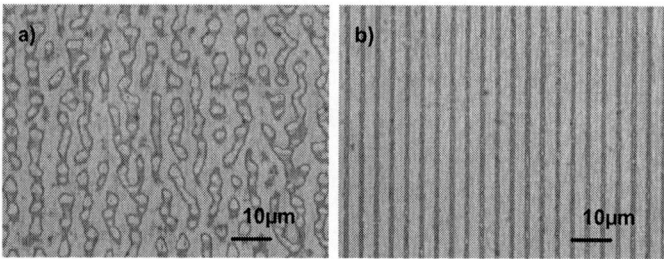

Figure 4. Typical optical-microscope surface images of crystallized Si films after Wright etching without (a) and with (b) preformed lines.

The crystallization process was improved by introducing the preformed lines prior to the deposition of amorphous silicon films. The addition of this extra process step resulted in a different crystallization behavior (see Figure 5). Although similar rapid melting occurred during the film irradiation, the interface between the solid and molten regions was not uniform and resembled that of a single amorphous layer. The periodically varied thickness locally resulted in non-molten regions deeply introduced into the molten silicon. These solid regions influenced the temperature gradient in the lateral direction perpendicular to the laser scan direction and served as the crystallization seeds where the super lateral crystal growth could initially start from (see Figure 5c-d). Thereby the dominate crystal orientation extend up to the grain boundaries with possible formation of the intragranular ridges and hillocks described in (17).

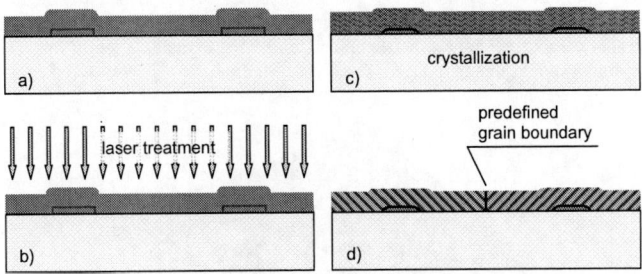

Figure 5. Crystallization process in silicon film with preformed lines (a) film formation, (b) laser treatment, melting of the silicon film, (c) crystallization process – super lateral growth, (d) final crystallized film with predefined grain boundary.

The average energy density and the area overlap for the subsequent laser pulses are critical parameters during crystallization process. The scan velocity is related to the laser pulse overlap (an overlap of 98% corresponds to a velocity of 1 mm/s, whereas an overlap of 90% corresponds to a velocity of 5 mm/s, both at 8.8 kHz repetition rate). The overlap area influences the crystallization mode, changing from lateral growth to homogeneous spontaneous nucleation in completely molten regions. During the lateral growth at the solid-liquid interface, the rest of liquid silicon is rapidly cooled and becomes a strongly undercooled liquid. Therefore the lateral crystal growth was stopped by homogeneous nucleation, formed a small-grained poly-Si region with poor electrical properties. Further re-crystallization with overlapping laser pulses was required, so the crystals created with one pulse are extended with the following one.

The laser treatment with an energy density of 1.00 J/cm^2, a laser pulse overlapping of 90% and repetition rate of 8.8 kHz provided the optimal crystallization process with predefined grain boundary location. Figure 4b shows the crystallized extremely long grains, visualized using Wright defect etching.

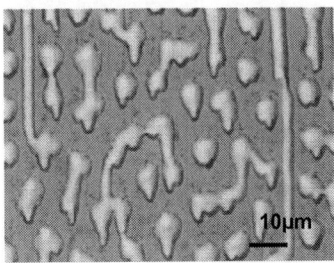

Figure 6. Optical image of silicon film agglomeration after laser treatment with a high energy density.

The irradiation with lower energies (< 0.6 J/cm^2 at 98% overlapping and < 0.8 J/cm^2 at 90% overlapping) resulted in partial melting of the film. For the lowest energy, only a part of the film was crystallized. The energy densities higher than 1.0 J/cm^2 at 98% overlapping and 1.2 J/cm^2 at 90% overlapping lead to undesirable film heating. It was observed that the agglomeration of silicon and following crystallization of microcrystals give discontinuous and inhomogeneous films (see Figure 6).

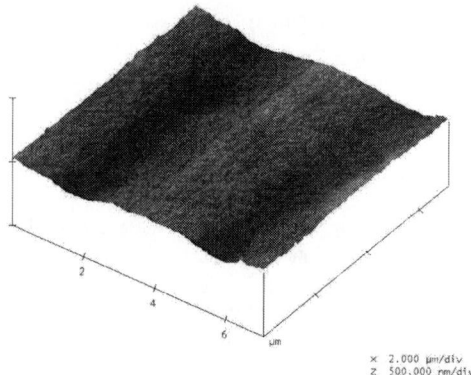

Figure 7. AFM image of crystallized Si film morphology (R_a = 19.9 nm, RMS = 24.2 nm).

As could be seen from analysis with optical microscopy (see Figure 4) and atomic force microscopy (see Figure 7), the surface of the silicon film even at optimum process conditions was not completely smoothed during laser crystallization. The crystallized film still exhibited a periodical height variation, remnant of the initial pre-formed lines. Therefore planarization (e.g. using chemical-mechanical polishing) is required before further device manufacturing.

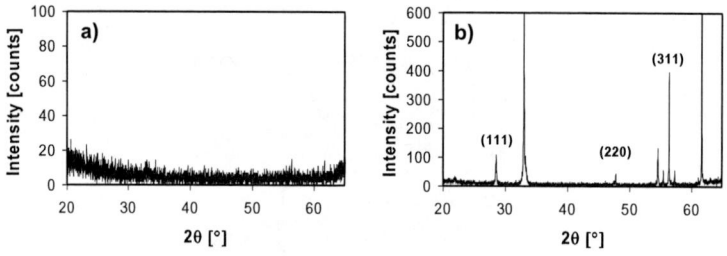

Figure 8. Measured XRD spectra (a) of 100 nm thick amorphous silicon film; (b) of crystallized silicon film with preformed lines. The XRD spectra were limited to $2\theta = 65°$ to exclude the (100) peak of the silicon substrate.

XRD measurements of the crystallized silicon film with preformed lines (Figure 8b) show peaks indicating the presence of (111), (220) and (311) oriented grains. As a reference, the sample with 100 nm thick a   us silicon film deposited by LPCVD at

550 °C on top of a 0.7-μm thick thermally grown SiO_2 layer was measured (Figure 8a). The shown measurement data confirmed the textured type of the laser-annealed silicon film.

Conclusions

Amorphous silicon films with underlying α-Si lines were deposited on thermally grown SiO_2. Laser crystallization with a green laser showed that the crystallization process is steered by the pre-patterned morphology. Through XRD analysis and optical inspection of the original, annealed and Wright-etched films, it was established that the films are indeed crystalline. Moreover, the grain boundary density was very low and the grain boundary position was controlled, allowing to design transistors in between grain boundaries.

Acknowledgments

This research was supported by the Dutch Technology Foundation (STW). Project number STW-TEL 6358.

References

1. K.C. Saraswat, S.J. Souri, V. Subramanian, A.R. Joshi, A.W. Wang, in *IEEE Int. SOI Conf.*, Proceedings p. 54 (1999).
2. S. Gu, et al., *J. Vac. Sci. Technol. B*, **23(5)**, 2184 (2005).
3. F. Li, X. Yang, A.T. Meeks, J.T. Shearer, K.Y. Le, *IEEE Trans. on Device and Material Reliability*, **4(3)**, 416 (2004).
4. S.B. Herner, et.al., *IEEE El. Dev. Lett.*, **25**, 271 (2004).
5. A.J. Walker, et al., in *2003 Symposium on VLSI Technology Digest of Technical Papers* (2003).
6. A.T. Voutsas, *Appl. Surf. Sci.*, **208-209**, 250 (2003).
7. S.D. Brotherton, D.J. McCulloch, J.B. Clegg, J.P. Gowers, *IEEE Trans. El. Dev.*, **40(2)**, 407 (1993).
8. S.-M. Han, M.-C. Lee, M.-Y. Shin, J.-H. Park, M.-K. Han, *Proc. of the IEEE*, **93(7)**, 1297 (2005).
9. P. Lengsfeld, N.H. Nickel, W. Fuhs, *Appl. Phys. Lett.*, **76(13)**, 1680 (2000).
10. M. Nerding, et al., *J. Appl. Phys.*, **91(3)**, 4125 (2002).
11. A. Hara, et al., *Jpn. J. Appl. Phys., Part 2*, **41(3B)**, L311 (2002).
12. C.-H. Kim, I.-H. Song, W.-J. Nam, and M.-K. Han, *IEEE El. Dev. Lett.*, **23**, 315 (2002).
13. L. Mariucci et al., *Thin Solid Films*, **383**, 39 (2001).
14. R. Ishihara, et al., *IEEE Trans. El. Dev.*, **51(3)**, 500 (2004).
15. G. Lubbert, B.C. Burkey, F. Moser, E.A. Trabka, *J. Appl. Phys.*, **52(11)**, 6870 (1981).
16. A. Hara, K. Yoshino, F. Takeuchi, N. Sasaki, *Jpn. J. Appl. Phys., Part 1*, **42(1)**, 23 (2003).
17. D.K. Fork, G.B. Anderson, J.B. Boyce, R.I. Johnson, P. Mei, *Appl. Phys. Lett.*, **68(15)**, 2138 (1996).

192

CHAPTER 5

POLY-Si TFTs FROM NON-LASER
CRYSTALLIZATION PROCESSES

ECS Transactions, 3 (8) 195-201 (2006)
10.1149/1.2356354, copyright The Electrochemical Society

Low-Temperature Crystallization of Amorphous Si Films using Ferritin Protein with Ni Nanoparticles

Yukiharu Uraoka[a], Hiroya Kirimura[a], Takashi Fuyuki[a] Mitsuhiro Okuda[b], and Ichiro Yamashita[a,b]

[a]Nara Institute of Science and Technology, 8916-5, Takayama, Ikoma, Nara 630-0192, Japan
[b]Matsushita Electric Industrial Co., Ltd., Hikaridai 3-4, Seika, Kyoto 619-0237, Japan

A poly-Si thin film with a high crystallinity was obtained using ferritin with a Ni core (7 nm), which enabled us to precisely control the density and position of the nucleus for crystal growth. The core density of ferritin adsorbed on the amorphous silicon surface was controlled in the range from $10^9 \, cm^{-2}$ to $10^{11} \, cm^{-2}$. Crystal growth was performed at 550°C in N_2. Crystallinity or grain size strongly depended on Ni core density. Poly-Si film with the average grain size of 3 μm and a high crystallinity was obtained at a low Ni atom density of $10^{12} \, cm^{-2}$.

Introduction

Polycrystalline silicon (poly-Si) films fabricated on glass or plastic substrates have attracted much attention due to their applications in thin-film transistors (TFT) for flat-panel displays or future displays, such as a system-on-panel. As low-temperature fabrication methods to obtain a large grain size, excimer laser annealing and metal-induced crystallization (MIC) methods[1] are widely investigated. The MIC method is carried out to increase crystallization speed utilizing the metal silicidation phenomena with an activation energy lower than that of silicon crystallization.

For the silicidation metal, Co, Cr and Pt are used, particularly Ni is regarded as a promising material due to the lattice constants of $NiSi_2$, which are comparable to those of Si lattice with less than 0.4 % mismatch. However, the remaining Ni atom increases the off-state current of TFTs. Therefore, in order to decrease the amount of Ni, metal-induced lateral crystallization (MILC) [2] or the introduction of a diffusion barrier of a SiN_x layer [3] is proposed. Furthermore, the metal imprinting method[4] employs a tip array coated with metal film for attachment on an a-Si surface. An extremely small amount of metal can be adsorbed on the Si surface as the crystallization nucleus. However, a method that enables the precise control of density and position has not been established yet. Recently we have succeeded in fabricating floating gate memory using bio-technology[5]. Effectiveness of Bio Nano technology was demonstrated by embedding the protein in the silicon oxide as electron confinement node.

In this study, we propose a new method utilizing biotechnology to obtain high-quality poly-Si thin films. As the nucleus of silicon crystallization, the Ni core of ferritin with a diameter of 7 nm was adsorbed on the surface of a-Si films by a bottom-up process. By adjusting the density of ferritin on the a-Si film, we controlled the Ni nucleus density and performed the solid-phase crystallization of the a-Si film.

Experimental

Thus far we have investigated a technique of accommodating various inorganic materials, such as Fe, Co, Ni, and Cr into recombinant ferritin on the nanoscale by biomineralization. We have successfully introduced Ni core into apoferritin which has no cores, by adjusting the pH value of ammonium nickel sulfate with a buffer solution (HEPES+CAPSO) and bubbling with CO_2 gas. This procedure is illustrated in Fig.1 (a). The yield of Ni core formation was confirmed to be approximately 95% from Transmission Electron Microscope (TEM) images shown in Fig.1 (b).

The fabrication procedure of poly-Si film proposed here is shown in Fig.2. The concentration of ferritin in Ni-Ferritin solution was adjusted by dilution with pure water and the solution was dropped onto the a-Si film with a thickness of 50 nm deposited by low-pressure chemical vapor deposition (LPCVD) method. After adsorbing ferritin for 10 min, the remainning solution was removed with a centrifugal machine. After the removal of the outer protein shell by UV-ozone treatment at 110℃ for 40 min, solid-phase crystallization was performed at 550℃ for 25 h in N_2 atmosphere.

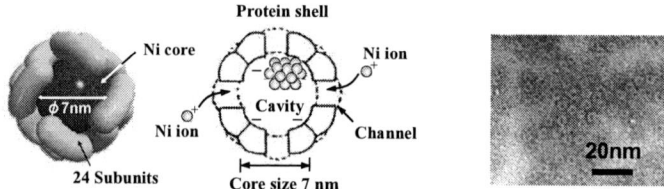

Fig.1 (a) Schematic of a apoferritin molecule with nickel ions.(b) TEM image of ferritin-formed nickel core in cavity. The core is approximately 7nm in diameter and is dependent by the cavity size.

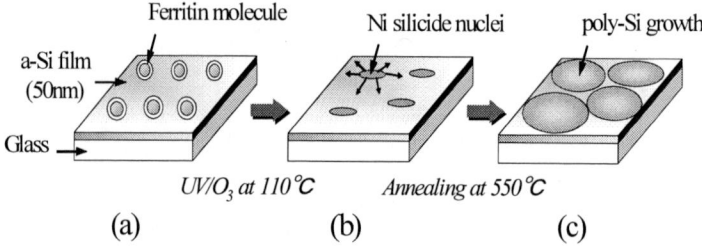

Fig. 2 Process flow of silicon crystallization using novel bio-nano process. (a) Adsorption of Ni-ferritin to a-Si films on glass. (b) Formation of nickel silicide nuclei after elimination protein with UV/O₃ treatment(c) Crystallization of silicon films by RTA.

Results and Discussion

A Ni-ferritin solution with Ni core concentrations of 0.5, 0.15 and 0.05 mg/ml was prepared by dilution with pure water. Figures.3 (a), (b) and (c) show SEM images of Ni- ferritin with various core densities adsorbed on the a-Si film. The relationship between the concentration of the ferritin solution and the density of the core determined for an area of 200×200 nm was investigated. A linear relationship was obtained and it was suggested that we can control Ni core density precisely to 2.5×10^{11}, 2.8×10^{10} and 2.6×10^{9} cm^{-2}.

The SEM images obtained after the UV-ozone treatment for removing the outer protein shell are shown in Figs. 3 (a'), (b') and (c'). Ni cores were not clearly observed, however, relatively uniform circular patterns with a diameter of approximately 50 to 150 nm were observed. Their size and density were decreased with decreasing Ni core density. It is reported that various stable Ni silicides such as Ni_3Si, Ni_2Si, Ni_3Si_2, $NiSi$, and $NiSi_2$ can be observed at different annealing temperatures and Ni_2Si is generated at a temperature lower than 200℃. With the increase in temperature, the crystal structure changes from Ni_2Si to $NiSi$, to $NiSi_2$. On the a-Si film, $NiSi_2$ is formed at less than 400℃. Therefore, the crystal growth of poly-Si from a-Si proceeds with $NiSi_2$ as the nucleus of the crystal. The patterns observed after UV-ozone treatment indicate the Ni_2Si generated during the reaction process between Ni and Si at a relatively low temperature.

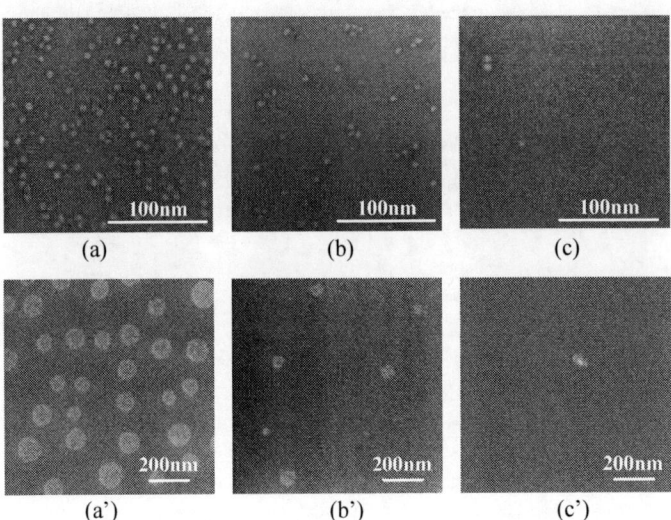

Fig. 3 SEM images (a-c) of Ni-ferritin cores adsorbed on a-Si films for various concentrations of Ni-ferritin and SEM images (a'-c') of nickel silicide nuclei after UV/O$_3$ treatment. (a, a') 2.5×10^{11} cm^{-2}, (b, b') 2.8×10^{10} cm^{-2}, (c, c') 2.6×10^{9} cm^{-2}

Under the condition of the highest Ni core density of 2.5×10^{11} cm^{-2}, the distance between cores was small, approximately 20 cores in the area of 100 nm square accumulated to form a circular Ni$_2$Si region with a diameter of 100 nm and with an interval of 200 to 300 nm. On the contrary, under the condition of the lowest Ni core density of 2.6×10^9 cm^{-2}, a few Ni cores gathered to form a small Ni$_2$Si pattern with a diameter of 50 nm or less, and with a long interval of $2 \sim 3\mu$m. Under the middle condition of a Ni core density of 2.8×10^{10} cm^{-2}, the interval of the Ni$_2$Si pattern was approximately 500 nm. These results suggest that we can control the position of the initial nickel silicide (Ni$_2$Si) as the nucleus of crystallization of Si films.

The a-Si film with a controlled crystal nucleus was heated up to 550°C in 10 min and then annealed for 25 h in N$_2$ ambient in an RTA furnace. Rapid heating suppresses the generation of a natural nucleus and poly-Si growth proceeds laterally from the controlled nucleus of NiSi$_2$. It is reported that the activation energies of Ni/a-Si $\rightarrow$ NiSi$_2$, a-Si $\rightarrow$ crystalline Si, and the generation of a natural nucleus are 1.45, 2.7, and 4.4eV, respectively. Therefore, it is estimated that the rapid heating and continuous annealing at 550°C enhanced the reaction of NiSi$_2$ dominantly, thus the crystal growth of poly-Si proceeded.

Crystallinity after the annealing was evaluated by XRD analysis, Raman spectroscopy and EBSD (electron backscattered diffraction) analysis. The XRD analysis indicated that peaks for <111>, <220> and <311> were observed under all conditions. Judging from peak intensity, crystallinity was high at densities on the order of 10^9, 10^{10} and 10^{11} cm^{-2}. The order of the network between Si-Si bonding was evaluated from the FWHM of the TO phonon peak at 520 cm^{-1} in Raman spectroscopy. As observed in the XRD analysis shown in Fig.4, the film with the core density of 10^9 cm^{-2} had the highest crystallinity, suggesting an enlarged grain size.

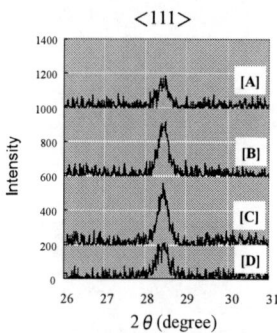 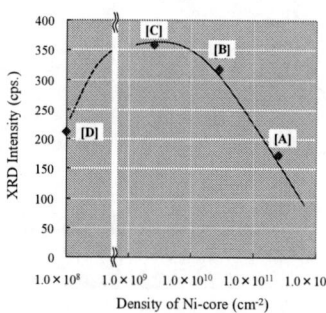

Fig.4 Relationship between intensity of XRD <111> peak and density of Ni-cores.[A] 2.5×10^{11} cm^{-2}, [B] 2.8×10^{10} cm^{-2} [C] 2.6×10^9 cm^{-2}, [D] without Ni-core.

Furthermore, to examine grain size and crystalline orientation, EBSD analysis was performed. The "grains" were defined as the regions with the same crystallographic orientations and the same phases. Grain and crystalline orientation mapping images are shown in Figs.5 (a) – 5 (c). As reference, a sample without cores is shown in Fig. 5(d). Figure 5(e) shows the color triangle coordinate, from which the orientation of the grain can be decided. Mapping images showed that definite grains were not observed at a density of 10^{11} cm^{-2}, therefore, the microcrystalline structure was dominant. For the film with a density of 10^{10} cm^{-2}, a mixture of grains and a microcrystalline structure was confirmed. However, for the film with a core density of 10^9 cm^{-2}, definite grains were confirmed and their size was uniform. The orientations of the grains were random in all samples.

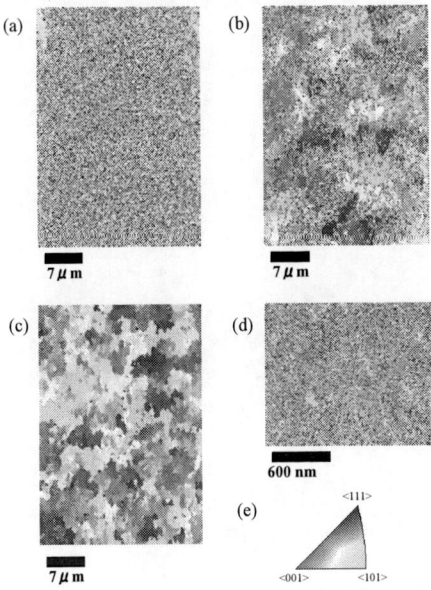

Fig. 5. EBSD grain images of poly-Si films after annealing for 25 h at 550 C as function of density of nickel cores. (a) 2.5×10^{11} cm^{-2}, (b) 2.8×10^{10} cm^{-2}, (c) 2.6×10^9 cm^{-2}, (d) without nickel cores.

The film without cores prepared as reference had no grains. Grain sizes was determined on the basis of a set of grains with different orientation angles within 5 degees. The average grain sizes estimated from the size distribution, were 0.36μm, 0.48μm and 3.06μm for the core densities of 10^{11}, 10^{10} and 10^9 cm^{-2}, respectively. Thus, grain size increased markedly with decreasing core density.

We also investigated the Ni contamination after the crystallization by using SIMS. Very low density of 5.2×10^{12} cm^{-2}, which is one order small compared with previous report[3], was obtained.

Here we discuss the crystallization mechanism in this method. When the UV ozone treatment is performed, ferritin cores aggregate to make the nucleus as shown in Fig. 6. Grains grow from the nucleus during the following annealing.

The interval of the Ni silicide estimated from the SEM image in Fig.3, was in good agreement with grain size, as shown in Fig.7. It can be easily surmised that the grains grew laterally until they met the nuclei of Ni silicide. Correspondingly, we could increase the grain size by decreasing core density and by increasing core interval. It is fairly definite that a few Ni cores with a diameter of 7 nm became a nucleus for crystallization.

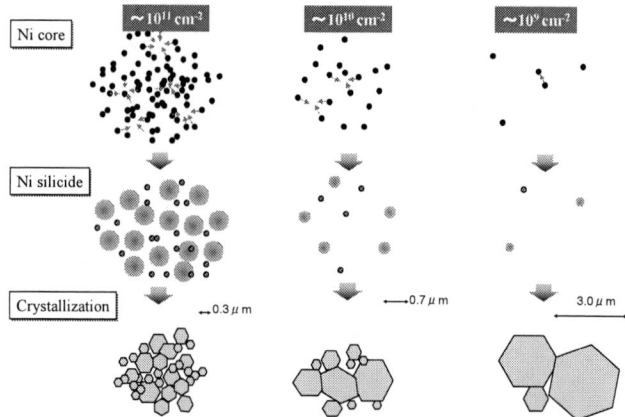

Fig. 6. Mechanism of grain growth in the method using bio-nano core.

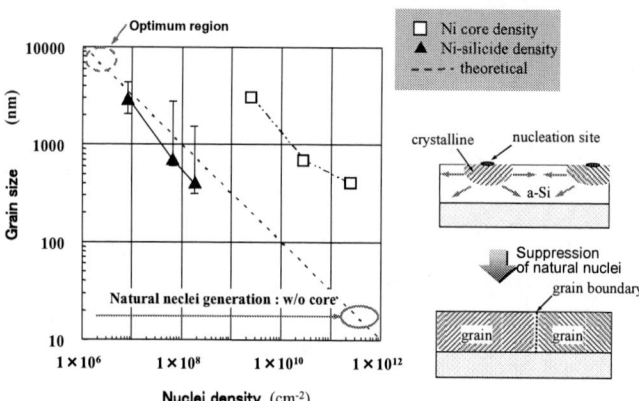

Fig 7. Relationship between nuclei density and grain size.

Feature of the Bio Nano process, ferritin has the ability to control the position on the silicon surface. Therefore, we can fabricate the poly-Si film with position controlled grain as shown in Fig.8.

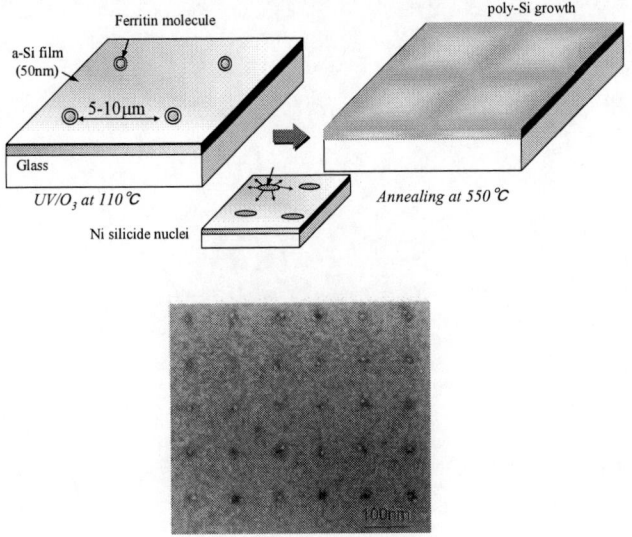

Fig. 8 Position control of poly crystalline silicon grain using the feature of "Bio Nano Technology".

Conclusion

Poly-Si film with a large grain size and a high crystallinity was obtained using ferritin accommodating nanosize Ni cores, which enabled us to precise control the density and position of nucleus for crystal growth. The poly-Si film with a large grain size of 3μm was obtained at a low Ni contamination of 5.2×10^{12} cm^{-2} and 550℃. This film will be very promising for the low-temperature poly-Si thin-film transistors. In this study, we could demonstrate that the quality of silicon thin film can be greatly improved by the introduction of bio-nanotechnology.

Acknowledgement

The authors would like to express their thanks to Prof.H. Harima of the Kyoto Institute of Technology for the Raman spectrometry measurements. They also thank Prof. S.Yokoyama in Hiroshima University for the experimental support.

This work was partially supported by Leading Project by MEXT, Japan.

References

1. R. C. Cammarata, *J. Mater Res* **5**, 2133 (1990).
2. J. Jang , *Nature* (London) **395**, 481 (1998).
3. J. H. Choi , *Electrochemical and Solid-State Letters*, **6**, G16-G18 (2003).
4. K. Makihara, T.Asano, *Appl. Phys. Lett.* **76**, 3774 (2000).
5. T.hikono, Y.Uraoka, T.Fuyuki, S.Yoshii, I.Yamashita, *Appl. Phys. Lett,* **88**, 023108, (2006).

A Simple Method for Gettering of Nickel within the NILC Polycrystalline Silicon Film Using a Gettering Substrate

Chih-Yuan Hou, Chi-Ching Lin and YewChung Sermon Wu

Department of Materials Science and Engineering, National Chiao Tung University, Hsinchu 300, Taiwan, Republic of China

Ni-metal-induced lateral crystallization (NILC) of amorphous silicon (α-Si) has been employed to fabricate high-performance low-temperature polycrystalline silicon (poly-Si) thin-film transistors (TFTs). However, the current crystallization technology often leads to trapped Ni and $NiSi_2$ precipitates, thus degrading the device performance. We proposed using α-Si-coated wafers as gettering substrates. By bonding the gettering substrate and NILC poly-Si film together, the Ni-metal impurity within the NILC poly-Si film was greatly reduced.

Introduction

Ni-metal-induced lateral crystallization (NILC) of amorphous silicon (α-Si) has been employed to fabricate high-performance low-temperature polycrystalline silicon (poly-Si) thin-film transistors (TFTs). In NILC, Ni islands are selectively deposited on top of α-Si films and allowed to crystallize at a temperature below 600°C. The polycrystalline silicon (poly-Si) formed under the metal film is called "Ni-metal-induced crystallization (NIC)"; whereas the poly-Si formed outside the metal layer is called "NILC". High-performance LTPS TFTs have been fabricated by the NILC method, and have been demonstrated to be suitable for systems in panel applications (1,2).

Unfortunately, the NIC area and poly-Si grain boundaries trap Ni and $NiSi_2$ precipitates which increases the leakage current and shifts the threshold voltage (3-8). Therefore, Ni contamination should be reduced to improve the device performance. In this study, an α-Si-coated Si wafer and an α-Si-coated quartz wafer were utilized as Ni-gettering substrates. Through a specialized wafer-bonding technique, the Ni concentration inside the NILC poly-Si film was significantly reduced.

Experimental

Two types of poly-Si films were investigated in this study. Samples designated as (1) "NILC POLY" were poly-Si films fabricated by traditional NILC methods, and (2) "GETR POLY", were poly-Si films fabricated by the same traditional NILC method with an additional Ni-gettering process. The basic NILC fabrication process of both poly-Si films began with four-inch Si(100) wafer substrates where wet oxide films of 500 nm were grown using a H_2/O_2 mixture and substrate temperature of 1050°C. Silane-based α-Si films with a thickness of 100 nm were deposited using low-pressure chemical vapor deposition (LPCVD) at 550°C and 100 mTorr. The photoresist was patterned to form the desired Ni lines, and a 2-nm-thick Ni film was deposited on the α-Si using an e-gun. The samples were then dipped into an acetone and ultrasonic bath for 5 min to remove the photoresist, and subsequently annealed at 550°C for 12 h to form the NILC poly-Si film.

To reduce Ni contamination, the unreacted Ni metal was removed by chemical etching. A lift-off process using photoresist was employed to form islands of poly-Si regions on the wafers. At this point, only 20% of the poly-Si film remained on the NILC POLY sample.

For the Ni-gettering substrate, 100-nm-thick α-Si films were directly deposited onto both sides (inner and outer sides) of the Si substrate. To form the GETR POLY film, the NILC POLY film was bonded to the Ni-gettering substrate and then annealed at 550°C for an additional 12 h (9). Following the gettering process, the Ni-gettering substrate was removed.

Results And Discussion

The NILC poly-Si was dipped into a silicide-etching solution ($HNO_3:NH_4F:H_2O=4:1:50$), numerous holes were observed at the NIC areas and at the boundaries where two poly-Si fronts intersected, as seen in Fig. 1. These holes were residues of the Ni silicides that had been etched away by the silicide-etching solution. Ni contamination in the NILC POLY film should be reduced, particularly in the NIC areas and grain boundaries, since they induce trap-charge carriers and build up potential barriers to the flow of carriers. After the gettering process, no silicide-etched holes were found on GETR POLY, with the exception of a very few etched holes in the NIC areas. This indicates that a substantial number of Ni atoms had diffused into the Ni-gettering substrate.

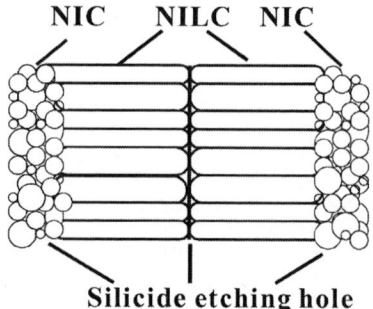

Silicide etching hole

Figure 1. Schematic illustration of the silicide-etching holes on the NIC areas and grain boundaries of the NILC POLY film.

Secondary-ion mass spectroscopy (SIMS) was also employed to measure the reduction of Ni residues in the NILC region of poly-Si films. The Ni concentration in the GETR POLY film has been reduced to 1/30 compared with that in NILC POLY films. This also demonstrates successful out-diffusion of Ni atoms to the Ni-gettering substrate.

The Ni-gettering substrate was also investigated. The OM and SEM images of the inner and outer Si shown that α-Si has been transformed into the poly-Si grains by NILC mechanism. The only difference was that the fraction of NILC poly-Si (determined by estimating the areas of dark (poly-Si) region vs. total area on the optical microscope micrograph) on the outer poly-Si film was 90%, which was smaller than that of the inner poly-Si film, 95%.

The performance of TFT devices was also used to examine the quality of poly-Si films, as shown in Figure 2.

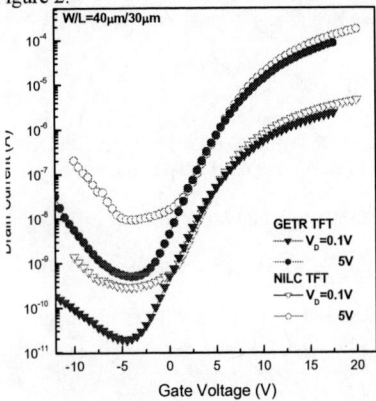

Figure 2. The Transfer characteristics of NILC TFT and GETR TFT

A quartz wafer coated with α-Si was also used as a Ni-gettering substrate. We also found a reduction of silicide-etched holes at the grain boundaries of the NILC POLY film, and NILC poly-Si grains formed on the inner Si film of the quartz wafer. However, no NILC poly-Si grains were found on the outer α-Si film because of the quartz diffusion barrier.

Summary

An investigation of the relationship between Ni-gettering substrates and Ni-metal impurity within the NILC poly-Si (NILC POLY) film has led to the development of a simple, effective Ni-gettering process for NILC poly-Si films. Ni-gettering substrates were fabricated by coating α-Si films on both sides of Si and quartz wafers. When the Si wafer was used as a Ni-gettering substrate, it was found that the silicide-etched holes at the NILC POLY grain boundaries were greatly reduced. The Ni concentration within the NILC POLY film was reduced to 1/30 after the Ni-gettering process. Both sides of the α-Si films were transformed into poly-Si by the NILC mechanism. The use of the quartz wafer also allowed to achieve the reduction of silicide-etched holes on the NILC POLY film. NILC poly-Si grains were located only on the inner Si film of the quartz Ni-gettering substrate. No NILC poly-Si grains were found on the outer side because of the quartz diffusion barrier. These improvements increased with annealing time and α-Si thickness.

Acknowledgments

This project was funded by Chunghwa Picture Tubes, Ltd. (CPT) and the National Science Council (NSC) of the Republic of China under No. 94-2216-E009-015.

References

1. S. W. Lee and S. K. Joo: IEEE Electron Device Lett. **17,** 160 (1996).
2. Z. Meng, M. Wang and M. Wong: IEEE Trans. Electron Devices **47,** 404 (2000).
3. P. J. van der Zaag, M.A. Verheijen, S. Y. Yoon and N. D. Young: Appl. Phys. Lett. **81,** 3404 (2002).
4. G. A. Bhat, Z. Jin, H. S. Kwok and M. Wong: IEEE Electron Device Lett. **20,** 97 (1999).
5. Z. Jin, K. Moulding, H. S. Kowk and M. Wong: IEEE Electron Device Lett. **20,** 167 (1999).
6. G. A. Bhat, H. S. Kwok and M. Wong: Solid-State Electron. **44** 1321 (2000).
7. S. W. Lee and S. K. Joo: IEEE Electron Device Lett. **17** 160 (1999).
8. D. Murley, N. Young, M. Trainor and D. McCulloch: IEEE Trans. Electron Devices **48** 1145 (2001).
9. P. C. Liu, C. Y. Hou and Y. S. Wu: Thin Solid Films **478** 280 (2005).

Molecular-Dynamics Simulations of Recrystallization Processes in Silicon:
Nucleation and Crystal Growth in The Solid-Phase and Melt

T. Motooka, S. Munetoh, R. Kishikawa, T. Kuranaga, T. Ogata and T. Mitani

Department of Materials Science and Engineering, Kyushu University
Motooka 744, Fukuoka 819-0395, Japan

It is of increasing importance to understand crystallization
processes of silicon (Si) in atomic scale for developing high-speed
thin film transistors. We have performed molecular-dynamics
(MD) simulations of nucleation, crystallization and defect
formation processes in amorphous Si (a-Si) as well as in liquid Si
(l-Si). Based on the MD simulation results combined with high-
resolution electron microscopy measurements, it is proposed that
the crystallization and defect formation processes are controlled by
the interface nanostructures between disordered and crystalline Si.
It is also suggested that nucleation and crystal growth in a-Si films
deposited on glass during excimer laser annealing occur in
amorphous-solid and low-density liquid phases rather than in
supercooled l-Si as has been generally accepted.

Introduction

The polysilicon (Si) thin film transistor (TFT) is a key component of liquid crystal
display (LCD) panels. Recently, much attention is paid to system-on-glass LCD in which
both LCD active-matrix switches and driving as well as controlling circuits are included
in TFT on glass for the development of mobile multimedia devices. In order to realize
such system-on-glass LCD, it is crucial to develop a method to grow high-quality
crystalline Si on glass. The most popular method is to use excimer-laser pulse (10-20 ns)
shots to anneal amorphous Si (a-Si) films deposited on glass where crystal growth is
considered to be initiated by rapid melting of a-Si and quenching of melted Si. Since
crystallization occurs in very thin (50-100 nm) a-Si films within a few 100 ns, almost all
experimental studies were carried out only after crystallization occurred. Although
theoretical studies have been carried out to analyze the crystallization processes based on
thermal diffusion equations[1], the detailed atomistic processes are not well understood.
In this paper, we have first reviewed our work[2-6] on the analysis of crystallization and
defect formation processes in amorphous/crystalline (a/c) Si as well as solid/liquid (s/l) Si
interfaces by using molecular-dynamics (MD) simulations combined with high-resolution
electron microscopy measurements. Then, preliminary results of MD simulations have
been presented for investigating recrystallization processes in a-Si thin films deposited on
glass during excimer laser annealing (ELA).

Method

MD simulations were carried out under constant NVT condition using a
rectangular MD cell including a/c or s/l interface as illustrated in Fig. 1. A typical number
of Si atoms in the MD cell was ~10000 and the interatomic forces were calculated using

the Tersoff potential [7] for crystalline Si. Although the Tersoff potential gives rise to high melting temperature near 2550 K, it can well reproduce crystallization processes in solid and liquid Si, and thus it is considered to be useful to describe nucleation and crystallization processes in Si. The MD cell was immersed in a thermal bath with various temperature gradients in the longitudinal direction. Temperature was controlled by using the ordinary Langevin equations. Periodic boundary conditions were employed in the lateral directions, while the atomic positions in the top two layers were fixed and an elastic hard wall was set at the bottom to prevent hot atoms from escaping out of the MD cell.

Solid Phase Epitaxy of Si

Solid phase epitaxy (SPE) of Si is one of the basic processes in Si microelectronics fabrication technologies. Various experimental works were carried out to determine the growth rate of thermal SPE by using furnace and laser annealing, and the activation energy was found to be ≈2.7 eV.[8] Recently, extensive investigations have been

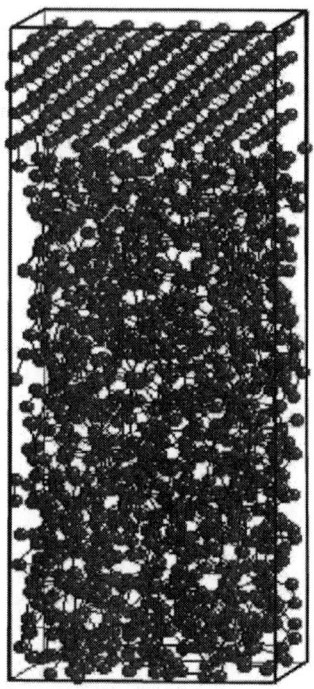

Fig. 1. MD cell with an a/c or s/l Si interface. (Taken from ref. [3].)

performed on ion-beam-induced epitaxial crystallization (IBIEC) where ion-beam-induced defects play an important role for crystallization, and the activation energy is found to be varied from 0.18 to 0.40 eV depending on ion mass and temperature. Based on these results, it is generally believed that crystallization occurs at amorphous/crystal (a/c) Si interfaces in both thermal SPE and IBIEC processes. However, the atomistic mechanism of crystallization at the a/c interface is not well understood. Figure 2 shows an example of cross-sectional transmission electron microscopy (XTEM) observation of the typical a/c interface btained by Si⁺ ion implantation on

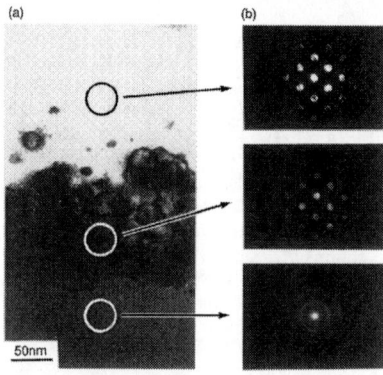

Fig. 2. (a) Bright-field image of the *a/c* interface region. (b) Diffraction patterns from the selected areas indicated by the circles.(Taken from ref. [2].)

Si(001) substrates. Figure 2(a) shows the detailed bright-field image near the a/c interface of the sample with a dose of 10^{17}cm^{-2}. Diffraction patterns from the three selected areas with a size of ~30 nm in diameter are also shown in Fig. 2(b). These diffraction patterns indicate that the a/c interface is very sharp and the diffraction spots corresponding to the (110) reciprocal-lattice plane of the (110) reciprocal-lattice plane of crystalline Si clearly appear even in the adjacent area to the completely amorphized region judged by the halo diffraction pattern. Figure 3 shows the high-resolution XTEM images taken in the <110> projection from the same sample described in Fig. 2. Near the surface region where the ion-beam induced damage is relatively weak, the observed image[Fig. 3(a)] reveals almost the perfect crystalline Si contrast (A). On the other hand, in the a/c interface region[Fig. 3(b)], the crystalline image (B, C) becomes deteriorated toward the amorphous region (D, E). Nevertheless, it should be noted that the crystalline image remains until the amorphous transition occurs within a few atomic layers which is consistent with the results obtained by the microdiffraction pattern described above.

Figure 4 shows typical snapshots of the SPE processes annealed at 1450 and 2000 K. These snapshots clearly suggest that layer-by-layer crystallization occurs along the (001) and (111) planes at 1450 and 2000 K, respectively. From these MD simulation results, we can envisage the SPE growth mechanisms in the lower and higher temperature regions as shown schematically in Fig. 5. In the lower temperature region, the a/c interface is essentially (001) with a small portion of (111) edges, and the rate-limiting step of crystallization can be considered to be two-dimensional nucleation on the (001)

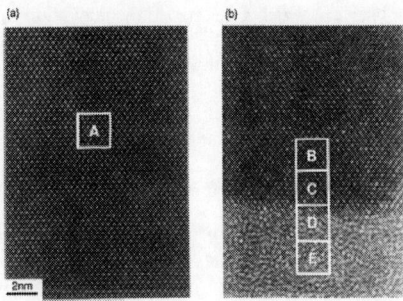

Fig. 3. High-resolution images:(a) Near the surface and (b) the a/c interface. (Taken from ref. [2].)

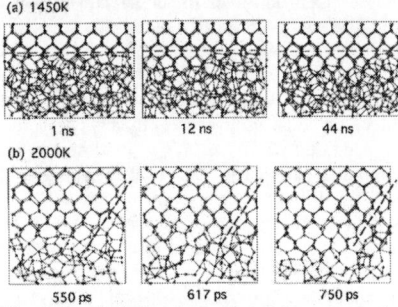

Fig. 4. Typical snapshots of atomic movements during SPE growth at (a) 1450 and (b) 2000 K obtained by MD simulations. (Taken from ref. [3].)

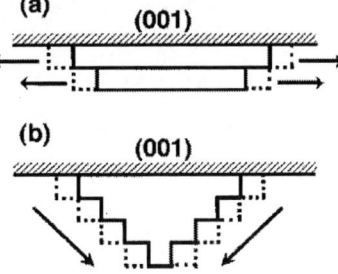

Fig. 5. Schematic diagram of the rate-limiting steps for (a) lower and (b) higher temperature SPE crystallization in the [001] direction. (Taken from ref. [3].)

a/c interface which gives rise to the activation energy, 2.6 eV. This is consistent with the experimental fact that the SPE rate in the [111] direction is much slower, since two-dimensional nucleation would be more difficult on the (111) interface than on the (001). On the other hand, in the higher temperature region, the a/c interface is predominantly composed of {111} facets and the rate-limiting step is a trapping process of Si at the kink sites associated with these facets.

Crystallization from Melted Si

Almost all crystal Si wafers for microelectronics device fabrication are produced by the Czochralski (CZ) method or crystal growth from Si melt. It is well known that there exist microdefects called grown-in defects in CZ grown crystal Si. Since these grown-in defects cause the deterioration of Si ULSI devices, the reduction of grown-in defects is increasingly important to obtain the required crystal quality. These grown-in defects are considered to be agglomerates of intrinsic point defects such as single vacancies and self-interstitials, which suggests that there is a strong relationship between the distribution of grown-in defects and the amount of excess point defects. It is empirically known that the type of grown-in defect can be controlled by the ratio V/G, where V is the growth rate and G the temperature gradient at the solid/liquid interface. When the V/G value is higher than $(V/G)_{critical}$; 0.1 mm^2/min K, vacancy type grown-in defects or voids are formed in the crystal, while interstitial type defects are formed under the condition of $V/G<(V/G)_{critical}$. Although many phenomenological models have been reported to discuss the physical meanings of the V/G rule for better prediction of the defect distribution, the atomistic mechanism is not well understood. As the large amount of intrinsic point defects are considered to be introduced at the solid/liquid interface, it is important to investigate the formation processes of point defects from microscopic point of view.

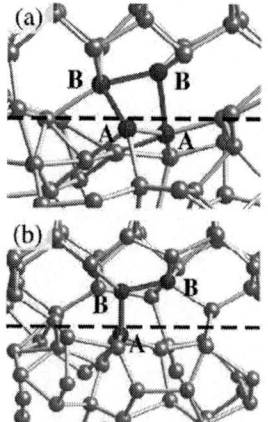

We propose [4,5] that the defect type is controlled by the structural difference in the five-membered ring at the s/l interface, while the crystallization mechanism is the same as that shown schematically in Fig. 5(b). Figures 6(a) and 6(b) show the typical five-membered rings to give rise to single interstitial and vacancy, respectively. In the former case, there exist two atoms indicated by **A** bonded to the **B** atoms in the five-membered ring, while one of these atoms is lost in the latter. Our MD simulations indicate that the former five-membered ring structure can always be seen for larger G, while the latter can only be seen for smaller G. This is consistent with the macroscopic V/G rule described above and can be microscopically attributed to a nature of atomic motions in a transition layer at the s/l interface as shown in Fig. 7. The nonequilibrium atomic

Fig. 6. Typical five-membered rings to initiate the formation of (a) single interstitial and (b) vacancy. The dashed lines indicate the s/l interfaces. (Taken from ref. [5].)

diffusion is found to be faster for larger G and thus the lower density region can be quickly filled by Si atoms from the liquid side which results in interstitial formation. On the other hand, the atomic diffusion constant becomes smaller for smaller G and space filling is not effective which increases opportunity to create vacancies.

Recrystallization Processes in a-Si Thin Films on Glass

We have investigated atomistic processes of nucleation and crystallization in bulk supercooled liquid Si (l-Si) as a first step to analyze excimer-laser induced growth of thin c-Si films based on MD simulations using the Tersoff potential. Figure 8 shows an example of nucleation processes under the temperature gradient shown in the figure.[6] Nucleation first occurs at the lower temperature region (Fig.8(b)) as expected and then crystallization proceeds toward the high temperature region (Fig.8(c)). The crystalline front-end hits the bottom of the MD cell and

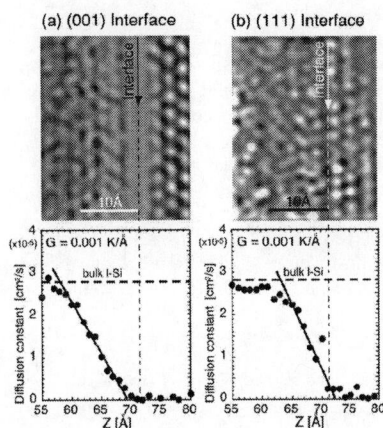

Fig. 7. Comparison of HRTEM measurements with the calculated atomic diffusion constants at the (001) and (111) s//l interfaces. The s//l interface position in the HRTEM data is determined based on the contrast change, while it is defined as the position where the diffusion constant becomes essentially zero from the liquidlike values in the lower panels. (Taken from ref. [5].)

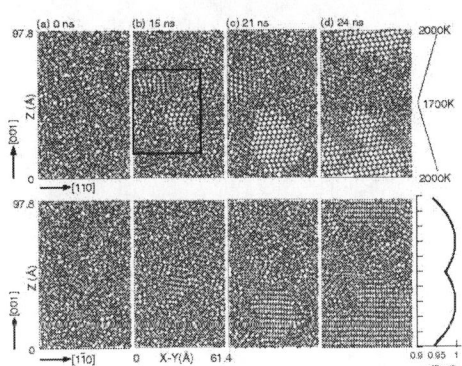

Fig. 8. Snapshots of atomic configurations during nucleation and crystal growth under a symmetric temperature gradient as shown in the top panel: (a) t=0 ns, (b) t=15 ns, (c) t=21 ns, and (d) t=24 ns. These snapshots were obtained by projecting the atomic coordinates in the central 10-Å thick regions including the diagonal $(1\bar{1}0)$ and (110) planes of the MD cell on the $(1\bar{1}0)$ and (110) planes, respectively. (Taken from ref. [6].)

then crystallization continues to the top region (Fig.8(d)) due to the periodic boundary condition. It is remarkable that the (111) surface of the crystallite is predominantly parallel to the z-axis or direction of the temperature gradient. This can be attributed to (1) the trend that the solid/liquid Si interface is stabilized by forming {111} facets and (2) temperature dependence of the crystallization velocity. The velocity of crystallization can be given as a function of temperature by $v(T) = Cexp(-Q/k_BT)[1-exp(-L\Delta T/k_BT_mT)]$ based on the transition-state theory. Here, C is a constant, Q the activation energy for the transformation of liquid to solid Si, k_B the Boltzmann constant, L the latent heat of fusion, and ΔT the amount of supercooling, $T_m - T$ (T_m: melting temperature). By assuming Q=0.42 eV/atom[9] and L =45.4 kJ/mol[10], the relative velocity $v(T)/C$ was calculated as illustrated in Fig. 8. Figure 9 shows a magnified view of the region enclosed by the thick rectangle in Fig.8(b). In this figure, well-ordered atoms are highlighted. Two nuclei

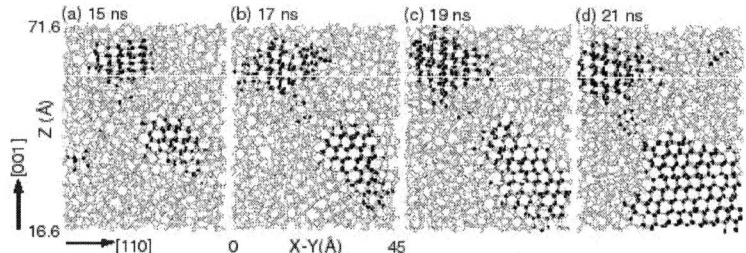

Fig. 9. Magnified view of the enclosed region in Fig. 8(b): (a) t=15 ns, (b) t=17 ns, (c) t=19 ns, and (d) t=21 ns. (Taken from ref. [6].)

can be clearly seen in the region. The right-side nucleus has the (111) plane parallel to the z-direction. Thus, this nucleus grows faster in the downward direction due to a relatively larger crystallization velocity. On the other hand, the (111) plane of the left-side nucleus is rather perpendicular to the z-direction at 15 ns (Fig. 9(a)), which results in slower growth of this nucleus as shown in Figs. 9(b), 9(c), and 9(d).

Figure 10 shows an example of MD simulations to observe nucleation and crystallization processes in a 15 nm a-Si layer on SiO$_2$. In this figure, well-ordered atoms are highlighted. By partially

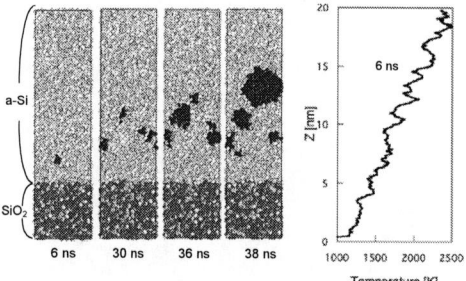

Fig. 10. Snapshots of atomic configurations observed in a 15 nm a-Si layer on SiO$_2$ during the thermal treatment described in the text. The right-hand panel indicates the steady-state temperature profile obtained at t=6 ns. Under this temperature condition, nucleation predominantly occurs in the lower temperature region (Z~5-12 nm) where Si is an amorphous-solid, while the nucleus observed at t=36 ns in the higher temperature region rapidly grows as can be seen at t=38 ns.

heating the a-Si surface region with 2 nm depth and removing thermal energy from the bottom of the SiO_2 substrate, a steady-state temperature profile was obtained after 6 ns as illustrated in Fig. 10. It is found that nucleation predominantly occurs in an amorphous-solid and rapid crystallization occurs toward the higher temperature region where Si is probably in a low-density liquid phase[11] as judged by the coordination numbers and diffusion constants of the atoms in the region.

Conclusion

We have investigated atomistic processes of crystallization in amorphous and liquid Si based on MD simulations combined with high-resolution electron microscopy measurements. It is proposed that crystallization and defect formation processes are controlled by the interface nanostructures between disordered and crystalline Si. It is also suggested that nucleation and crystal growth in a-Si films deposited on glass during ELA occur in amorphous-solid and low-density liquid phases rather than in supercooled liquid Si as has been generally accepted.

Acknowledgments

This work was supported by JSPS Research for the Future Program in the Area of Atomic Scale Surface and Interface Dynamics. This work was also partly supported by the Japanese Ministry of Economy, Trade and Industry, New Energy and Industrial Technology Development Organization and Advanced LCD Technologies Development Center Co. Ltd.

References

1. K. Ishikawa, M. Ozawa, CH Oh, and M. Matsumura, *Jpn. J. Appl. Phys.*, **37**, 731 (1998).
2. T. Motooka, S. Harada, and M. Ishimaru, *Phys. Rev. Lett.*, **78**, 2980 (1997).
3. T. Motooka, K. Nisihira, S. Munetoh, K. Moriguchi, and A. Shintani, *Phys. Rev.* B**61**, 8537 (2000).
4. Teruaki Motooka, Ken Nishihira, Ryuichiro Oshima, Hirosato Nishizawa, and Fuminobu Hori, *Phys. Rev.* B**65**, 081304(2002).
5. K. Nishihira and T. Motooka, *Phys. Rev.* B**66**, 2333101(2002).
6. T. Motooka and S. Munetoh, *Phys. Rev.* B**69**, 073307 (2004).
7. J. Tersoff, *Phys. Rev.* B **38**, 9902 (1988).
8. G. L. Olson and J. A. Roth, *Mat. Sci. Rep.* **3** , 1(1988).
9. Mark D. Kluge and John R. Ray, *Phys. Rev.* B **39**, 1738 (1989).
10. M. Ishimaru, K. Yoshida, T. Kumamoto, and T. Motooka, *Phys. Rev.* B **54**, 4638 (1996).
11. A. Hedler et al., *Nature Mater.* Vol.3, 804(2004).

214

ECS Transactions, 3 (8) 215-225 (2006)
10.1149/1.2356357, copyright The Electrochemical Society

Ge Nuclei for Fabrication of Poly-Si Thin Films on Glass Substrates

K. Yasutake, H. Watanabe, H. Ohmi, and H. Kakiuchi

Department of Precision Science and Technology, Graduate School of Engineering,
Osaka University, 2-1 Yamadaoka, Suita, Osaka 565-0871, JAPAN

As nucleation sites for solid-phase crystallization (SPC) of a-Si
thin films, crystalline Ge islands controlled in size and density
were fabricated on glass and SiO_2 substrates. Self-assembled
crystalline Ge islands were formed by the SPC of a-Ge thin films
on glass substrates. By annealing the samples under a reduced
pressure of oxygen above 350°C, Ge islands were etched to form
volatile GeO species (oxygen etching). Single-crystalline Ge
islands with good crystallinity and controlled in size and density
were fabricated on glass substrates by a simple process combining
SPC and oxygen etching. The crystallization times by SPC
annealing of a-Si thin films deposited on SiO_2 substrates with and
without Ge nuclei were studied by Raman spectroscopy and
transmission electron microscopy. Marked reductions of the
crystallization time and temperature were observed in the SPC of
a-Si:H thin films deposited by atmospheric pressure plasma CVD
on SiO_2 substrates with crystalline Ge islands.

Introduction

Large-grained polycrystalline silicon (poly-Si) thin films on glass substrates are of
interest for the fabrication of high-performance thin-film transistors (TFT) and cost-
effective solar cells. To obtain large-grained poly-Si thin films, it is important to control
the formation of nucleation sites for Si crystal growth on a glass or SiO_2 surface prior to
film deposition. Glass substrates having nucleation sites with tailored properties, such as
size and spacing, are particularly desirable for the solid-phase crystallization (SPC) of
amorphous silicon (a-Si) (1) or the selective epitaxial growth of Si from crystalline nuclei
by chemical vapor deposition (CVD) (2). Crystalline Ge islands formed on SiO_2 are
promising candidate nucleation sites for Si crystal growth.

For TFT or solar cell application of poly-Si thin films, a grain size significantly larger
than 1 μm is needed. To use the nucleation site spacing in defining the characteristic
grain size in poly-Si films, the average spacing of Ge islands on glass or SiO_2 surface
should be larger than 1 μm. Then, the density of Ge islands should be lower than
approximately 10^8 cm^{-2}. Other requirements for Ge islands as nucleation sites are i) the
enough smallness to be single crystals, and ii) the harmlessness for the electrical
properties of Si thin films. Since germanium oxide is not stable at high temperatures
(3,4), Ge islands can be etched in oxygen ambient at elevated temperatures (oxygen
etching). The oxygen etching of Ge is a simple but effective process for reducing the size
and density of Ge islands as has been recently reported by us (5).

215

In this work, we have proposed a new process to fabricate poly-Si thin films using crystalline Ge islands as nucleation sites for Si crystallization. To study the effect of Ge islands on the SPC of *a*-Si thin films, we have compared the crystallization times of *a*-Si thin films deposited on SiO₂ substrates with and without Ge islands.

Experimental

Figure 1 schematically illustrates the proposed process to grow poly-Si thin films on glass substrates. The process consists of several steps, namely i) deposition of amorphous Ge (*a*-Ge) thin films on glass substrates, ii) vacuum annealing to form self-assembled crystalline Ge islands, iii) oxygen etching of Ge islands to reduce their size and density, iv) deposition of *a*-Si thin films, and v) SPC annealing of *a*-Si thin films. Instead of processes iv) and v), selective heteroepitaxial growth of Si from crystalline Ge islands vi) is also possible because the CVD growth rate of Si on Ge (6, 7) is much higher than that of Si on SiO₂ (2, 8). Experimental procedure in each step is described below.

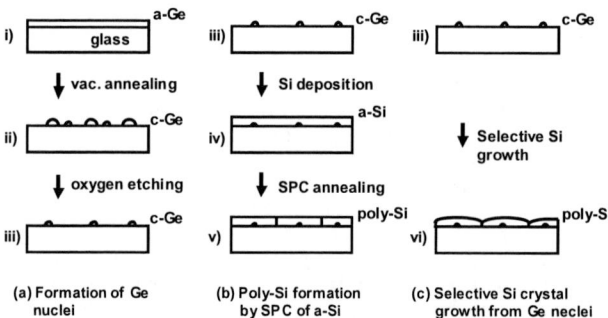

Figure 1. Formation processes of (a) crystalline Ge islands on glass substrate, (b) crystallization of *a*-Si using Ge nuclei by SPC annealing, and (c) selective hetero-epitaxial growth of Si from Ge nuclei.

i) *a*-Ge thin films were deposited on Corning#1737 glass or SiO₂ substrates at room temperature by vacuum evaporation using a p-BN crucible. SiO₂ substrates were prepared by thermal oxidation of Si wafers with the oxide thickness of approximately 150 nm. The thickness of the *a*-Ge layer deposited was in the range from 10 to 100 nm.

ii) The samples were annealed for SPC of *a*-Ge in the chamber under a vacuum of 1×10^{-6} Pa at various temperatures ranging from 250 to 500°C and for various annealing durations from 0.1 to 12 h.

iii) After the vacuum annealing, oxygen etching of Ge islands was performed in the same chamber under a controlled oxygen flow. The oxygen partial pressure was 10^{-4} - 10^{-2} Pa, and the oxygen etching temperature ranged from 250 to 500°C.

iv) For the experiment to study the effect of Ge islands on the SPC of a-Si, a-Si thin films with the thickness of 50-150 nm were deposited on SiO_2 substrates with and without crystalline Ge islands. a-Si films were prepared by electron beam (E-beam) evaporation at room temperature with a deposition rate of 0.03 nm/s. Hydrogenated a-Si films (a-Si:H) were also prepared by atmospheric pressure plasma CVD (AP-PCVD) using gas mixtures containing $SiH_4(0.1\%)$, $H_2(1\%)$ and $He(98.9\%)$ at a total pressure of 760 Torr. 150 MHz VHF power of 50 W was applied to the gap (0.6 mm) between the substrate and the cylindrical electrode rotating at 1000 rpm. a-Si:H films were deposited at 200°C with a deposition rate of 2.5 nm/s. Details on the AP-PCVD method have been described elsewhere (9, 10). The total concentration of hydrogen atoms in a-Si:H films prepared by AP-PCVD was approximately 30%. Before the SPC annealing, both of the a-Si films were annealed at 400°C for 1 or 2 h in vacuum for densification or to avoid peeling of the film from the substrate during SPC annealing, respectively.

v) SPC annealing of a-Si thin films was carried out in the temperature range from 450 to 600°C under a vacuum of $(1 - 5) \times 10^{-6}$ Pa. To compare the crystallization times of a-Si thin films on the SiO_2 substrates with Ge islands and those on the reference SiO_2 substrates without Ge islands, both the samples were annealed simultaneously in the same chamber.

vi) Selective heteroepitaxial growth of Si from Ge nuclei may be possible by using AP-PCVD at the lower temperatures than 600°C (11, 12). The experimental results will be presented elesewhere.

Structures of Ge islands and Si films were characterized by various methods including atomic force microscopy (AFM) (SEIKO Instruments, SPA400), selective etching using a diluted Secco etchant (13), transmission electron microscopy (TEM) (JEOL, JEM-2000FX) operated at an acceleration voltage of 200 kV, scanning electron microscopy (SEM, Hitachi S-800 or S-4200R), reflection high energy electron diffraction using a highly focused electron beam from field emission gun (μ-RHEED, Hitachi S-4200R), and Raman spectroscopy (JASCO, R-1100) using Ar^+ laser at 514.5 nm.

Formation of Crystalline Ge Islands on Glass Substrate

Solid Phase Crystallization of a-Ge Thin Films

Crystalline Ge islands can be formed in a self-assembling manner through SPC and agglomeration from a-Ge thin films on SiO_2 by thermal annealing (14, 15). Figure 2 is a series of AFM and TEM micrographs showing time evolution of Ge island formation by the vacuum annealing at 400°C. The initial thickness of the a-Ge thin film was 50 nm. Surfaces of the glass substrate and the a-Ge thin film (Fig. 2(a)) were smooth with root mean square (RMS) roughness of 0.57 nm and 0.25 nm, respectively. From Fig. 2, it is confirmed that Ge crystal islands are formed through crystallization and agglomeration of the a-Ge thin film by the vacuum annealing. The growth of Ge islands stops when the glass surface was uncovered because of the limited surface diffusion lengths of Ge atoms on a glass surface.

(a)   (b)   (c)

Figure 2. AFM and TEM micrographs showing time evolution of Ge island formation on glass substrate by vacuum annealing at 400°C for (a) 0 h, (b) 2 h, and (c) 4 h.

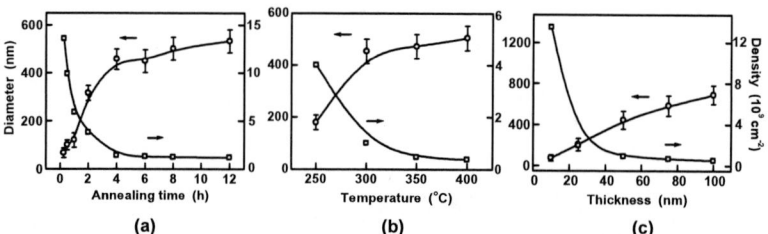

(a)   (b)   (c)

Figure 3. Average diameter and density of Ge islands as functions of (a) annealing time at 400°C, (b) annealing temperature for 30-min-annealing, and (c) initial a-Ge thickness.

Figure 3 shows the average diameter and density of Ge islands measured by AFM as functions of annealing time, temperature and initial a-Ge thickness. It is demonstrated that the average diameter of Ge islands becomes large with increasing annealing time, temperature and initial a-Ge thickness. Correspondingly, the density of Ge islands becomes lower with increasing these parameters. It is natural to conclude that to form small Ge islands in nano-scale, the annealing time should be short, the annealing temperature low, and the initial thickness of a-Ge film small, and vice versa. In the present range of experimental conditions, however, the minimum island density is around 10^{10} cm^{-2} when the Ge islands are smaller than about 100 nm.

Size and Density Control of Ge islands by Oxygen Etching

We performed oxygen etching (5) to reduce both the size and density of Ge islands using the samples with relatively large crystalline Ge islands. Figure 4(a) shows AFM and TEM images of the crystalline Ge islands on the glass substrate formed by the SPC annealing of a 50-nm-thick a-Ge film at 400°C for 12 h. Under this annealing condition, the a-Ge thin film fully crystallizes to form Ge islands as shown in the TEM image with the corresponding diffraction pattern (Fig. 4(a)). The average diameter of Ge islands is 530 nm and the average density is 4.0×10^8 cm^{-2}.

Figures 4(b) and 4(c) show AFM and TEM images of the crystalline Ge islands after oxygen etching at 400°C for 15 and 30 min, respectively. The oxygen partial pressure was 1×10^{-3} Pa. It is clear that the Ge islands become smaller as a result of the oxygen etching at 400°C. The relatively small islands seen in Fig. 4(a) disappear in Figs. 4(b) and 4(c). The density of the Ge islands decreases as a result of the oxygen etching. From TEM observations, it is confirmed that Ge islands smaller than approximately 300 nm are single crystals with no preferred orientation. By comparing the Ge islands seen in Fig. 4, the smaller Ge island in 4(c) has a lower density of extended defects. The smaller Ge island shows a well-facetted feature, as can be seen in the magnified AFM image in Fig. 4(c). It is speculated that defect or amorphous portions of Ge are easily removed by oxygen etching and stable surfaces of Ge crystals appear in the advanced stage of oxygen etching. This suggests possible reduction in the SPC annealing time required for the Ge island formation. In the experiments in the later section, we prepared crystalline Ge islands by the SPC annealing at 500°C for 2 h followed by the oxygen etching at 400°C for 1 h.

Figure 5(a) shows the temperature dependences of the average diameter and density of Ge islands after isochronal oxygen etching for 30 min. At temperatures lower than 300°C, the size and density of Ge islands are almost unchanged. This is in accordance with the experimental result of thermal desorption spectroscopy showing that GeO

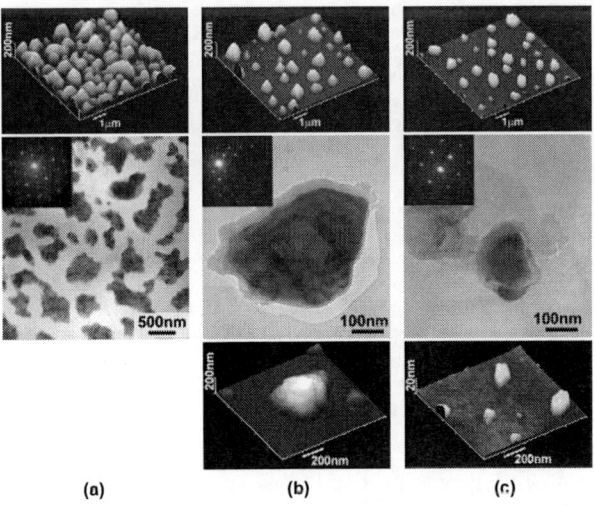

(a)   (b)   (c)

Figure 4. AFM and TEM images of crystalline Ge islands formed on glass substrate. (a) Crystalline Ge islands were formed by the SPC annealing of a 50-nm-thick a-Ge thin film at 400°C for 12 h in vacuum. The size and density of Ge islands were reduced by the oxygen etching at 400°C for 15 min (b) or 30 min (c).

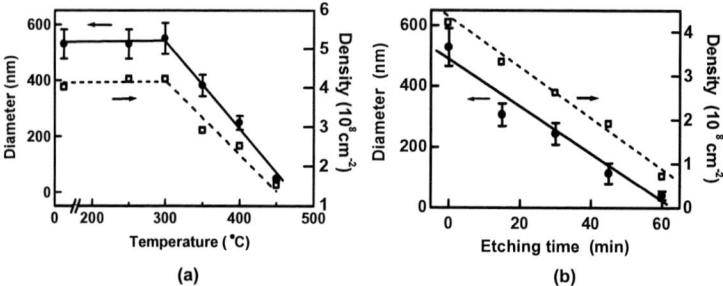

Figure 5. Average diameter and density of Ge islands as functions of oxygen etching temperature for 30-min-annealing (a), and oxygen etching time at 400°C (b). The oxygen partial pressure was 1×10^{-3} Pa. The preparation conditions of the samples before oxygen etching were the same as those in Fig. 4(a).

desorption peaks appear at 300 - 350°C (16). At temperatures higher than 350°C, the oxygen-etching rate increases with increasing temperature. Figure 5(b) shows the average diameter and density of Ge islands as a function of oxygen etching time at 400°C. An etching rate at 400°C is estimated to be approximately 4.3 nm/min from the time dependence of the average island size. From Fig. 5, we can control the diameter and density of Ge islands by combining the temperature and time of oxygen etching. We have obtained low-density (1.0×10^{-7} cm^{-2}) crystalline Ge islands with a small average diameter (40 nm) and a large average spacing (3.5 μm) on the glass substrate by oxygen etching at 400°C for 1 h (5).

Effects of Ge Islands on Crystallization of a-Si Thin Films

Ge Islands as Nuclei for Crystallization of a-Si

To study the effect of Ge islands on the SPC of a-Si thin films, we have deposited 150-nm-thick a-Si thin films on two kinds of SiO$_2$ substrate by E-beam evaporation. One has crystalline Ge islands on it and the other is the reference substrate without Ge nuclei. As a preliminary experiment to detect the effect of Ge islands on Si crystallization, Ge islands with a relatively large diameter (approximately 300 nm) were used in this experiment. The average density was approximately 3×10^8 cm^{-2} and the corresponding average spacing between Ge islands was approximately 0.6 μm. Figure 6 shows RHEED patterns of the Si thin films after the SPC annealing at 520°C for 16 h. The Si thin film on the reference substrate shows a halo pattern (Fig. 6(a)), indicating that no crystallization occurs at 520°C for 16 h. On the other hand, in the Si thin film on the SiO$_2$ substrate with Ge nuclei, we can see diffraction rings of polycrystalline Si (Fig. 6(b)). The ring pattern first appeared at the SPC annealing time of 1 h.

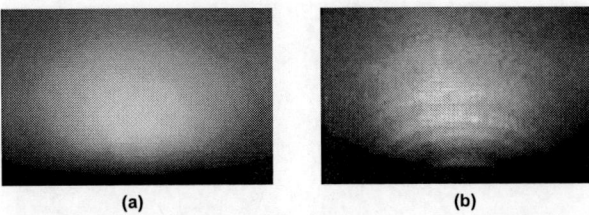

(a)   (b)

Figure 6.   RHEED patterns of a-Si thin films after SPC annealing at 520°C for 16 h.
(a) Reference a-Si sample on SiO_2 substrate without Ge nuclei. (b) Si sample deposited on SiO_2 substrate with Ge nuclei.

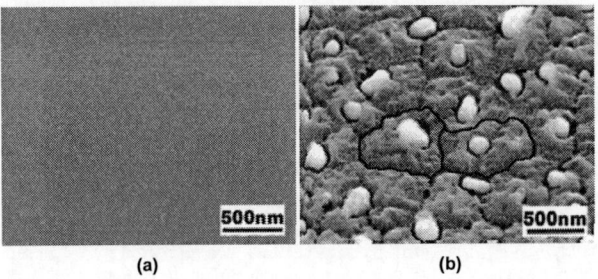

(a)   (b)

Figure 7.   SEM photographs of Secco-etched specimens after SPC annealing at 520°C for 16 h. (a) Reference a-Si sample on SiO_2 substrate without Ge nuclei.   (b) Si sample deposited on SiO_2 substrate with Ge nuclei.

To delineate grain boundaries in the annealed Si thin films shown in Fig. 6, the samples were etched in the diluted Secco etchant. Figure 7 shows SEM photographs of the etched specimens after the SPC annealing at 520°C for 16 h. The reference sample shows featureless appearance corresponding to the amorphous state. On the other hand, the Si sample on the SiO_2 substrate with Ge nuclei shows grain boundaries and Ge islands in the Si grain. It appears that peripheries of Ge islands are deeply etched like Si grain boundaries. Probably the lattice misfit strain between Si and Ge crystals generates defects at the peripheries of Ge islands.

In order to confirm that the crystallization of Si starts at the Ge islands, we performed μ-RHEED observation on the partially crystallized sample that was annealed at 480°C for 6 h using the smaller Ge islands than those in Fig. 7.   Figure 8(a) shows a SEM micrograph of the surface of a Si film on SiO_2 substrates with Ge islands. Two ellipses indicate the irradiated areas for μ-RHEED observation by a focused electron beam (6.25 nm in diameter) at an incidence angle of 2°. Since Ge islands are relatively large, the Si surface on Ge islands shows protrusions. A μ-RHEED pattern of the Si film on a Ge island shows crystalline diffraction spots as shown in Fig. 8(b) On the other hand, the Si film on SiO_2 surface shows an amorphous pattern (Fig. 8(c)). From these results, we can conclude that crystalline Ge islands act as   ion sites for Si crystallization.

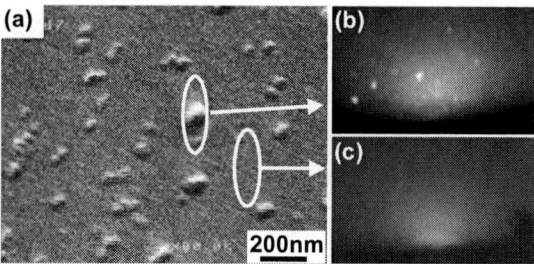

Figure 8. SEM photograph of Si film surface on SiO_2 substrate with Ge islands (a). The sample was partially crystallized by annealing at 480°C for 6 h. Right hand panels show μ-RHEED patterns of the Si film on a Ge island (b), and on a SiO_2 surface (c).

Reduction of Crystallization Time of a-Si Thin Film Using Ge Nuclei

Effects of relatively small Ge islands on SPC of a-Si films were studied by Raman spectroscopy. Ge islands were formed on SiO_2 substrates by the vacuum annealing of 50-nm-thick a-Ge thin films at 500°C for 2 h followed by the oxygen etching at 400°C for 1 h. The diameter of Ge islands was 50 - 70 nm and their average density was approximately $(1 - 3) \times 10^8$ cm^{-2}. The a-Si films were prepared by E-beam evaporation and AP-PCVD to see the difference between the preparation methods of a-Si films for SPC. Figure 9 shows Raman spectra of as-deposited and annealed Si films prepared by AP-PCVD on SiO_2 substrates. A small peak at about 520 cm^{-1} for the as-deposited film in Fig. 9(b) indicates that crystalline Si grows on Ge nuclei during the deposition, while a-Si grows on the bare SiO_2 surface. The reduction of crystallization time by using Ge nuclei is confirmed by SPC annealing experiments at 520°C as shown in the Raman spectra. After the annealing at 520°C for 2 h, Raman band at around 480 cm^{-1} of a-Si disappears in Fig. 9(b) for the sample with Ge nuclei. On the other hand, the Raman band of a-Si dominates in the sample without Ge nuclei even after the annealing at 520°C for 8 h (Fig. 9(a)).

Interestingly, it is found that the crystallization time of a-Si:H deposited by AP-PCVD using Ge nuclei is much shorter than that of a-Si deposited by E-beam evaporation. Figure 10 shows Raman spectra of E-beam deposited Si films on SiO_2 substrates with and without Ge nuclei. Comparing the crystal peaks at 520 cm^{-1} in as-deposited Si films in Figs. 10(b) and 9(b), the crystal peak in the E-beam deposited sample (Fig. 10(b)) is larger than that by AP-PCVD. This may be related with the difference in the deposition rate of a-Si films between E-beam evaporation and AP-PCVD. While the deposition rate by the former method was 0.03 nm/s, that by the latter was 2.5 nm/s. Therefore, the migration time of Si precursors during deposition was much longer in the E-beam evaporation than that by AP-PCVD. On the other hand, although the post-deposition annealing at 520°C for 8 h has little effect on the crystallization of a-Si films deposited by E-beam evaporation (Fig. 10(b)), the crystallization of a-Si:H films deposited by AP-PCVD is greatly enhanced by 520°C annealing for 2h (Fig. 9(b)).

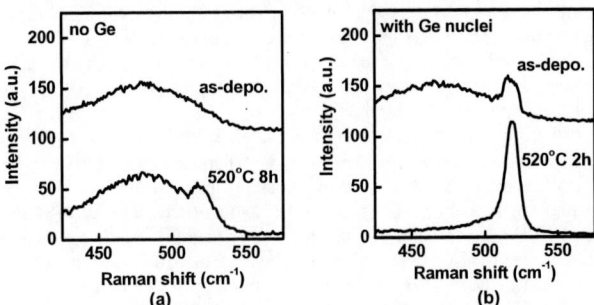

Figure 9. Raman spectra of as-deposited and annealed Si films at 520°C. The Si films were deposited by AP-PCVD at 200°C on SiO$_2$ substrates without Ge (a), and with Ge nuclei (b). Raman bands at 480 and 520 cm^{-1} are related with amorphous and crystalline Si phase, respectively.

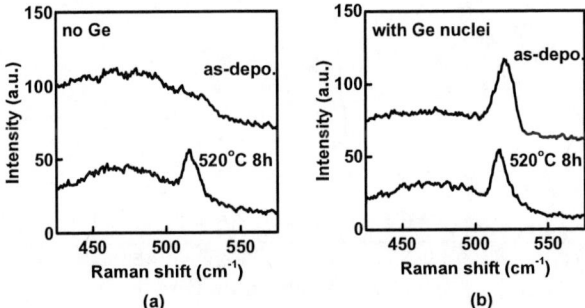

Figure 10. Raman spectra of as-deposited and annealed Si films at 520°C. The Si films were deposited by E-beam evaporation at room temperature on SiO$_2$ substrates without Ge (a), and with Ge nuclei (b).

Figure 11 shows the initiation of crystallization of a-Si:H by annealing at various temperatures measured by Raman spectroscopy. The a-Si:H films were deposited by AP-PCVD on SiO$_2$ substrates with Ge nuclei. Since Si films are thin (50 nm), the Raman peak at 520 cm^{-1} from the Si substrate appears when the a-Si film begins to crystallize because of the low absorption coefficient for 514.5 nm laser light in the crystalline state. Therefore, only the time when the spectral change starts can be determined from Fig. 11. A broken curve in Fig. 11 shows a crystallization curve at 600°C for a-Si:H prepared by a conventional low-pressure PECVD without Ge nuclei deduced from ref. 17. From Fig. 11, it is concluded that the crystallization time is much shorter in the present sample using Ge nuclei than that reported for usual a-Si:H films.

As shown in Fig. 11, although the Ge islands act as nuclei for Si crystallization, there exist incubation times for crystallization of a-Si:H films. Two possibilities for the cause of the incubation time are conceivable. One is the thin oxide layer present on the surface of Ge islands before a-Si deposition and the other is the misfit strain between Ge and Si crystals. Both of them should retard the crystallization reaction of a-Si from Ge nuclei. In the present experiments, E-beam deposition of a-Si was performed in the same chamber as that for the oxygen etching of Ge islands, and AP-PCVD of a-Si:H was performed after a flash anneal at 400°C in vacuum. Therefore, the presence of thin oxide film on Ge surface may be excluded from the possible causes of the incubation time. The observations of Raman peaks at 520 cm^{-1} in the as-deposited states of a-Si and a-Si:H (Figs. 9(b) and 10(b)) are in favor of this argument. Therefore, lattice strains in the Si crystal regions in the vicinity of Ge islands may be the cause of the incubation time. It is considered that since the a-Si:H film deposited by AP-PCVD initially contains high concentration of SiH and SiH$_2$ bonds, misfit strains between the Ge and Si crystals may be easily accommodated in the a-Si:H region around Ge nuclei resulting in the decrease of incubation time for Si crystallization.

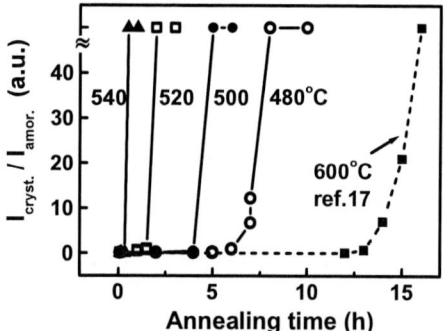

Figure 11. Crystallization processes of a-Si:H with Ge nuclei by annealing at various temperatures. $I_{cryst.}/I_{amor.}$ denotes the ratio of integrated intensities of the Raman band at 520 cm^{-1} for crystal Si to that at 480 cm^{-1} for a-Si. The Si films were grown by AP-PCVD at 200°C on SiO$_2$ substrates with Ge nuclei. A broken curve shows a crystallization curve for a-Si:H without Ge nuclei deduced from ref. 17.

Summary

We have proposed a new process to fabricate crystalline Ge islands with controlled size and density in a tailored range to form nuclei for solid-phase crystallization of a-Si thin films. The process consists of the deposition of a-Ge thin films on glass or SiO$_2$ substrates at room temperature, the subsequent vacuum annealing for SPC of the deposited a-Ge thin films to form self-assembled crystalline Ge islands, and the oxygen etching for controlling the size and density of Ge islands. By combining vacuum annealing conditions for the SPC of a-Ge thin films and oxygen etching conditions, we can control the size and spacing of Ge islands on glass substrates over a wide range. From the μ-RHEED observation of the partially crystallized a-Si thin film using Ge nuclei, we have confirmed that Ge islands act as nuclei for Si crystallization. Marked reductions of the crystallization time and temperature are observed in the annealing experiments of a-Si:H thin films deposited on the SiO_2 substrates with crystalline Ge islands. Moreover, it is found that the crystallization time of a-Si:H deposited by AP-PCVD is much shorter than that of a-Si deposited by E-beam evaporation.

Acknowledgments

This work was partially supported by a Grant-in-Aid for Scientific Research (No. 16360070) and a Grant-in-Aid for the 21st Century COE Program from the Ministry of Education, Culture, Sports, Science and Technology. The authors wish to thank Professor H. Mori, Dr. K. Yoshida, Dr. T. Sakata and E. Taguchi of the Research Center for Ultra-High Voltage Electron Microscopy for their kind support in the TEM observations. The technical assistance of A. Takeuchi, N. Nishimoto, H. Mishima, D. Nakajima, S. Koyama, K. Minami and C. Yoshimoto of Osaka University is also greatly appreciated.

References

1. A. T. Voutsas, Appl. Surf. Sci. **208-209**, 250 (2003).
2. K. Yamagata and T. Yonehara, Appl. Phys. Lett. **61**, 2557 (1992).
3. A. A. Frantsuzov and N. I. Makrushin, Surf. Sci. **40**, 320 (1973).
4. K. Prabhakaran and T. Ogino, Surf. Sci. **325**, 263 (1995).
5. K. Yasutake, H. Ohmi, H. Kakiuchi, H. Watanabe, K. Yoshii and Y. Mori, Jpn. J. Appl. Phys. **43**, L1552 (2004).
6. K. Fujinaga, Y. Takahashi, H. Ishii, I. Kawashima and S. Hirota: J. Vac. Sci. & Technol. **B5**, 1551 (1987).
7. K. Fujinaga, Y. Takahashi, H. Ishii, S. Hirota and I. Kawashima: J. Vac. Sci. & Technol. **B7**, 225 (1989).
8. W. A. P. Claassen and J. Bloem: J. Electrochem. Soc. **195**, 194 (1980).
9. H. Kakiuchi, M. Matsumoto, Y. Ebata, H. Ohmi, K. Yasutake, K. Yoshii and Y. Mori, J. Non-Cryst. Solids **351**, 741 (2005).
10. H. Kakiuchi, H. Ohmi, Y. Kuwahara, M. Matsumoto, Y. Ebata, K. Yasutake, K. Yoshii and Y. Mori, Jpn. J. Appl. Phys. **45**, 3587 (2006).
11. K. Yasutake, H. Kakiuchi, H. Ohmi, K. Yoshii and Y. Mori, Appl. Phys. A **81**, 1139 (2005).
12. H. Ohmi, H. Kakiuchi, N. Tawara, T. Wakamiya, T. Shimura, H. Watanabe and K. Yasutake, Jpn. J. Appl. Phys. **45** [10B] (2006) in press.
13. F. Secco d'Aragona, J. Electrochem. Soc. **119**, 948 (1972).
14. Y. Wakayama, T. Tagami and S. Tanaka, Thin Solid Films **350**, 300 (1999).
15. Y. Wakayama, T. Tagami and S. Tanaka, J. Appl. Phys. **85**, 8492 (1999).
16. K. Yasutake, H. Watanabe, H. Ohmi, H. Kakiuchi, S. Koyama, D. Nakajima and K. Minami, in *Proc. Thin Film Materials & Device Meeting, Nara, 2004*, p. 19.
17. K. Pangal, J. C. Sturm, S. Wagner and T. H. Büyüklimanli, J. Appl. Phys. **85**, 1900 (1999).

226

CHAPTER 6

TFTs ON FLEXIBLE SUBSTRATES

ECS Transactions, 3 (8) 229-236 (2006)
10.1149/1.2356358, copyright The Electrochemical Society

The Road towards Large-Area Electronics without Vacuum Tools

J.H. Daniel[*], A.C. Arias, B. Krusor, R. Lujan, R.A. Street

Palo Alto Research Center, 3333 Coyote Hill Road, Palo Alto, CA 94304, USA
[*]daniel@parc.com

The fabrication of conventional electronic circuits relies on vacuum deposition and etching steps. However, for large-area and low-cost electronics other processes may be advantageous. We have explored the fabrication of electronic pixel circuits with a low-temperature all-additive process using jet-printing technology. These low-cost pixel backplanes are expected to find application in active-matrix addressed paper-like electronic displays.

Introduction

Photolithographic patterning, vacuum deposition and etching are typically employed in the manufacture of conventional electronic circuits. Significant progress in these technologies has made it possible to shrink the feature size far into the sub-micron range. Small features are important for highly integrated circuits, but other large-area electronic circuits do not require them. The feature size in active-matrix pixel circuits for displays, for example, has not significantly changed. However, the size of the displays has increased. Moreover, the substrate size on which multiple display backplanes are fabricated has continuously increased, with current 8th generation glass substrates reaching a size of 2160×2400 mm. The manufacturing equipment, such as photolithography systems and vacuum tools, is becoming increasingly complex and expensive (1).

Alternative fabrication processes based on inkjet-, offset-, flexography-, gravure- and microcontact-printing have been recently explored (2-6). The goal is to simplify the fabrication process by reducing process steps, to use less expensive manufacturing equipment and to save on processing materials. Many printing techniques are also compatible with roll-to-roll processing which has proven to be an inexpensive process in the printing industry. However, significant research and development is required with regards to processes and materials in order to evaluate these novel fabrication approaches.

Our research focuses on jet-printing processes because it is a non-contact printing method and it allows a rapid modification of the printed patterns. We are using piezoelectric printheads because they allow a wide variety of materials to be ejected as long as their viscosity is low enough. A typical drop placement accuracy of +/-5 μm is currently achieved with our multi-ejector (~1000 nozzles) printheads and the drop size is ~20-40 μm (7). For materials evaluation and process development we also use single-ejector printheads (MicroFab Technologies).

Previously, jet-printing has been shown to simplify conventional photolithography (8,9). In this method, termed 'digital lithography', a phase-change material, such as a wax, is printed onto a substrate as an etch mask. Using this method, pixel circuits based on organic thin film transistors (TFTs) were fabricated with a combination of additive and subtractive processing steps (10,11).

Here, we report on printed TFT backplanes fabricated in an all-additive, low-temperature process without employin~ ~~~ photolithography systems, vacuum

229

deposition or etching tools. An all-additive fabrication method is expected to have the lowest cost since it employs a minimum number of processing steps and because hardly any material is wasted. Active-matrix backplanes for paper-like displays are a potential first application for such low-cost processes. In these displays the performance of the pixel circuit can be slightly lower because high refresh rates are usually not required. Particularly promising electronic paper displays are based on bistable display media such as electrophoretic ink (12,13).

In our all-additive printed backplanes, the metal conductors for the gate/data lines and for the pixel pads were jet-printed with a colloidal ink based on silver nanoparticles. The dielectric layers were spin coated and the organic semiconductor for the active region of the TFTs was again jet-printed. Below, we will describe the individual steps and the challenges for fabricating all-printed pixel circuits. Initial results for TFT performance and pixel response will be discussed.

Components of an All-additive Printed Circuit

Insulators

Insulating or dielectric layers play an important role in electronic circuits because they significantly determine the circuit or device performance. In most silicon-based circuits, thermal silicon dioxide is used as a dielectric because of its outstanding quality. Good dielectric properties are particularly important for the gate dielectric in transistors. Silicon dioxide gate dielectrics are also often used in the evaluation of TFTs with organic semiconductors. An additional advantage of silicon dioxide is the fact that the surface can be functionalized with silanes, a method which often leads to significant improvement in the transistor performance.

For large-area electronic circuits on low-temperature substrates, other dielectric materials such as solution processable polymers are desirable. These materials can be deposited by spin-coating, dip-coating, spray coating, printing or similar large-area compatible processes.

However, there are many requirements facing these dielectric materials (14-16). Due to the relatively low dielectric constant of most polymers, the thickness of the film has to be ideally sub-micron in order to assure relatively high values for the gate capacitance. Such ultra-thin polymer films are difficult to achieve without any pinholes or defects. Moreover, the polymer has to be a material with low charge leakage and it has to perform well in conjunction with the organic semiconductor in a TFT. A further requirement is the polymer's compatibility with the subsequent processing steps. In our bottom gate TFT structures, the gate dielectric must withstand the solvents used for printing the source/drain layer as well as the solvent in the organic semiconductor solution. Several commercial polymer dielectrics require curing temperatures above 200°C. However, our goal was to limit the maximum process temperatures to ~150°C in order to develop a process that is compatible with a wide variety of low-temperature substrates. Table I lists a number of polymer dielectrics which we evaluated in comparison to thermal silicon dioxide. The dielectric properties were measured on simple capacitor structures and the contact angles were determined using a sessile drop method. From the data it is evident that the leakage currents in polymer dielectrics are generally higher. Moreover, it is challenging to achieve a gate capacitance comparable to a ~110-120 nm thick silicon dioxide layer. Regarding the printing of continuous lines for the source-drain layer, we also had to evaluate the wetting properties of the dielectrics. On highly hydrophobic

TABLE I. Potential low-temperature gate-dielectrics in comparison to thermal silicon dioxide.

material	ε @ 1kHz	capacitance @1kHz (nF/cm2)	leakage current @ 0.5 MV/cm (A/mm2)	water contact angle (deg)
SiO2	3.6-3.9	28-29 (110-120nm)	3.4×10^{-14}	~62-72
Parylene C	1.9-2.0	3.3-3.5 (500nm)	7.6×10^{-12}	~80
Avatrel	2.4-2.6	1.9-2.1 (1.1micron)	1.6×10^{-11}	~93
Techneglas GR-650UP	3.1-3.2	2.3-2.4 (1.2micron)	8.8×10^{-11}	~92
Cytop	1.7-1.8	2.7-3.1 (540nm)	2×10^{-11}	~106
PVA	7	17-18 (340-350nm)	1.7×10^{-9}	~20
SU-8	3.2-3.9	14-17 (200nm)	5×10^{-11}	~75

surfaces, the inks would form discontinuous lines and on hydrophilic surfaces, the printed lines would exhibit excessive spreading. We found that the epoxy-based photopolymer SU-8 (MicroChem Corp.) was a good compromise. SU-8 can be diluted and spin coated as a thin layer. Thereafter, SU-8 is cross-linked by UV exposure and subsequent bake at 150°C. Cross-linked SU-8 is chemically inert to most chemicals due to its high cross-link density and the measured leakage currents were reasonably low. Most importantly, the surface energy of SU-8 was in a range that allowed us to print continuous lines and features with good definition using our water-based inks.

Conductors

All-additive printed circuits require conductors that can be jet-printed. There are several requirements for the conductors. In a TFT, the source and drain electrodes should be electronically matched to the semiconductor. Here the work function of the printed conductor is important. A high mismatch causes an excessive contact resistance. The gate and the pixel pad have to be conductive, but the conductivity does not have to be extremely high because of the small size or short length of the structures. High conductivity of the conductor can however become important for the gate and data lines. Particularly if the address lines are long, a high resistivity will cause long RC pixel charging times. In today's large-format LCD displays for TV applications, the conductors have been replaced by copper conductors to achieve a low resistance.

Organic TFTs with printed solutions of organic conductors based on PEDOT/PSS (polyethylenedioxythiophene doped with polystyrene sulfonic acid) have been reported (17). PEDOT/PSS is a promising material for the source/drain electrodes due to the efficient hole-carrier injection into organic semiconductors, but the conductivity is not high enough to replace long metal conductors (16).

Recently, dispersions of metal nanoparticles have become available. We have used nano-silver which consists of silver particles with a diameter of ~30-50 nm dispersed in a water-based solution. Others have reported gold nanoparticle solutions which they used for the source and drain contacts in TFTs (18). However gold is rather expensive for low-cost displays. Silver nanoparticles are less expensive, they provide a good conductivity and they also do not easily oxidize. Nanoparticles from copper or nickel are more problematic because of their propensity for oxidation.

Once the nanosilver is deposited and the solvent has dried, the film is not yet conducting. Annealing is required to sinter the silver nanoparticles. In metal nanoparticles, the melting temperature is greatly reduced compared to bulk material and silver films with low resistivity were obtained at annealing temperatures as low as 150°C (19). The resistivity value depends not only on the annealing temperature but also on the annealing time. The lowest resistivity value we observed was $\sim 5 \times 10^{-6}$ Ωcm which is about three times higher than the bulk value for silver.

The nanosilver solution was printed with a single ejector piezo-printhead and a ~ 40 μm nozzle aperture. In order to print continuous lines or square patterns, various parameters have to be adjusted. If the conditions are not optimized, various effects can happen. The lines may break up into discontinuous lines and the edges of the lines or structures may not be straight. Moreover, bulges may form in the lines, or the lines may spread excessively if the ink wets the surface too much. When writing larger continuous areas which consist of several parallel lines, the ink may pull towards the center or concentrate in some areas which leads to strong thickness variations. Thick silver layers tend to show cracking after annealing and ideally the film thickness should stay below ~ 200 nm. This also avoids a strong topography variation which can negatively affect the deposition of subsequent layers. Another problematic topography often occurs due to the coffee ring effect observed in printed features (20).

The parameters that affect the printed pattern are viscosity, surface tension and the composition of the metal ink, as well as the surface energy of the substrate. Moreover, the printing conditions are important. The overlap between adjacent drops has to be adjusted so that narrow continuous lines without bulges form.

As shown in Fig. 1, we have printed continuous silver lines with a width of ~ 100 μm using optimized conditions. Fig. 1a shows the gate layer for a 60x60 pixel array printed on a SU-8 coated glass substrate. Fig. 1b shows a close-up of the well-defined silver features for the gates, the bottom electrodes of the pixel storage capacitor and the gate lines. In Fig. 2, images of multiple patterned layers are shown. Fig. 2a is a photograph of a completed pixel structure. After depositing the SU-8 gate dielectric, the data lines and the pixel pads were jet-printed from nanosilver solution. Fig. 2b is an SEM image of a stack of SU-8 layers and printed nanosilver. The thickness of the gate dielectric is ~ 300 nm.

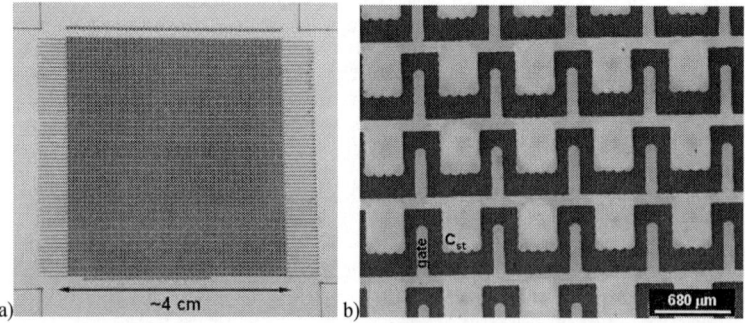

Figure 1. Jet-printed gate-level of a 60x60 pixel array on a glass substrate (a). In (b) a close-up photograph of the printed silver lines is shown. The lines are ~ 100 μm wide and the pixel pitch is 680 μm.

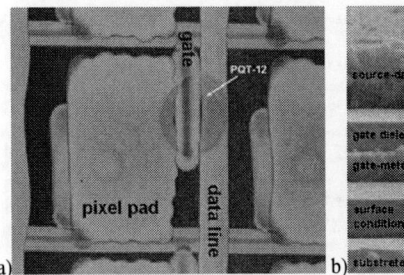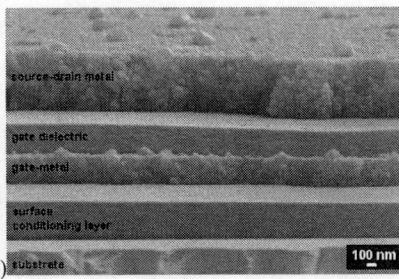

Figure 2. Pixel structures (a) fabricated in an all-additive process (~680 μm pixel pitch). The gate and data layers were patterned with jet-printed nano-silver and the polymeric semiconductor was jet-printed PQT-12. The SU-8 gate-dielectric was spin-coated. In b) an SEM image is shown of a cross-sectioned circuit with jet-printed nano-silver metal layers and SU-8 dielectric layers. The gate-dielectric is ~300 nm thick.

Semiconductors

In order to jet-print the semiconductor it has to be in the form of a solution. A variety of solution processable polymeric organic semiconductors have been reported and many of them are from the class of polythiophenes. Others have developed organic semiconductors which are deposited from a precursor (21). Recently, even solution processable liquid silicon has been reported, although this material still requires high annealing temperatures (22).

We are currently focusing our work on the polythiophene PQT-12 (23). It is jet-printed from a solution of nanoparticles in dichlorobenzene. A mobility of greater than 0.1 cm^2/Vs and high on-to-off ratios ($>10^7$) have been reported for TFTs fabricated with this material and the high mobility was achieved on a silicon dioxide dielectric that was functionalized with a silane (OTS - octyltrichlorosilane) (23,24). These values are in the range needed to drive a display. Moreover, PQT-12 has good stability in air.

Here, we jet-printed the material from a single-ejector piezo-printhead (from MicroFab Technologies) with an aperture of ~40 microns. After solvent evaporation, the semiconductor was heated. This heating step has been reported to increase the ordering and the alignment of the PQT-12 molecules (23).

All-additive Printed TFTs

Fig. 3 shows a measurement of a bottom-gate TFT fabricated in an all-additive process with SU-8 gate dielectric. Jet-printed silver pads defined the source and drain of the TFT with PQT-12 semiconductor. In this device, the SU-8 gate dielectric was spin-coated and then cross-linked by UV illumination and a post-exposure bake. The source and drain contacts were rendered conductive by an annealing step. Then, the PQT-12 was dispensed and the solvent dried off. Afterwards, the TFTs were annealed to improve the organic semiconductor. The shown TFT measurements were performed in the dark. From the transfer curves we extracted a saturation mobility of 1.7x10-3 cm^2/Vs and a turn-on voltage of ~10-15 V. The output curves show a linear behaviour at low V_{ds} which

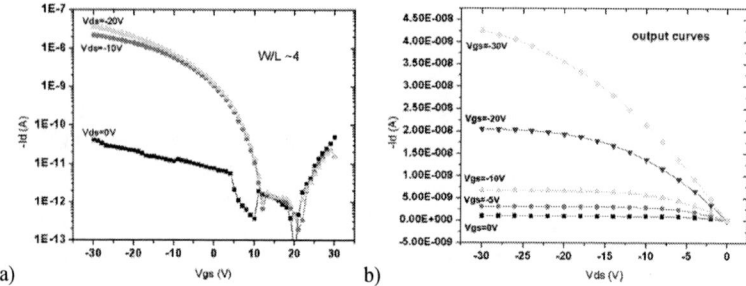

a)   b)

Figure 3. Transfer characteristic (a) and output curves (b) of a TFT structure with SU-8 gate dielectric (230-240 nm thick), printed silver source and drain pads and solution deposited PQT-12 semiconductor. The TFT structure had a channel length of 80 μm and a channel width of ~300 μm. A saturation mobility of ~1.7x10-3 cm^2/Vs was extracted from the graphs.

indicates ohmic contacts. However, further measurements on TFTs with shorter transistor channel lengths are required to give definitive answers to the contact issue. In TFTs with silver nanoparticles, others have reported a dependence of the contact resistance on the kind of additives used to stabilize the nanoparticles in solution (25). If necessary, different materials, such as PEDOT or gold nanoparticles could be locally jet-printed onto the source-drain regions. The measured carrier mobility with SU-8 gate dielectric was rather low compared to TFTs with OTS functionalized silicon dioxide dielectrics. Other organic dielectrics will have to be evaluated and promising results with PQT-12 have been already reported (18).

All-additive Printed Pixel Circuits

Fig. 2a shows a photograph of a pixel of an active-matrix backplane with a pixel pitch of ~680 μm. The PQT-12 semiconductor was deposited by jet-printing. Fig. 4 shows a measurement of the pixel response of a slightly larger pixel (~1.1 mm pixel pitch). The pixel signal was measured with a high-impedance picoprobe. The measurement shows a typical pixel response in which the pixel charges up to the data potential when the gate voltage switches the TFT into the ON-state. After switching off the TFT, the charge on the pixel remains stored within the limits of the leakage currents. When the gate potential changes, a feedthrough voltage is measured on the pixel pad. For the pixel in Fig. 4, this feedthough voltage is ~2 V for a 20 V gate voltage swing. This is due to the parasitic capacitance C_p caused by the overlap of the pixel pad with the gate electrode (indicated in Fig. 4b) and the contribution of the channel capacitance. The goal is to minimize the overlap of the printed lines in an improved printing process and also to optimize the ratio of C_p to the storage capacitance C_{st} without excessively increasing the pixel charging time. The measurement in Fig. 4 shows a relatively long pixel charging time of ~35 ms to reach 90% of the data voltage with a gate-ON potential of -10 V. This time is shortened to ~27 ms with a more negative gate-ON potential of -15 V. For driving displays, these rather long pixel charging times may be acceptable if an image does not need to be updated at a fast rate. However, an improved TFT performance (higher mobility) and an optimized pixel design will result in fast   ·l response. The pixel signal in Fig. 4

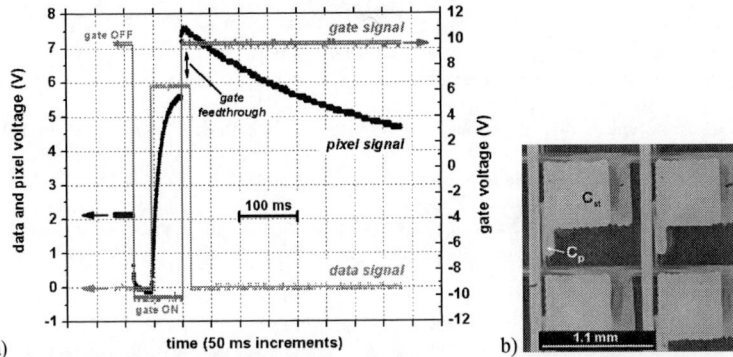

a)

b)

Figure 4. Picoprobe measurement (a) of a pixel which was fabricated in an all-additive process with SU-8 gate-dielectric and jet-printed gate, source-drain metal and PQT-12 semiconductor. The gate-feedthrough voltage (20 V gate-swing) is ~2 V and the pixel charging time to ~90% of the data voltage (5.4 V) is ~35 ms. In the off-state the pixel shows good charge storage. In (b), a photograph of the measured pixel structure is shown. Indicated are the storage capacitor (C_{st}) and the parasitic capacitance (C_p) due to the overlap of gate and pixel-pad.

shows good charge storage when the TFT is switched off (gate: +10 V) and at higher OFF-voltages, a further improvement was seen. However, the larger gate-swing also causes a stronger voltage feedthrough.

Summary and Conclusions

We reported on the fabrication of all-additive pixel circuits using jet-printing. A careful choice of materials and printing conditions is important to achieve well defined printed structures. However, the materials also have to perform well electronically, which means, they must provide a sufficiently high conductivity for the gate and data lines and they must show good performance in a thin film transistor. In the described pixel circuits we used jet-printed nanosilver for defining conductors, SU-8 photo polymer for the insulating layers and printed PQT-12 as the organic semiconductor. The maximum process temperature was 150°C and therefore the process is compatible with many substrates. The pixel circuits showed good switching characteristics which are promising for driving a bistable display medium such as electrophoretic ink. Future research will investigate other solution processable dielectrics to improve the transistor performance with regard to mobility and charge leakage. Moreover, the interaction of PQT-12 with the new dielectrics in terms of bias stress will have to be evaluated.

Acknowledgments

We would like to thank Beng Ong, et al. at the Xerox Research Centre of Canada for collaboration on materials and NIST for partial funding. At PARC, we also would like to thank Michael Chabinyc for helpful discussions and Steve Ready for building the printing system.

References

1. T. Takehara, *Digest AM-FPD 06*, 5-1.95-98 (2006)
2. B.-J. de Gans, P.C. Duineveld, U.S. Schubert, *Adv. Mater.*, **16**, No. 3, 203-213 (2004)
3. W. Clemens, W. Fix, J. Ficker, A. Knobloch, A. Ullmann, *J. Mater. Res.*, **19**, No. 7, 1963-1973 (2004)
4. T. Kawase, H. Sirringhaus, R.H. Friend, T. Shimoda, *SID 01 Digest*, 6.1, 40-43
5. G. Blanchet, J. Rogers, *J. Imaging Sci. Techn.*, **47**, 296-303 (2003)
6. M.L. Chabinyc, W.S. Wong, A.C. Arias, S. Ready, R. Lujan, J.H. Daniel, B. Krusor, R.B. Apte, A. Salleo, R.A. Street, *Proc. IEEE*, **93**, No. 8, 1491-1499 (2005)
7. R.A. Street, W.S. Wong, S.E. Ready, M.L. Chabinyc, A.C. Arias, S. Limb, A. Salleo, R. Lujan, *Materialstoday*, **9**, 32-37 (2006)
8. W.S. Wong, S. Ready, R. Matusiak, S.D. White, J.-P. Lu, J. Ho, R.A. Street, *Appl. Phys. Lett.*, **80**, 610-612 (2002)
9. W.S. Wong, S.E. Ready, J.-P. Lu, R.A. Street, *IEEE Electr. Dev. Lett.*, **24**, No. 9, 577-579 (2003)
10. K.E. Paul, W.S. Wong, S.E. Ready, R.A. Street, *Appl. Phys. Lett.*, **83**, 2070-2072 (2003)
11. A.C. Arias, S.E. Ready, R. Lujan, W.S. Wong, K.E. Paul, A. Salleo, M.L. Chabinyc, R. Apte, R.A. Street, Y. Wu, P. Liu, B. Ong, *App. Phys. Lett.*, **85**, 3304-3306 (2004)
12. B. Comiskey, J.D. Albert, H. Yoshizawa, J. Jacobson, *Nature*, **394**, 254-255 (1998)
13. J. Daniel, A.C. Arias, W.S. Wong, R. Lujan, B.S. Krusor, R.B. Apte, M.L. Chabinyc, A. Salleo, R.A. Street, *SID 05 Digest*, 54.1, 1631-1633
14. A. Facchetti, M-H. Yoon, T. Marks, *Adv. Mater.*, **17**,1705-1725 (2005)
15. J. Veres, S. Ogier, G. Lloyd, D. de Leeuw, *Chem. Mater.*, **16**, 4543-4555 (2004)
16. S. Lee, B. Koo, J-G. Park, H. Moon, J. Hahn, J.M. Kim, *MRS Bulletin*, **31**, 455-459 (2006)
17. S.E. Burns, P. Cain, J. Mills, J. Wang, H. Sirringhaus, *MRS Bulletin*, Nov., 829-834 (2003)
18. P. Liu, Y. Wu, Y. Li, B. Ong, S. Zhu, *J. Am. Chem. Soc.*, **128**, 4554-4555 (2006)
19. Ph. Buffat, J-P. Borel, *Phys Rev. A*, **13**, No. 9, 2287-2298 (1976)
20. H. Hua, R.G. Larson, *J. Phys. Chem. B*, **110**, No 14, 7090-7094 (2006)
21. S. Aramaki, A. Ohno, Y. Sakai, M. Tazoe, S. Sugimoto, G. Yip, A. Ikeda, R. Hattori, *AM-FPD 06*, 57-58
22. T. Shimoda, et al., *Nature*, **440**, 783 (2006)
23. B. Ong, Y. Wu, P. Liu, S. Gardner, *Adv. Mater.*, **17**, 1141-1144 (2005)
24. B. Ong, Y. Wu, P. Liu, S. Gardner, *J. Am. Chem. Soc.*, **126**, 3378-3379 (2004)
25. Y. Wu, Y. Li, B. Ong, *J. Am. Chem. Soc.*, **128**, 4202-4203 (2006)

ECS Transactions, 3 (8) 237-247 (2006)
10.1149/1.2356359, copyright The Electrochemical Society

230 DPI High Resolution AMPLED Display on Flexible Metal Foils with Integrated Row Drivers

Matias N. Troccoli[a], Ta-Ko Chuang[a], Abbas Jamshidi Roudbari[a], Po-Chin Kuo[a], Jeffery A. Spirko[a], M. K. Hatalis[a], A. T. Voutsas[b], T. Afentakis[b], and John W. Hartzell[b]

[a]Display Research Lab, Lehigh University, Bethlehem, PA 18015
[b]Sharp Laboratories of America, Camas, WA 98607

This paper presents a 230 dpi Active Matrix Polymer Light Emitting Diode (AMPLED) display on flexible stainless steel substrates fabricated with laser annealed poly-silicon TFT technology. The high resolution display is based on the standard 2 TFT pixel circuitry to form a full VGA array. Polysilicon TFT based shift registers for integrated row driving are also demonstrated. Display fabrication is compatible with standard polysilicon TFT CMOS processing temperatures and conditions. This paper discusses backplane fabrication, PLED deposition and display encapsulation; all performed in-house.

Introduction

Attention on large area electronics on flexible substrates has risen considerably in recent years. Mechanically flexible electronics have the potential to realize novel applications with physical and mechanical restrictions that do not permit the use of rigid substrates. The main applications targeted are flat panel displays, but other matrix-based electronics such as active sensor arrays are being developed as well. A large degree of system integration that includes control and processing electronics in a single substrate is also being targeted. A higher integration level will increase system yield by reducing external connections and substantially reduce manufacturing cost, but most importantly, it will facilitate a large reduction in a system's physical size and weight. This requires the incorporation of high performance digital and analog circuits. The implementation of such circuits with poly-silicon TFTs in non-conventional substrates presents challenges not usually encountered when designing standard IC's. Some of these factors are: TFT performance considerations (threshold, mobility, electrical stability, etc.); and limitations presented by large flexible substrates such as dimensional instability, large feature sizes and poor yield. These elements must be accounted for in the electrical circuit design, the physical layout of circuits, and the fabrication of systems.

Polymers substrates have been the basic contender for these applications (1,2) and have received a growing research interest in the past few years. However, they are found to be the reason for a number of drawbacks, the most important of which are their reduced compatibility with standard CMOS processing, the resulting low device stability, and large device features. Although NMOS devices with high mobility are feasible, the lack of high temperature steps compromises device stability and requires new low temperature processes and materials. All inorganic, thin metal foil backplane substrates represent an excellent alternative to polymers or hybrid backplanes (inorganic substrates with polymer/organic coatings) for use as flexible systems. They offer superior chemical resistance in a number of environments compared to plastics, and they are compatible with high temperature processing. In the case of TFTs on stainless steel (2,3) for example, high temperature steps such as thermal oxide growth, thermal dopant activation, silicide growth etc. pose no problem. Furthermore, the greater dimensional stability that metal foil substrates offer allows for the implementation of circuit designs with minimum feature size of 1μm or less. High performance, high speed circuits are then feasible as gate lengths are significantly reduced.

In this work, we present the compilation of research efforts initiated with the development of polysilicon TFT's on metal substrates (3,4), implementation of high resolution active matrix backplanes (5), and integration with a suitable PLED process. Furthermore, we have demonstrated in our previous work (4,6), that poly-Si TFT technology on metal foil is a suitable platform for the fabrication of high performance large area systems onto flexible substrates; in part due to their high mobility, high stability and small design rules. Polysilicon TFT backplanes on stainless steel foils are then suitable for high performance digital and analog large area circuit applications, in particular, high resolution, low power, AM-PLED displays.

Poly-Silicon Thin Film Transistor

The device technology presented here is based on laser annealed poly-silicon Thin Film Transistors (TFTs) as it can produce circuits with dramatically better performance than amorphous silicon while considerably reducing costs over the silicon-on-insulator technology. The high carrier mobility values that can be obtained with this approach and the ability to realize stable, low power CMOS circuitry are the two strongest assets of poly-silicon TFT technology.

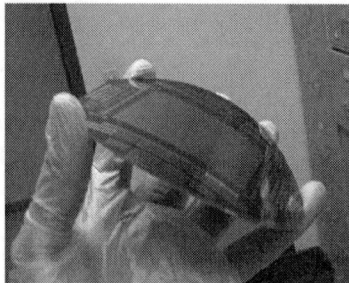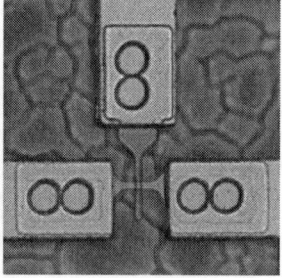

Fig. 1 VGA AMPLED backplane on flexible stainless steel metal foil 6 inch wafer. Display consists of 640 x 480 2-TFT PMOS pixels of 105 x 110 microns pitch (230dpi) with a 3.3 inch diagonal. Also shown is a Polysilicon TFT on Metal foil of width and length of 1 micron

All polysilicon TFT devices and circuits in this work were fabricated on 150 mm flexible stainless steel type 304 substrates that were 100 micron thick (figure 1). The surface of the steel foil was polished to a final surface roughness (Ra) around 3 nm. Then, a passivation layer of PECVD SiO_2 was deposited on top of the substrate to isolate the devices from the conductive substrate. The polysilicon film was 50 nm thick and was excimer-laser annealed. The silicon dioxide gate dielectric was 100 nm thick and was deposited by plasma enhanced chemical vapor deposition (PECVD) at 300 °C. After deposition and n+ doping of a 200 nm thick polysilicon layer, patterning and dry etching of the gate electrodes were performed. Afterwards doping of source and drain regions was performed by ion implantation. Activation of the dopant was accomplished by thermal annealing at 700 °C and this was the highest processing temperature. After the activation anneal silicidation of source, drain and gate regions was performed by first depositing 10 nm of nickel followed by a thermal anneal at 400 °C to complete the silicidation process. A 10 minutes plasma hydrogenation was then performed followed by deposition of silicon dioxide by PECVD, to serve as the device passivation layer, and opening of the contact windows. Finally lift-off process was used to define the final aluminum metallization layer followed by a post metallization anneal at 300 °C.

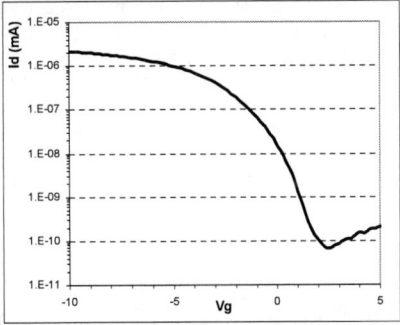

Fig. 2 Transfer characteristic of PMOS on stainless steel foil at Vds = -0.1V. The PMOS transistors had mobility of 42 cm^2/Vs, threshold voltage of -1.5V and sub threshold slope of 1.35V/dec.

As mentioned above, stainless steel foils offer great dimensional stability during the thermal processing steps encountered in TFT fabrication. This allows for minimum fabricated channel length of 1μm (minimum feature size) as shown in figure 1. The technology presented here has produced in previous runs (6) TFTs of W/L = 1/1μm which recorded a mobility value of 358 cm^2/Vs (this high value was achieved with SLS laser annealing techniques and with grain sizes approaching channel lengths). A typical PMOS transistor used in the display presented here showed mobility's of 42 cm^2/Vs, threshold voltage of -1.5V and sub threshold slope of 1.35V/dec. The transfer curve for this PMOS TFT is shown in figure 2 at |Vsd|=0.1V. Line fidelity is also extremely good, as devices of length 1 μm and widths up to 2 mm were fabricated as well, further increasing the capabilities of this technology. The combination of all these assets; high mobility, high stability and small design rules makes these devices suitable for high performance digital and analog large area circuit applications; in particular, high resolution, low power, AMPLED displays.

Active TFT Pixel

The VGA active matrix array presented in this work (5) consists of a full VGA array of 480 by 640 pixels with a 3.3 inch diagonal. The standard 2 TFT pixel circuit is implemented with PMOS transistors (figure 3). In this circuit, M1 behaves as a voltage controlled current source, and M2 is the address switch. This means that there are three lines going into each pixel (Vdata, Vadd and Vdd). This circuit works as follows: an analog voltage is programmed in the data line during addressing time (when M2 is turned ON by the address line.) This voltage activates M1 allowing a controlled current to flow through it. When the addressing period is over, M2 is turned OFF and C stores the data voltage. This allows M1 to keep supplying the correct current during the non-addressing time. This pixel architecture was chosen over other more complex circuit topologies in order to increase yield. It is a simple proven benchmark design which can be implemented with only 2 transistors and can potentially produce displays with high aperture ratio. In our current design (5), a mirrored pixel pair was utilized in order to obtain a more efficient use of the area. As shown in figure 3, the power supply is implemented with vertical aluminum lines shared by two adjacent pixels.

The standard CMOS transistor (7) equation [i] was used to calculate the width to length ratio of the drive transistor 'M1'. Maximum current levels of approximately 10μA were used in the calculation, as well as mobility values of 50 cm²/Vs and threshold voltage of -1V. This calculation yielded values of W/L between ½ and 2 depending on the power supply used. This leaves layout considerations such as space available in the pixel and aperture ratio as the final factors to consider when sizing 'M1'.

$$I_d = \frac{1}{2} \cdot \mu \cdot C_{OX}^* \cdot \frac{W}{L} \cdot \left(V_{gs} - V_{th} \right) \qquad [i]$$

In order to size the storage capacitor, the parallel plate charge equation [ii] was used. By setting the change in voltage equal to a greyscale data level of 0.1V, the time period equal to the display frame time (based on 60 Hertz) and the discharging current equal to the leakage current of transistor 'M2' we obtained an approximate capacitor size of 1pF. In order to keep this capacitance low, we have implemented the switch transistor 'M2' as a triple gate device. This type of gate configuration, where the total length of the transistor gate is split into 3 segments spaced by a few microns, has shown to reduce leakage current. Furthermore, the capacitor was implemented with the active Polysilicon layer as the bottom plate and the silicided polysilicon material used for transistor gates as the top plate leaving only the high quality gate oxide as the capacitor dielectric.

$$I_{off} = C \frac{\partial V}{\partial t} \qquad [ii]$$

$$C = I_{off} \cdot \frac{FrameTime}{GreyLevel} \qquad [iii]$$

Another important factor taken into account in the layout of the pixel was the final aperture ratio of the emissive area. When using metal foils as a substrate, top emissive structures are required. Therefore, if planarization techniques are used, the lighting emission area of the PLED can occupy a large percentage of the pixel. However, in order to simplify processing of our panels, we have limited the emissive area to flat areas of the pixel, mainly the space occupied by the storage capacitor. High mobility Polysilicon TFT's and high dimensional stability of metal foil make possible the implementation of small size transistors, hence occupying a small fraction of the pixel. This resulted in an emissive area of 70 μm by 74 μm, which amounts to a 45% aperture ratio.

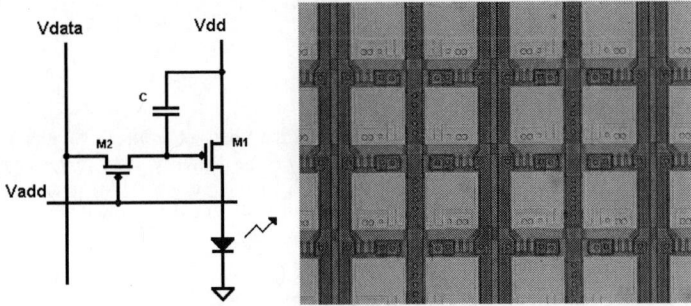

Fig. 3 PMOS 2-TFT Active Pixel Circuit consisting of a transistor switch 'M2', a current source 'M1' a storage capacitor and a Polymer Light Emitting Diode. Also shown is the AMPLED backplane close up formed of mirrored pairs of pixels and an aperture ratio of 45% (emissive are of 70 x 74 microns).

One of the main disadvantages of the pixel architecture used here is the dependence of pixel current on the properties of the pixel transistors. As discussed above, the current that flows through the pixel is programmed by a data voltage stored at the gate of 'M1'. The trans-conductance of 'M1', its threshold voltage and other TFT properties such as kink effects can vary by large amounts over the area of the display, resulting in different pixel currents for the same data voltages. In order to compensate for these mismatches, complex pixel structures have been proposed (8,9,10) that use extra number of transistors to perform a correction algorithm. However, in order to increase yield and maintain aperture ratio high we have opted to implement the standard 2-TFT pixel with some alterations in an attempt to keep mismatches to a minimum. One such modification was the implementation of a double gate transistor for 'M1'. This was observed to reduce TFT kink effects and therefore limiting the number of mismatch factors to just mobility and threshold voltage.

We can obtain the theoretical variation in pixel currents by solving the difference equations for mobility 'iv' and threshold voltage 'v' at different gate bias. (First level approximation computed with TFT's working in saturation and assuming the CMOS current equation [i].)

$$\partial I = \partial Vt \cdot 2 \cdot k \cdot \mu \cdot (Vgs - Vt) \qquad \text{[iv]}$$

$$\partial I = \partial \mu \cdot k \cdot (Vgs - Vt)^2 \qquad \text{[v]}$$

We can note here that the mobility deviation can be written as $\partial \mu = \overline{\mu} \cdot \xi_\mu$ and the threshold voltage deviation as, $\partial Vt = \overline{Vt} \cdot \xi_{Vt}$ where $\overline{\mu}$ and $\overline{Vt}$ are the average recorded values, while ξ_μ and ξ_{Vt} are the percentage mobility and threshold voltage variations respectively. We can then solve for the normalized changes with respect to the total current as follows:

$$\frac{\partial I}{I} = \xi_\mu \qquad \text{[vi]}$$

$$\frac{\partial I}{I} = 2 \cdot \frac{\overline{Vt} \cdot \xi_{Vt}}{Vgs - \overline{Vt}} \qquad \text{[vii]}$$

From these equations, we can clearly see that threshold voltage variations are more dominant at low data levels, whereas mobility is the dominant cause of mismatches at high light intensity levels. Figure 4 shows the distribution of threshold voltage values for a Polysilicon TFT wafer on stainless steel foil with an average of -0.96V and an average deviation of 0.16V.

Fig. 4 Threshold voltage distribution on steel wafer. Average of -0.96V and variation of 0.16 V

Display Drivers

One of the advantages of the 2-TFT pixel architecture is that it can easily be controlled with simple voltage signals. This means that standard LCD display drivers can be used with minor modifications. In order to have maximum control over the display signals we used National Instruments NI-6323 Analog Output computer cards to generate all the signals needed for display operation. However, because the number of control signals was limited to the number of channels in the computer cards, pixel rows and columns were grouped at the wafer level in sets of 20. Therefore, a total of 24 row blocks and 32 column blocks were externally driven.

For the row drivers implemented on-panel, we have accommodated Static Half Bit Shift Registers that were selected based on stability, reliability and yield from a variety of tested designs. This is one of the key technology advantages made possible by high performance Polysilicon TFT circuits on metal foils. High mobility values, dimensional stability and device stability due to high temperature process are fundamental aspects in the implementation of display drivers.

When implementing such circuits, it is generally recommended that in order to reduce charge up transients wide current buffers be added at the end of each shift register stage. The reasoning behind this practice is that long display lines form large capacitive loads for the row drivers and therefore the voltage change is limited by the current sourcing/sinking capabilities of the buffers. However, the implementation of row lines with Polysilicon lines also present a resistive load to each shift register stage, and for the displays presented here, this resistance becomes the limiting factor. The transient times are then determined by the simple 'RC' network formed by the line resistance and line capacitance. For this reason, the size of the buffers of the shift registers were reduced to 0.75 mm CMOS inverters, further increasing yield.

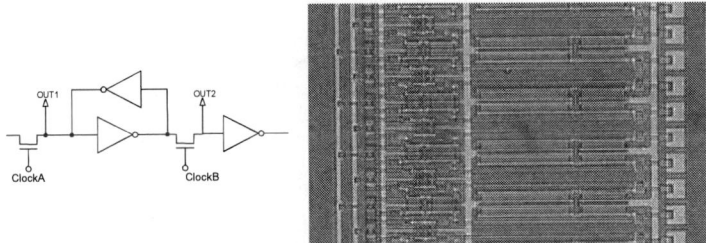

Fig. 5 Static Half Bit Shift Register stage block circuit schematic with 0.75mm CMOS inverter as current buffer. Functional Shift Register on stainless steel foil with operating supply voltages as low as 3V and frequency operation as high as 10MHz

The operation speed of the shift registers ranges from a few Hertz to several MegaHertz, which puts them well within the operational range for VGA column driving speeds of approximately 50Khz. Operation at 1 MHz of test circuit on stainless steel foil is shown in figure 6. Transition times under no load conditions were analyzed, in particular, the effects of the conductive substrate studied as they present capacitive loads to all high speed circuits. Several factors are considered to affect stage transition times (7): firstly; the average resistance from drain to source of the TFTs during switching (as input is swept from ground to Vdd.) A first order approximation (6,7) of this number is obtained through equation 'viii' and is a function of device mobility, threshold voltage and channel size. Secondly; the output and input capacitances of the TFTs as defined by equations [ix] and [x]. These capacitances are added to the parasitic capacitance to the substrate, Cp, of the key nodes in order to obtain a total node capacitance seen by the inverters; see equation [xi]. Finally, equation [xii] computes rise (or fall) times for the inverters by multiplying average PMOS (or NMOS) resistance and total capacitance.

$$R_{avg} = \frac{V_{dd}}{\frac{1}{2}\mu \cdot C_{OX} \cdot (V_{dd} - V_{th})^2} \cdot \frac{L}{W} \qquad [viii]$$

$$C_o = C_{OX} \cdot W \cdot L \qquad [ix]$$

$$C_{in} = \frac{3}{2} \cdot C_{OX} \cdot W \cdot L \qquad [x]$$

$$C_T = C_o(NMOS) + C_o(PMOS) + C_{in}(NMOS) + C_{in}(PMOS) + C_p \qquad [xi]$$

In equation [xi], we note a parasitic capacitance $C_p = A \cdot \frac{\varepsilon}{h}$ where h is the substrate oxide thickness over a node area A. The total time constant is then:

$$\tau = R_{avg} \cdot C_T \qquad [xii]$$

This work has produced shift registers with transition times under no loading conditions of approximately 50 ns. In the case of the circuits used for the displays, the dominant factor in equation 'xi' is the input capacitance of the inverters as they were implemented with lengths of 6 μm. However, when connecting the buffers to the row lines, the parasitic capacitance and resistance of the line node starts to dominate as discussed above. This was corroborated with circuit simulations where we can see how the transient times of the buffer are affected by the resistive load and not buffer size.

On panel shift register operation is shown in figure 6. Even though the row lines were grouped together on one end to allow for block external driving, the very resistance of the line made possible the implementation of shift registers on the other side of the display. The resistance allowed each line to swing appropriately in one end, even though they were at the same potential at the other. Some flicker was observed in part due to the overlapping nature of half bit shift registers.

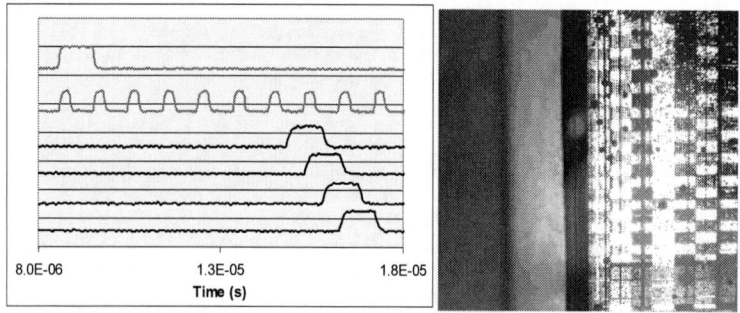

Fig. 6 Operation of shift registers on steel foil. Input signal, clock and 4 successive stages shown. Notice the overlapping output typical of half bit registers. Also shown is AMPLED driven by an integrated row driver consisting of a polysilicon TFT shift register.

Fabrication of Polymer Light-Emitting Diodes

Due to the opaque nature of the stainless steel substrates, the top-emission OLED's is adopted (figure 7). Instead of small-molecular OLED, polymeric OLED (PLED) is applied in this study (11,12). There are several basic requirements to meet for a top-emission OLED in order to achieve the high efficiency of illumination, and they are the reflective anode, the low work-function and transparent cathode, and a smooth anode surface to prevent shorts. In this study, the anode is gold, which provides 5.1 eV of work-function and a very smooth surface. On top of the anode, a conductive hole-transporting polymer, PEDOT, was spin-coated to make the good metal/poymer junction and smootherize the anode surface even further. Sequentially, a light-emitting polymer, poly-phenylenevinylene (PPV), was spun. A thin layer of calcium with 2.8 eV of work-function, was then thermally evaporated and deposited onto PPV, being capped by a thick transparent ITO cathode. The combined Ca/ITO cathode gives the greatest transmittance of 85% at wavelength of 550 nm, where the peak of PLED's electroluminescence locates, so that the light output was maximized. Figure 6 shows the I(V) curves for this structure and the exponential fit used for circuit analysis. The light turn-on voltage (V_L) of the PLED devices is 4.0 V and they become visible at 5.0 V under ambient lighting conditions.

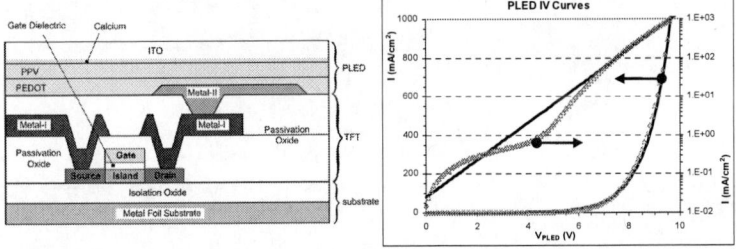

Fig. 7 Active pixel cross section with top emitting PLED. Highlights of the technology are 1.3 micron thick metal patterning through lift of process and nickel silicides for reduced contact resistance. PLED IV curves in log and linear scales with fitted exponential function used for circuit analysis. Turn ON voltage of 4V.

Results and Discussion

Figure 8 shows a completed AMPLED on stainless steel foil under operation. It is being driven by external drivers and operated at a power supply of 15 V. The images were clearly visible under ambient light conditions. The best performance was observed when non-overlapping clocks were used for the row signals (roughly 30% duty cycle). Furthermore, an added time shift was implemented between the row signal and the data signal that allowed for transients to settle. Under these conditions, no flicker was observed and image quality remained acceptable for as low as 10 Hz frame rate.

As discussed above, the variation of light intensity across the array can be attributed to mismatches in TFT parameters, although over a large area as shown in figure 8, variations in the PLED films can also introduce further non-uniformities. A close up image of pixels can be seen in figure 8 where localized mismatches of intensity that are most probably due to device parameter variations are not very noticeable.

Figure 9 shows the current for an individual pixel as a function of the data voltage. Full ON operation consumes roughly 11.5 μA. We can also see in this figure the voltage across the PLED as a function of data voltage. This voltage consumption is at most 8V which means the rest of the voltage is lost across the TFT 'M1'. This is in part due to the long channel double gate implementation used to reduce non-uniformities. For this reason, exists here a trade off between performance and power consumption. Small W/L ratios as implemented in this array can dissipate a lot of the power; for instance, at full ON conditions 99 μW were consumed by the emissive element while 82 μW were consumed by the TFT (55% of total power).

Fig. 8 Encapsulated AMPLED display on flexible stainless steel foil in operation driven by external drivers. Close up of Active pixels in operation; notice slight variations in pixel brightness attributed to TFT mismatches

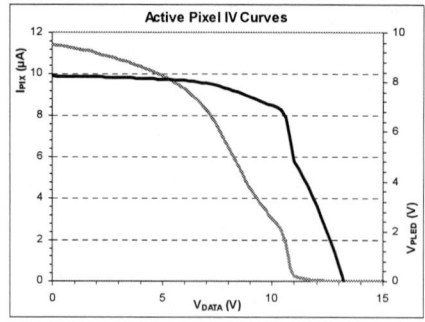

Fig. 9 Current of active pixel with integrated PLED as a function of data voltage, and PLED voltage as a function of data voltage

Conclusion

Positive results in Organic/Polymer Light Emitting Diode (O/PLED) science and technology as well as higher performance thin film devices such as poly-silicon TFT's have propelled the development of active TFT backplanes for O/PLED display applications. Large area, high resolution, slim displays are now the focus of attention for many consumer electronics applications and Active Matrix O/PLED Displays have the potential of overtaking a portion of the market currently being dominated by the LCD industry. To this end, O/PLED displays are one of the preferred flexible electronic applications because of the mechanical and space saving advantages a conformal or foldable display would have (robust, lightweight). Furthermore, poly-silicon TFT's make it possible to include in the same substrate all or many of the components needed to drive a flat panel O/PLED display.

We have implemented a 230 dpi Active Matrix Polymer Light Emitting Diode (AMPLED) display on flexible stainless steel substrates with laser annealed poly-silicon TFT technology. The high resolution displays consist of a full VGA array. This is the highest resolution AMPLED display demonstrated to date on a flexible metal foil, and processed with CMOS compatible techniques.

References

1. S. D. Theiss and S. Wagner, *IEEE Electron Device Lett.* vol. 17, pp. 578-580 (1996).
2. S. Wagner et al., 1997 *Symposium on Amorphous and Microcrystaline Silicon Technology*, pp.843-849, (1997)
3. R. Howell, M. Stewart, S. Karnik, S. Saha, and M. Hatalis, *IEEE Electron Device Letters*, Vol. 21, pp. 70-72, 2000.
4. T. Afentakis, M. Hatalis, T. Voutsas, and J. Hartzell, *Proceedings of the SPIE*, Vol. 5004, pp. 122-126, 2003.
5. M. Troccoli, M.; Afentakis, T.; Chuang, T.K.; Chang, Y.L.; Hatalis, M.; Voutsas, Apostolos T.; Hartzell, John W.; Chouvardas, Vasilios *Proc. ECS, Thin Film Transistor Technologies VII*, 2004
6. M. Troccoli, A. J. Roudbari, T.K. Chuang, and M. Hatalis, *Journal of SSE*, 2006.
7. CMOS Circuit Design and Simulation, R. Baker. *IEEE Press*, (1998)
8. R. Hattory, *IEICE Transactions on Electronics*, Vol. E83-C, pp. 779 (2000)
9. J. Goh, *SID Digest* 2003 P.72 No.1 p. 494 (2003)
10. R. Dawson, *IEDM* 1998, Vol. 32 p. 875 (1998)
11. T.K. Chuang, A. J. Roudbari, et. al. *Proceedings of the SPIE*, v 5801, n 1, 2005, p 234-48.
12. H. Najafov, I. Biaggio, T.-K. Chuang, and M. K. Hatalis, *Phys. Rev.* B 73, (2006) 125202.

248

ECS Transactions, 3 (8) 249-253 (2006)
10.1149/1.2356360, copyright The Electrochemical Society

Overlay Alignment in a-Si:H TFTs Fabricated on Foil Substrates

H. Gleskova[a], I-C. Cheng[a], A. Kattamis[a], S. Wagner[a], and Z. Suo[b]

[a] Department of Electrical Engineering and Princeton Institute for the Science and Technology of Materials, Princeton University, Princeton, New Jersey 08544, USA
[b] Division of Engineering and Applied Science, Harvard University, Cambridge, Massachusetts 02139, USA

> Thermo-mechanical theory of a device film-on-foil structure reveals that the layer-to-layer alignment accuracy and the radius of curvature of the structure are both controlled by the mismatch strain between the deposited films and the substrate. Amorphous silicon thin-film transistors fabricated on thin foils of steel or plastic must be grown with tensile built-in stress to make the structure as flat as possible, and for accurate layer-to-layer alignment.

Introduction

Flexible electronics based on amorphous silicon (a-Si:H) thin film transistors (TFTs) is a field of rapidly growing interest, to the display industry and for large-area electronics in general. Thus building transistor electronics that are dimensionally stable and that allow for high design tolerances has become very important. Recent experimental results show that the alignment accuracy between source/drain and gate electrodes in a-Si:H TFTs on Kapton changes with the deposition power for the gate dielectric (1). When a-Si:H TFTs are grown on free-standing foil substrates, the structure changes its in-plane dimensions with the growth of each TFT layer. Consequently, misalignment is produced during the patterning of TFT layers, leading to poor layer-to-layer alignment accuracy. We analyze the correlation between the layer-to-layer alignment accuracy and the radius of curvature of the free-standing film-on-foil structure after plasma enhanced chemical vapor deposition (PECVD) of the film at elevated temperature.

Thermo-mechanical behavior of film-on-foil structure

Curvature of the film-on-foil structure

a-Si:H TFTs are built on substrates layer-by-layer at elevated temperature. Strain develops in the structure due to built-in stresses in the deposited layers, or, upon cooling down, due to the differences in the thermal expansion coefficients between the deposited film and the substrate, or between different films. The mechanics of the film-on-substrate structure depends strongly on the elastic moduli and thicknesses of the substrate Y_s, d_s and the thin film Y_f, d_f (see figure1). A stiff film and a compliant substrate, like amorphous silicon TFT film on an organic polymer foil, have similar mechanical strength, i.e., products of elastic modulus and thickness, $Y_f \cdot d_f \approx Y_s \cdot d_s$. The coefficient of thermal expansion (CTE) of the organic polymer foil typically is much larger than that of a-Si:H TFT films. Upon cooling the substrate tends to shrink more than the TFT film that is bonded to it, putting the film into compression. When the structure is taken off the

249

substrate holder used for deposition, it rolls into a cylinder that partially relieves the compressive strain in the film in the bending direction. If the Poisson ratios v of the film and the substrate are identical, the classical theory of the bimetallic strip leads to the following radius of curvature of the film-on-foil structure (2):

$$R = \frac{d_s}{6\frac{Y_f d_f}{Y_s d_s}\left(e_f - e_s\right)(1+v)} \cdot \frac{\left(1 - \frac{Y_f d_f^2}{Y_s d_s^2}\right)^2 + 4\frac{Y_f d_f}{Y_s d_s}\left(1 + \frac{d_f}{d_s}\right)^2}{\left(1 + \frac{d_f}{d_s}\right)} \qquad [1]$$

The first factor is the Stoney formula divided by $(1+v)$, a factor arising from the generalized plane strain condition (cylindrical shape). The second fraction constitutes the deviation from the Stoney formula for thin and/or compliant substrates. $\left(e_f - e_s\right)$ is the mismatch strain between the film and the substrate when they are not bonded to each other. Note that a large radius R (small curvature $1/R$) reflects a small mismatch strain $\left(e_f - e_s\right)$. The mismatch strain has two dominant components. One is the thermal mismatch strain caused by the difference between the coefficient of thermal expansion of the substrate, α_s, and that of the film, α_f. The other is the built-in strain ε_{bi} in the deposited film. Therefore,

$$e_f - e_s = \alpha_f \cdot \Delta T + \varepsilon_{bi} - \alpha_s \cdot \Delta T = \left(\alpha_f - \alpha_s\right) \cdot \Delta T + \varepsilon_{bi} \qquad [2]$$

where $\Delta T = T_r - T_d$ is the difference between the room and deposition temperatures. The built-in strain ε_{bi} is produced by the built-in stress σ_{bi} that develops in the film during its growth at elevated temperature. For a-Si:H circuits on polymer substrates a negative ε_{bi} is desired.

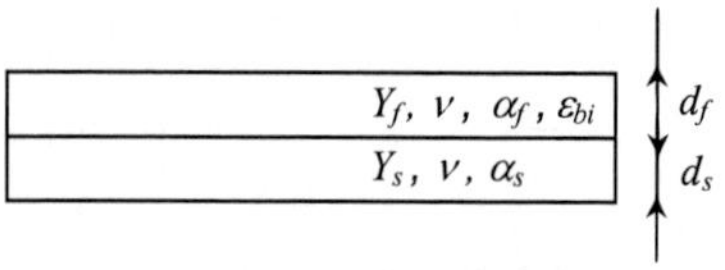

Figure 1. Film-on-substrate structure.

σ_{bi} is a function of growth conditions. In some materials σ_{bi} can be easily controlled, for example in silicon nitride deposited by PECVD (1). Since in thin films the built-in stress is typically reported, we will replace the built-in strain in equation [1] with the built-in stress. Assuming that a (i) thin and/or compliant substrate is held flat during PECVD growth, (ii) σ_{bi} is constant, and (iii) no net force is applied at any time, the built-in strain ε_{bi} is given by:

$$\varepsilon_{bi} = -\frac{(1-v)\ \sigma_{bi}}{Y_f} \cdot \left(1 + \frac{Y_f d_f}{Y_s d_s}\right) \qquad [3]$$

The second factor on the right-hand side is important when film and substrate have comparable mechanical strength Yd. Substitution of Eqs. [2] and [3] into Eq. [1] leads to the radius of curvature of the film-substrate couple at room temperature T_r following film growth at an elevated temperature T_d. The radius of curvature R is defined positive when the film is on the outside (compressive stress in the film) and negative when the film is on the inside (tensile stress in the film).

Overlay alignment in film-on-foil structure

Every time an alignment is performed, the structure must be flattened. However, when compared to the original size of the flat, bare substrate the film-substrate couple is typically elongated in both in-plane directions because the a-Si:H film deposited at elevated temperature does not tend to contract upon cooling as much as the polymeric substrate. Mechanical modeling of the film-on-substrate structure gives (3):

$$\varepsilon(T_r) = \frac{(\alpha_s - \alpha_f) \cdot (T_d - T_r)}{1 + \dfrac{Y_s \, d_s}{Y_f \, d_f}} - \frac{(1-\nu) \; \sigma_{bi} d_f}{Y_s d_s} \qquad [4]$$

Here $\varepsilon(T_r)$ is the in-plane strain of the substrate at room temperature, after film deposition at elevated temperature. $\varepsilon(T_r)$ compares the in-plane dimensions of the substrate between after and before film growth, i.e., it specifies the elongation or shrinkage of the substrate after film growth. The first term on the right-hand side is the strain caused by the $\Delta\alpha \cdot \Delta T$ mismatch between the film and the substrate. The second term is the strain resulting from the built-in stress in the film.

Correlation between the radius of curvature and overlay alignment in film-on-foil structure

Figure 2 shows the film-on-foil curvature $1/R$ and the strain of the substrate $\varepsilon(T_r)$ at room temperature as a function of the built-in stress σ_{bi} in a 1-μm-thick PECVD silicon nitride film deposited on a 100-μm-thick steel or Kapton E foils at temperatures of $T_d =$ 150, 200, and 250°C. The following parameters were used in the calculations: $d_f = 1$ μm, $d_s = 100$ μm, $Y_f = 194(1-\nu)$ GPa, $Y_s = 200$ GPa (steel) (4) or 5.3 GPa (Kapton E) (5), $\nu = 0.3$, $\alpha_f = 2.7 \times 10^{-6}$ K^{-1} (6), $\alpha_s = 18 \times 10^{-6}$ K^{-1} (steel) (4) or 16×10^{-6} K^{-1} (Kapton E) (5), and $T_r = 20$°C.

Both the curvature $1/R$ of the work piece and the strain of the substrate $\varepsilon(T_r)$ become zero if the silicon nitride film is grown with a built-in tensile strain equal to the $\Delta\alpha \cdot \Delta T$ mismatch between the film and the substrate. Thus, the radius of curvature of the film-on-foil structure may be controlled by the built-in stress in the TFT films. If $\Delta\alpha \cdot \Delta T$ and ε_{bi} lie within certain ranges one can achieve flat and dimensionally stable structures.

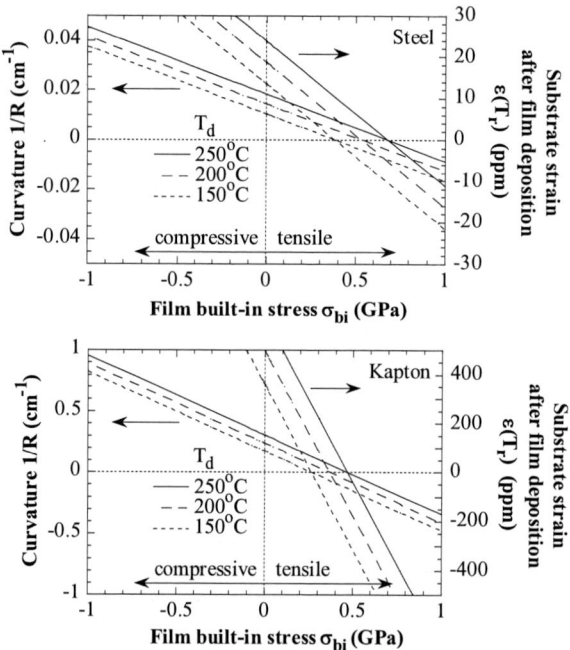

Figure 2. Film-on-foil curvature $1/R$, and strain of the substrate $\varepsilon(T_r)$ at room temperature, as a function of the built-in stress in a 1-µm-thick PECVD silicon nitride film. Substrates are a 100-µm-thick steel (upper frame) or Kapton E (lower frame) foils. Deposition temperatures are $T_d = 250$, 200, and 150°C.

Conclusions

Mechanical analysis of a device-film-on-substrate-foil structure reveals that the magnitude of the layer-to-layer misalignment, caused by the in-plane change in size of the TFT structure, and the curvature $1/R$ are both caused by the mismatch strain between the deposited films and the substrate. The layer-to-layer overlay accuracy can become particularly poor in (stiff) a-Si:H / (soft) plastic couples.

Acknowledgments

This work is supported by the United States Display Consortium.

References

1. I-C. Cheng, A. Kattamis, K. Long, J.C. Sturm, and S. Wagner, *Journal SID*, **13**, 563 (2005).
2. Z. Suo, E.Y. Ma, H. Gleskova, and S. Wagner, *Appl. Phys. Lett.*, **74**, 1177 (1999).
3. H. Gleskova, I-C. Cheng, S. Wagner, and Z. Suo, *Appl. Phys. Lett.*, **88**, 011905 (2006).
4. Stainless steel AISI 304 grade, Goodfellow.
   http://www.goodfellow.com/csp/active/gfHome.csp
5. Kapton E, DuPont.
   http://www2.dupont.com/Kapton/en_US/products/E/index.html
6. M. Maeda and K. Ikeda, *J. Appl. Phys.*, **83**, 3865 (1998).

254

ECS Transactions, 3 (8) 255-259 (2006)
10.1149/1.2356361, copyright The Electrochemical Society

Mist Deposition for TFT Technology

K. Shanmugasundaram, S. C. Price, K. Chang,*   D.O. Lee, and J. Ruzyllo

Department of Electrical Engineering, Penn State University, University Park, PA 16802
*Currently at Freescale Semiconductor, Austin, Texas

Abstract

In this work the process of mist deposition is explored as a method used to deposit organic semiconductors and gate dielectrics for TFTs. With an expanding use of TFTs in both electronic and photonic applications mist deposition offers advantages in terms of versatility, throughput and process cost. The method of mist deposition is first introduced and then example of results obtained with mist deposited dielectric and semiconductors thin films are discussed.

Introduction

Due to the rapid growth of active matrix display technology, flexible electronics and photonics, as well as various large-area electronics and photonic systems there is a pressing need to continue mastering Thin-Film Transistor (TFT) manufacturing technology toward process simplification and cost reduction. In particular, deposition and patterning of active semiconductor layer and gate dielectric components of TFT's structure are of interest as quality of the processes and materials used predetermines performance of the transistor.

At present, there is a growing interest in organic semiconductors which, due to their properties, will very likely open up new applications for semiconductor electronics and photonics. An important precondition to the success of organic semiconductors in these applications is their compatibility with the requirements of low-cost, large-scale TFT technology in terms of conditions of deposition and film patternability. Depending on chemical composition, organic semiconductors are either vacuum deposited or deposited from the liquid solution by spin coating. The former is not suitable for low-cost, high throughput atmospheric pressure processing while the latter can not used in the case of large or oddly shaped substrates.

The mist deposition method [1] allows deposition of organic semiconductors from liquid solutions at atmospheric pressure and room temperature independently of the shape and size of the substrate. Hence, it is considered to be an attractive alternative in TFT processing in large electronic and photonic system manufacturing. This study was undertaken to contribute to the development of low-cost TFT technology by studying mist deposition of thin-film high-k dielectrics and semiconductors.

Mist Deposition

The principle of mist deposition is to convert the liquid source material into a fine mist with droplet size in the order of 0.25 micron, which is then carried into the deposition chamber in a pressurized stream of N_2 gas (Fig.1), and the droplets are allowed to coalesce on the substrate at room temperature at atmospheric pressure. This forms a uniform film of liquid on the substrate, which is then thermally treated to burn off solvent and leave a thin film of solid on the surface. In order to control deposition rate

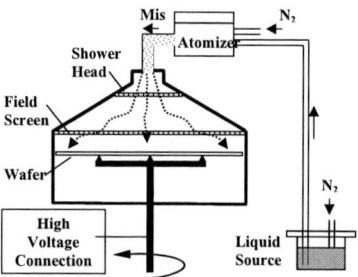

Figure 1 Schematic representation of the mist deposition apparatus.

beyond gravitational interactions, which in the case of sub-micron sized droplets are very weak, an electric field is created between the grounded field screen and a wafer (Fig.1). After deposition the film is thermally cured at the temperature of 160-300°C in ambient air or in the controlled ambient of either O_2 or N_2 at the atmospheric pressure. In the case of select inorganic materials, wafers may also be subjected to an additional anneal typically in the temperature range from 600°C to 800°C either in nitrogen or in nitrogen with some oxygen added.

Experimental

In this experiment mist deposition of high-k dielectrics, Hf(Si,O) based in particular, as well polymeric semiconductors was carried out using a commercial stand alone mist deposition module. As mentioned earlier, both are of potential use in TFT manufacturing, but in the reported experiment no attempt to make a working thin-film transistor was made. Substrates used were mainly Si wafers bare or oxidized and on occasion an ITO covered glass plate. Film thickness in the case of both semiconductors and dielectrics was measured using an ellipsometer (in some cases was determined using TEM). Film morphology was characterized by means of AFM and its composition by means of angle resolved XPS. Liquid precursors used were obtained from commercial vendors and modified as needed. Wetting angle measurements were carried out to characterize surface wettability with various precursors. Fundamental electrical characteristics of mist deposited films were determined base on current-voltage and capacitance-voltage measurements. Depending on material, either Pt (e-gun evaporation) or Al (thermal evaporation) contacts were used

Results and Discussion

An advantage in using high-k dielectric in TFT is that at the thickness sufficient to prevent excessive gate leakage current, gate including dielectric featuring k higher than that of CVD SiO_2 or Si_3N_4 provide superior capacitive coupling, and hence, control of the drive current. Fig. 1 shows TEM pictures of $HfSiO_4$ films mist deposited on oxide covered Si surface. As seen in this figure continuous films can be mist deposited even at the thickness below 3 nm, although, in TFT processing thickness in the range of 20-40 nm is more adequate for gate dielectric applications. About 2.4 nm thick interfacial oxide SiOx seen in the figure is specific to the Si substrate will not be present when high-k dielectric is formed on the surface of an organic semiconductor. Electrical characterization of mist deposited gate dielectric films in 20-40 nm thickness regimes have shown that electrical integrity is sufficient for those films to act as effective gate dielectrics in TFTs [2, 3].

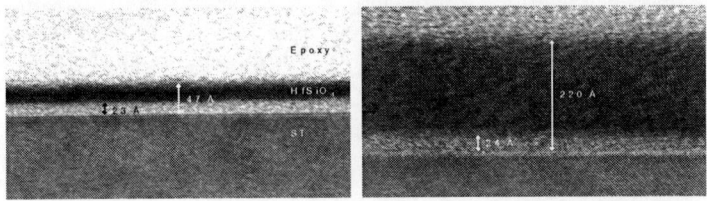

Figure 2 TEM cross-section of mist deposited 2.4 nm and 19.6 nm thick $HfSiO_4$.

Figure 3 shows current density - voltage characteristics of MOS structures with three different mist deposited dielectrics. As seen in this figure even at the thickness on 9 nm mist deposited gate insulators display structural integrity that prevents excessive leakage

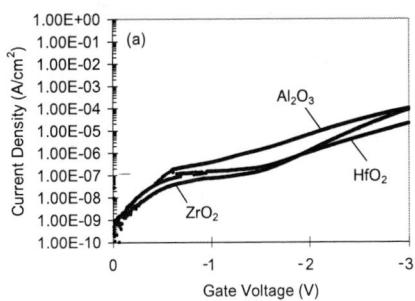

Figure 3   Current-voltage characteristics of about 9 nm thick high-k dielectrics mist deposited on Si substrate.

current. Considering the fact that in typical TFT gate insulators are significantly thicker than 9 nm it is expected that mist deposited high-k dielectric films should readily meet gate leakage requirements of standard TFTs.

In the second part of this exploratory investigation mist deposition was tested as a method of depositing organic semiconductors. Organic semiconductors are of two types oligomeric and polymeric. Oligomers are often referred to as "small molecule" semiconductors and consist of benzene rings with delocalized π bonds [4]. Polymeric semiconductors are long chains of carbon atoms with alternating single and double bonds which are functionalized with other atoms and molecules to form the delocalized π bonds, which make the material semiconducting [4]. The film quality and uniformity are very important to allow the charges to move freely across the material [5]. Most polymers are deposited from liquid precursors using common techniques such as spin coating. In this study a silanized tetraphenyldiaminobiphenyl (TPD) semiconductor was deposited by mist deposition with the intention of using it as a channel material in future TFTs.

Fully reproducible and controllable deposition of polymeric semiconductor films in the 10-400 nm thickness regime was accomplished using mist deposition, although, outcome of the process was dependant on the solvent used. Results of preliminary electrical characterization of electrical properties of these films show promise for mist deposited polymer semiconductors in TFT technology.

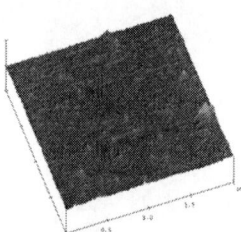

Figure 4 AFM images of mist deposited semiconducting polymer film on ITO substrates

Shown in Fig. 4 is the AFM image of a polymeric semiconductor film after a 10 minute deposition at 4 kV potential applied to the substrate on an ITO substrate. The RMS roughness of the films was measured to be 0.492 nm on the glass substrate and 0.571 nm on the ITO substrate. Preliminary results show that the concentration of the polymeric semiconductor in the liquid precursor plays an important role in the film nucleation and hence the final film quality or surface roughness.

Figure 5 shows the growth in thickness of the films with deposition time. The films studied here were all deposited at a substrate potential of 4 kV. The films display a linear growth with time which is typical of most liquid precursors deposited by mist deposition. The AFM analysis performed on these films indicated that the material nucleates to form islands before a complete film is formed. It was also seen that islands were stacked on top of each other before the underlying layer is completely formed.

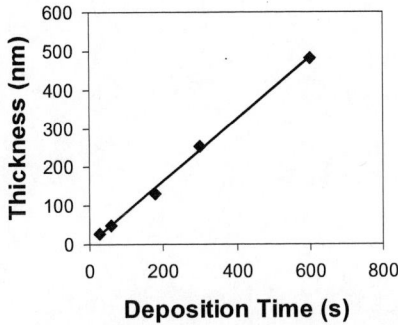

Figure 5 Thickness vs. deposition time of mist deposited polymeric semiconductor films.

Summary

In this experiment the characteristics of mist deposited thin-film high-k gate dielectrics and long-chain organic semiconductors that in combination could be used in TFT technology were studied. In both cases properties of deposited films appear to be sufficient to assure adequate operation of the transistor. However, final determination in this regard can be only done when both mist deposited materials are tested in the TFT structure.

Acknowledgements

Authors would like to thank Primaxx, Inc. for providing deposition module used in this study and Agiltron, Inc. for supplying organic semiconductor precursors used.

References

[1]. P. Mumbauer, K. Chang, W. Mahoney, D.-O. Lee, P. K. Shanmugasundaram, P. Roman, M. Brubaker, R. Grant, and J. Ruzyllo, *Semiconductor International*, **27**, 12 (2004).
[2]. D.-O. Lee, D.-O., P. Roman, C.-T. Wu, P. Mumbauer, M. Brubaker, R. Grant, and J. Ruzyllo, *Solid-State Electronics*, **46**, 1671 (2002).
[3]. D.-O. Lee, Ph.D. Dissertation, Penn State University, (2002).
[4]. K.Waragai, H. Akimichi, S. Hotta., H. Kano, and H. Sakaki, *Physical Review B*, **52**, 3 (1995).
[5]. C.D. Dimitrikapoulos and D.J. Mascaro, *IBM Journal of Research and Development*, **45**, 1 (2001).

260

CHAPTER 7
NON-SILICON TFTs

262

ECS Transactions, 3 (8) 263-272 (2006)
10.1149/1.2356362, copyright The Electrochemical Society

Hole Mobility in Structurally-Different Pentacene Field-Effect Transistors

H.L. Kwok[a]

[a]Dept. of ECE and Center for Advanced Materials & Related Technology,
University of Victoria, Victoria, B.C., Canada

This work examined pentacene field-effect transistors (OFETs)
made on polycrystalline, single-grain, and single-crystal active
layers. Mobility data obtained in the literature were matched to
equations developed from charge hopping and the Gaussian
disorder model (GDM). Making use of a logical sequence of
extraction steps, we were able to obtain materials parameters
relevant to the charge transport (only one fitting parameter was
needed). Our results indicated that there were major differences in
the transport site distribution ranging from a very narrow rms
width of the density of states (DOS) in the case of the single-
crystal OFET to broader values in the single-grain and
polycrystalline OFETs. Our model suggested an escape process
dominated by correlated hopping at short distance and one limited
by site energy distribution and re-organization energy as the active
layer became more crystalline.

Introduction

Recently there has been an increasing interest in the study of hole mobility in
pentacene organic field-effect transistors (OFETs) (1-4). One of the problems related to
the performance of these devices has been the low hole mobility which as reported (5,6)
rarely exceeds 10 cm^2/Vs. From an engineering perspective, hole mobility in pentacene
OFETs has to increase substantially to compete in applications such as in the construction
of backplane circuits. In addition, there is the scientific curiosity to find out why the
field-effect mobility is low even for the single-crystal device. These questions cannot be
answered in a straightforward manner and one method to resolve them would be to
examine the different device characteristics. Presently, only a few theoretical models
exist that quantify hole mobility in a disordered semiconductor in terms of the
"microscopic" materials parameters, and few researchers have applied these models
directly to the mobility data of pentacene OFETs of different crystal structures. This is
understandable from an experimental perspective since the properties of pentacene
OFETs can be difficult to control and variations in device properties are frequently
observed (1). Furthermore, field effect due to the gate bias during transistor operation
can complicate the charge transfer and escape process even in the absence of undesirable
contact resistances (7).

This work attempted to model hole mobility in pentacene thin films of different
crystal structures. Our approach was to examine theoretically how the OFET
characteristics depended on the materials parameters. For the polycrystalline thin films,
we assumed they are composed of crystal grains with traps residing in the inter-granular
region. Within each crystal grain, normal charge hopping prevailed and the gate voltage
bias chiefly acted on the trap states. In the limit, assuming that the trap states had a longer
retention time, charge transport would be mainly limited by the filling and emptying of
the traps. The presence of trap states would therefore generate a lower hole mobility.

Both the trap states and the transport states were assumed to be localized and their energy distribution was Gaussian. Charge transport was therefore governed by the Gaussian Disorder Model (GDM) (8) within the crystal grains and barrier-height limited elsewhere. In general, "disorder" in the polycrystalline thin films was assumed to have arisen from the charged states in the inter-granular region as well as the random orientation of the crystal grains. Such structural variations would be collectively characterized by a "disorder" parameter, α ($\alpha = 0$ in the single-grain and single-crystal devices). In addition to the derivation of the field-effect mobility equations, we also studied the effect of the gate voltage on the different materials parameters.

Theory

The theory of charge transport in conventional field-effect transistors is well documented in the literature (9) and the charge mobility can be extracted from the *I-V* characteristics. A similar scheme ought to be applicable to the OFETs. We will first examine the polycrystalline thin film transistor. For this device, we assume the current density to be due to holes and is given by:

$$J = n_{eff}\, q\, v. \tag{1}$$

n_{eff} is the effective hole density, q is the electron charge, and v is the velocity. Since by definition $v = \mu\acute{E}$, where $\acute{E}$ is the electric field, Eqn.(1) can be combined with the field-dependent mobility (10) (i.e., $\mu = \mu'\exp(\alpha\acute{E}^{0.5})$) for a disordered semiconductor to give:

$$J = n_{eff}\, q\mu_0 \exp(\alpha E^{0.5})\, E \tag{2}$$

μ' is the zero-field mobility, and α is the "disorder" parameter. As mentioned earlier, charge transport through the trap states is assumed to be barrier-height limited and this changes the zero-field mobility to a form: $\mu' = \mu_0 \exp[-\Phi_B/kT]$, where Φ_B is the barrier-height associated with the traps, k is the Boltzmann constant, and T is the absolute temperature. Eqn.(2) becomes:

$$J = n_{eff}\, q\mu'\exp(-\Phi_B/kT)\, E \tag{3}$$

Charge injection into the inter-granular region will result in space charge accumulation and barrier height modulation. According to the Poole-Frenkel effect, Φ_B can be replaced by $\Phi_B - \Delta\Phi$ in the presence of $\acute{E}$, where $\Delta\Phi = q(q\acute{E}/4\pi\varepsilon_s)^{0.5}$. Taking the barrier height lowering effect into account gives:

$$J = n_{eff}\, q\mu'\exp(-\Phi_B/E_\infty)\exp[q(qE/4\pi\varepsilon_s)^{0.5}/E_\infty]\, E = n_{eff}\, q\mu_0 \exp(\alpha E^{0.5})\, E \tag{4}$$

$\alpha = q(q/4\pi\varepsilon_s)^{0.5}/kT$, and $\mu_1 = \mu_0 \exp(-\Phi_B/kT)$. Eqn.(4) can now be combined with the microscopic expression of the low-field hopping mobility (11) ($T \rightarrow \infty$) given by:

$$\mu_\infty = q\acute{a}^2\, v \exp(-2\acute{a}/L)/\sigma \tag{5}$$

$\acute{a}$ is the "average" hopping distance, v is the escape frequency, L is the localization length, and σ is the rms width of the density of (transport) states (DOS). Note that $\mu_1 = \mu_\infty$ when $T \rightarrow \infty$. Comparing Eqn.(4) to Eqn. (5) gives:

$$\Phi_B/E_\infty = 2\acute{a}/L \text{ and } q\acute{a}^2 v/\sigma = \mu' \qquad [6]$$

Eqn.(6) harmonizes the charge transfer and hopping processes. The low-field carrier mobility therefore becomes:

$$\mu_\infty = q\acute{a}^2 v/\sigma \exp[- \Phi_B/kT] \qquad [7]$$

Furthermore, according to the GDM (8), α and σ in an "amorphous" organic semiconductor are related by the following expression:

$$\alpha = 2.9 \times 10^{-5} [(\sigma/kT)^2 - 2.25] \qquad [8]$$

Eqn.(8) can be combined with Eqn.(4) to give the following field-effect mobility:

$$\mu = \mu_\infty \exp\{\{2.9 \times 10^{-5} [(\sigma/kT)^2 - 2.25]\}E^{0.5}\} \qquad [9]$$

Eqn.(9) is only applicable to the polycrystalline OFET. For the single-grain and the single-crystal devices, $\alpha = 0$ and $\mu = \mu_\infty$ (Eqn.(7)) To compute μ, we need to know á, L, σ, v and Φ_B.

Parameter Extraction

In general, there would be limitations in matching the *I-V* data to an analytical current transport model. We first assumed the condition of quasi-static conductance, meaning that there was no time rate of change in the materials parameters. Furthermore, we also assumed no alternative degree of freedom in the fitting process. This would be the situation when the mobility data (12,13) were taken from the "linear regime" of the *I-V* curves. In the original data, contact resistance had been accounted for either through corrections and/or using four-terminal guarded measurements. Non-linear gate voltage dependent charge injection nonetheless could exist and would limit the applicability of our analyses to within the voltage range under consideration. As anticipated, parameter extraction linked to the *I-V* data in the OFETs could be complicated by the presence of the gate voltage bias. Gate voltage bias had previously been shown to affect the carrier density as discussed in (5). The same effect in principle could analogously be viewed as a change in the carrier mobility through a modulation in the barrier height for the trap states (see Eqn.(3)). Physically, what this meant was that the gate voltage bias instead of changing the occupancy of the trap states modulated the hopping process. We first examined the gate voltage bias effect on the hole mobility. In the presence of a gate voltage V_g, we replaced Φ_B in Eqn.(7) by $\Phi_{B0} - \gamma V_g$, where γ was a barrier-height modulation factor (14) and Φ_{B0} was the equilibrium barrier height. To deduce the gate voltage dependence, we examined the mobility data in (12) which showed how the activation energy E_a varied with the gate voltage for the polycrystalline OFET and the single-grain OFET. From these data, we obtained the following parameters: $E_a/q = -0.00375 |V_g| + 0.147$ V for the polycrystalline OFET, and $E_a/q = -0.001 |V_g| + 0.075$ V for the single-grain OFET. This suggested $\gamma - 0.00375$ and 0.001 respectively. Physically, γ denoted the extent the gate voltage changed Φ_B. A large γ generally implied the presence of significant trap states. These parameters also gave the equilibrium barrier-heights Φ_{B0} which were 147 meV and 75 meV, respectively. Using the same procedure applied to the *I-V* data of the single-crystal OFET in (13) gave $\gamma = 0.000033$ and $\Phi_{B0} =$

24.7 meV. The reduced gate voltage dependence in the single-grain and the single-crystal OFETs was in line with the fact that the densities of the trap states in these devices ought to be small. Similarly, the barrier-heights should also be low. Knowing Φ_{B0} in the thin films and the gate voltage dependence, we could now proceed to evaluate á, L, σ and v.

Rms Width of the Density of the (Transport) States

The rms width of the DOS σ in the polycrystalline OFET could be determined using Eqn.(8) if α was known. α was computed from Eqn.(4) as a function of temperature assuming $\varepsilon_s = 3\varepsilon_0$. The results were plotted in Figure 1a. As shown, α decreased with increasing temperature and this implied that "disorder" only dominated at low temperature. This was in agreement with similar observations found in other organic semiconductors (15). σ (shown in Figure 1b) on the other hand increased somewhat with increasing temperature. It would be quite tempting to suggest that such broadening in σ might be due to thermal fluctuations in the molecular geometry (16) which tended to increase with increasing temperature.

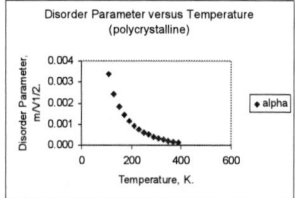 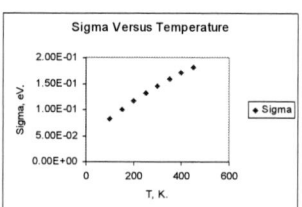

Figure 1: Plots of (a) α and (b) σ vs temperature for the polycrystalline OFET.

The computation of σ for the single-grain and the single-crystal OFETs had to come directly from the values of the sheet charge density. Assuming the transport site density N_T to be appreciably larger than the free hole density, we computed N_T using the relationship: $N_T \approx C_i V_g/q$, where C_i is the oxide capacitance per unit area. As given in (12,13), $C_i = 18$ pF.cm^{-2} in the single-grain OFET and $C_i = 3.5$ nF.cm^{-2} in the single-crystal OFET. Using the computed values of N_T at two different gate voltages (12 V and 20 V), we fitted them to the Gaussian distribution function and extracted the standard deviations (note that given the limited number of data points, these widths ought to be treated with care; however, they served as guiding parameters that described the influence of the gate voltage bias). Figure 2 shows the deduced σ as a function of temperature for the polycrystalline, the single-grain and the single-crystal OFETs. As observed, σ was relatively independent of temperature and the values ranged from ~ 1 – 100 meV. The upper limit was in line with the value of 0.1 eV +/- 0.01 eV evaluated from other thin film transistors using Kelvin probe force microscopy (17). The narrower DOS in the case of the single-crystal OFET was interesting and might imply in our opinion the presence of low-dimensional charge transport as in the quantum well devices. This had indeed been suggested to be what happened at high field in some organic polymers (15). In addition, according to (18) this also implied potentially an increase in non-local coupling between the charged states (i.e., a reduction in the polaron bandwidth). The fact that a small σ would give rise to a large hole mobility in an OFET had also been suggested (1). With σ evaluated, á, L, and v were the remaining parameters to be determined.

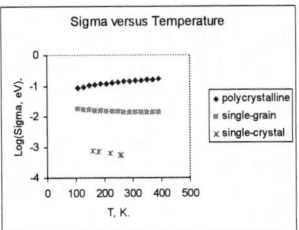

Figure 2: A plot of σ vs temperature for the three different types of OFETs.

"Average" Hopping Distance

According to (19), the average hopping distance á ought to be inversely linked to the site density N_T. Since $N_T \approx C_i V_g/q$, we could deduce that the density of the transport states which had the form: $\sim N_T/kT \exp(\gamma V_g/kT)$, where the exponential factor came from the distribution statistics. Making use of Eqn.(4) in (19), we arrived at the following relationship: $á \sim 1/(\pi N_T)^{0.5} \exp(-\gamma V_g/2kT)$. Fig.3 shows the computed values of á for the three different OFETs. The peak value for the single-crystal OFET was ~ 8 nm which was appreciably larger than similar values found in the case of hopping ($\sim 1 - 3$ nm). All of the OFETs showed increasing values of á with increasing temperature suggesting the favoring of charge transport between more distant hopping sites as the thermal energy increased. The moderate decrease in á with increasing gate voltage could be explained in terms of an increase in the trap density which caused a downward shift in the (quasi-) Fermi level towards the HOMO band edge. To a lesser extent, this might also be attributed to an increase in energy correlation amongst the more distant transport sites in the presence of the gate voltage bias. The most significant increase in the "average" hopping distance nevertheless occurred in the single-crystal OFET. This appeared to come naturally as the semiconductor became more crystalline. L and v were the remaining parameters to evaluate.

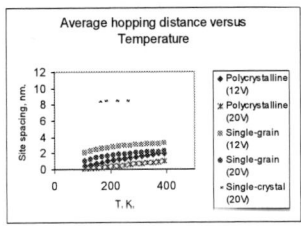

Figure 3: A plot of á vs temperature at two different gate biases for the OFETs.

Localization Length

Using the values of á and Φ_B, we determined the localization length L in the polycrystalline OFET based on Eqn.(6) and the fact that $\Phi_B = \Phi_{B0} - \gamma V_g$. Figure 4 shows a plot of the localization length for the different OFETs as a function of temperature. As shown, L increased with increasing temperature for all of the OFETs, which implied a spatial broadening of the hole wave function with increasing temperature. The range of L in the polycrystalline and the single-grain OFETs were in general agreement with values reported for the other organic semiconductors (~ 1 – 2 nm) as L tended to increase in conjunction with an increase in the carrier mobility. The larger values of L in the case of the single-crystal OFET could be related to the increase in á and partially due to the much reduced value of Φ_{B0}. The difference in L between the polycrystalline OFET and single-grain OFET was appreciable only at low temperature, apparently due to an increase in "disorder" in the former (less obvious when plotted in the linear scale). As expected, the observed effect due to the gate voltage bias was insignificant. We were left with ν as the remaining unknown.

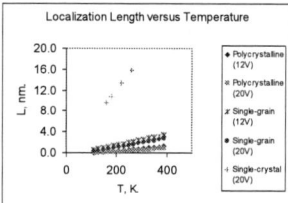

Figure 4: A plot of L vs temperature at two different gate biases for the OFETs.

Escape Frequency

Using the deduced values of á, L, σ, α, and Φ_B, we matched the mobility data (12,13) to our mobility equations using the escape frequency ν as the fitting parameter. Figure 5 show the simulated and reported experimental values of log(μ) plotted as a function of inverse temperature for the three OFETs. A reasonably good fit was observed in all cases. In addition, the deduced values of log(ν) plotted as a function of temperature are shown in Figure 6. The three escape frequency curves behaved very differently with significantly lower values in the case of the single-grain and the single-crystal OFETs (such values were nevertheless of the same order of magnitude (~ 10^{12} s^{-1}) as those reported (11) near room temperature for a different OFET). The highest values of ν in our case were found in the polycrystalline OFET suggesting (quite contrary to conventional thinking) that the presence of "disorder" appeared to assist in the charge escape process. The fact that ν increased with increasing gate bias in the polycrystalline OFET was in essential agreement with the existence of a gate voltage reduction due to barrier-height lowering effect. We were however unsure of the temperature dependence of ν in the single-grain and the single-crystal devices. In general, if the escape process was thermally activated, ν ought to increase logarithmically with increasing temperature and the results of the single-grain OFET appeared to be quite plausible. This could also imply that the temperature dependence of ν was affected by unknown external factors.

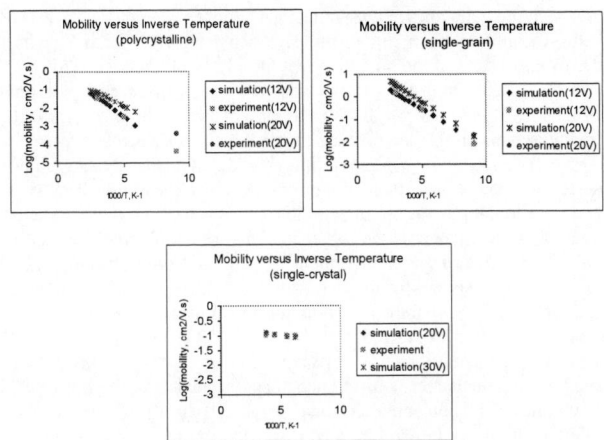

Figure 5: Simulated and experimental values of the hole mobility at different biases for the OFETs.

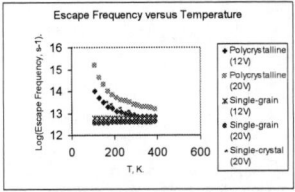

Figure 6: A plot ν vs temperature at different biases for the OFETs.

Discussion

Our model appeared to work well in matching simulations to the measured hole mobility. The smaller values of σ, the rms widths of the DOS in the single-grain and the single-crystal OFETs (see Figure 2) suggested a reduced level of "disorder" due to the "absence" of trap states. "Disorder" nevertheless prevailed in the polycrystalline OFET particularly at low temperature. This was shown in Fig.1a. Correlating such observations with the mobility data, it was tempting to suggest that the low hole mobility in the polycrystalline device could be due to the crystal morphology. The fact that mobility increased with increasing gate voltage bias in the polycrystalline OFET could be explained by a lowering in the barrier heights for the trap states in the inter-granular region, while the higher hole mobilities in the single-grain and the single-crystal OFETs were associated with the reduced values of σ. Figures 3 and 4 showed increased values of á and L in the single-crystal OFET. A large á normally could be the result of either a decrease in the transport site density or an increase in energy correlation between the more distant sites. The latter appeared to 丨 ·· : probable in this case, even though the effect of an increase in the gate voltage bias and hence a higher charge density was noted. Similar observations could be found in the localization length L which increased by almost one order of magnitude in the single-crystal OFET (see Figure 4). This could be explained as a changeover in the transport mechanism from localized charge hopping to extended band-like wave functions. In all three types of OFETs, the increases in L as a function of temperature appeared to support the fact that there had been an increase in the spread of the wave function at high temperature. We did not observe any significant effect on L due to the gate voltage bias. Figure 5 showed a close match between the mobility data and the simulations for the three OFETs at two different biases. The lack of sensitivity to a temperature change in the *I-V* characteristics of the single-crystal OFET was in our opinion the result of an improvement in structural order accompanied by a reduction in σ and/or energy spread for the transport states. This narrowing in the energy spread lowered the transition probability for hopping to occur as illustrated in Figure 7. In the figure, the joint density of states for transition would diminish in a "transport" band with a small σ. The re-organization energy between the charged states and the uncharged states was denoted by Δε. Such energy difference in pentacene has been estimated (20) to be about 60 -100 meV, a value smaller than σ (~ 150 meV) in the case of the polycrystalline OFET, but much larger than σ (~ 1 meV) in the single-crystal device. This could have suppressed the near-neighbor transitions and resulted in a larger á.

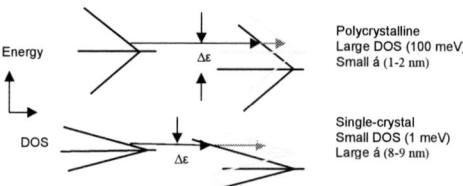

Figure 7: Hole transition between charged and uncharged transport sites.

According to Eqn.(6), changes in the escape frequency ν would have a direct effect on the hole mobility. The observed behavior of ν with changing temperature for the three types of OFETs suggested that different transport mechanisms were involved. The values of ν at room temperature however did not differ significantly reflecting that the escape process in these devices could be similar (probably dominated by the effects of the bulk/surface states). The distributed transport states in the polycrystalline OFET apparently facilitated the charge escape process at low temperature when the Fermi level was further away from the HOMO band edge where there would be a higher degree of charge occupancy. The smaller σ in the single-grain and the single-crystal OFETs probably did not enhance the charge escape process. An important observation was the smaller values of ν in these devices, an observation that appeared to play an important role in limiting the values of the hole mobility. Nevertheless, the increase in the average hopping distance and localization length in the single-crystal OFET could be to a demarcation separating transfer processes from near-neighbor molecular-type interactions to distant coulomb interactions, a concept frequently mentioned in considering charge relaxation for disordered solids (21). The slight increase in ν with increasing gate voltage bias could be explained as a decrease in the "average" hopping distance as the transport site density increased. Finally, it might be useful to mention the potential contribution of polymorphism to changes in the transport properties. This issue had not been considered in (12,13) even though two distinct bulk crystalline phases of pentacene had been reported (22). It was however doubtful that a small conductance change and/or anisotropy due to polymorphism could have affected the mobility values in the devices we examined. A second concern was the relatively large localization length extracted from the mobility data of the single-crystal OFET (see Figure 4). This contradicted the conventional thinking (23) that in the case of polaron related charge transport, the localization length would be limited to the length of the molecule (1-2 nm in the case of a pentacene molecule). Such restriction would be lifted if one considered both local and non-local electron-phonon coupling on equal footing and allowed for a progressive transition towards free carrier charge transport in the case of the single-crystal OFET.

Conclusions

We have examined pentacene OFETs prepared under different conditions to produce polycrystalline, single-grain and single-crystal active layers. Hole mobility data reported in the literature are compared with simulations based on a model developed to account for the differences in the materials properties, gate voltage bias and temperature. By systematically comparing data from the I-V characteristics with theory, we have been able to extract materials parameters in our models under different physical conditions. The observed hole mobility is explained in terms of structural changes such as a change in the rms width of the DOS, the "average" hopping distance, the localization length, as well as the escape frequency. In our opinion, these parameters to a larger or lesser extent are inter-related through the structural properties. In general, the single-crystal OFET behaves more like the crystalline semiconductor FET, while the polycrystalline OFET has the greatest effect linked to "disorder", a parameter closely associated with the site energy distribution and the barrier height. Apparently, the combination of materials parameters offering the highest value of the hole mobility in an OFET can not be achieved easily due to mutually offsetting effects. For instance, it is interesting to note that the escape frequency in the polycrystalline OFET is sensitive to the gate voltage bias and is higher than what is observed in the single-grain and the single-crystal OFETs. In our opinion, such low escape frequency is primarily due to the intrinsic properties of pentacene (as well as other organic semiconductors) being more akin to those of an insulator. There are indications that such behavior may be due to low-dimensional charge transport. The observation that the localization length increases to the tens of nanometer range in the single-crystal OFET is encouraging in the search for a higher mobility material. A proper means to control the properties of the pentacene thin film transistors with an increase in the escape frequency can be an interesting proposition. In our opinion, until a mechanism that offers an improved carrier escape process is found, the mobility of the single-crystal pentacene OFET will unlikely be significantly better that their polycrystalline counterparts.

Acknowledgments

This work is supported in part by NSERC, Canada.

References

1. S.F. Nelson, Y.Y. Lin, D.J. Gundlach, and T.N. Jackson, Appl. Phys. Lett. 72,1854 (1998).
2. C.D. Dimitrakopoulos, and P.R.L. Malenfant, Adv. Mater. 14, 99 (2002).
3. G.H. Gelinck, T.C.T. Geuns, and D.M. de Leeuw, Appl. Phys. Letts. 1487, (2002).
4. R.W. I. de Boer, M.E. gersgenson, A.F. Morpurgo, and V. Podzorov, phy. Stat. sol. (a), 201, 1302 (2004).
5. M.C.J.M. Vissenberg, and M. Matters, Phy. Rev. B57, 12964 (1998).
6. V.Y. Butko, X, Chi, D.V. Lang, and A.P. Pamirez, Appl. Phys. Letts. 4 4733, (2003).
7. P.V. Necliudov, M.S. Shur, D.J. Gunlach, and T.N. Jackson, Solid-State Electron. 47, 259 (2003).
8. H. Bassler, phys. stat. sol. (b) 175, 15 (1993).
9. S.M. Sze, *Physics of Semiconductor devices*, 2nd Ed., p.313, J. Wiley & Sons, New York, (1981).
10. H.Meyer, D. Haarer, H. Naarmann, and H.H. Horhold, Phys. Rev. B52, 2587 (1995).
11. H.C.F. Martens, P.W.M. Blom, H.F.M. Schoo, Phys. Rev. B61, 7489 (2000).
12. T. Minari, T. Nemoto, and S. Isoda, J. Appl. Phys. 96, 769 (2004).
13. J. Takeya, C. Goldmann, S. Haas, P.K. Pernstich, B. Ketterer, and B. Batlogg, J. Appl. Phys. 94, 5800 (2003).
14. Gate voltage reduction in the OFET channel has previously been proposed. See for example: H.L. Kwok, IEE Proc. Part G, 147, 125 (2000).
15. P.W.M. Blom, M.J.M de Jong, and M.G. van Munster, Phys. Rev. B55, R656 (1997).
16. A. Troisi, and G. Orlandi, Phys. Rev. Letts. 96, 86601 (2006).
17. O. Tal, Y. Rosenewaks, Y. Preezant, N. Tessler, C.K. Chan, and A. Kahn, Phys. Rev. Letts. 95, 256405 (2005).
18. K. Hannewald, V.M. Stojanovic, J.M.T. Schellekens, P.A. Bobbert, G. Kresse, and J. Hafner, Phys. Rev. B69, 075211 (2004).
19. S. Nakasuji, V. Dobrosavljevic, D. Tanaskovic, M. Minakata, H. Fukazawa, and Y. Maeno, Phys. Rev. Letts. 93, 146401 (2004).
20. N.E. Gruhn, D.A. da Silva Dilbo, T.G. Bill, M. Malagoli, V. Coropceanu, A. Kahn, and J-L. Bredas, J. Am. Chem. Soc., 7918 (2002).
21. J.C. Phillips, Rep. Prog. Phys. 59, 1133 (1996).
22. E. Venuti, R. G. D. Valle, A. Brillante, M. Masino, and A. Girlando, J. Am. Chem. Soc. 124, 2128 (2002).
23. M. Pope and C. Swenberg, *Electronic Processes in Organic Crystals and Polymers*, 2nd ed., p.1328, Oxford University Press, New York, (1999).

ECS Transactions, 3 (8) 273-278 (2006)
10.1149/1.2356363, copyright The Electrochemical Society

Design, Fabrication and Characterization of Parylene-Packaged Thin-Film Transistors

Hsi-wen Lo and Yu-Chong Tai

Micromaching Laboratory, California Institute of Technology, California, 91125, USA

A micro-fabricated parylene-packaged flexible pentacene thin film transistor is presented. Different from preceding devices that have been reported, this thin film transistor employs parylene as the substrate, the gate insulator and also the encapsulation layer. Also, this thin film transistor uses pentacene, an organic semiconductor with high mobility, as the active material. The transistor consists of Au/Cr gates and Au source and drain electrodes and takes a bottom-contact configuration. The freshly made thin film transistor shows a hole mobility of 0.084809 cm^2/V-s with an on-off ratio of 10^4.

Introduction

Vision loss due to retinitis pigmentosa (RP) and age-related macular degeneration (AMD) has troubled millions of people around the world. Recently, a retinal prosthesis has been developed for the treatment of aged-related blindness. This technology is based on the concept of replacing photoreceptor function with an electronic device (1). For this technology, a huge amount of electrodes are needed to achieve reasonable or high resolutions. A biocompatible and scalable high lead count electrode array for retinal prosthesis has been successfully fabricated (2). These electrodes are directly connected to the implanted control electronics through metal interconnects. As the resolution increase, the number of electrodes and interconnects increase, too. So is the volume of the implanted device. One way to reduce the number of metal interconnects and to satisfy the small volume constraint is to introduce a multiplexer into the system. This multiplexer has to overcome such difficulties as the corrosive environment and integration with the metal interconnects and so on.

One revolutionary approach to solve the problem is to explore biocompatible electronics that do not require conventional hermetic packaging and, at the same time, flexible enough for implantation use.

The combination of organic semiconductor and polymer substrate can serve this purpose. Pentacene ($C_{14}H_{22}$) thin film transistors (TFT) have been fabricated and possess a hole mobility up to 2.59 cm^2/V-s (3), which is comparable to the popular a-Si:H TFT technology.

Pentacene, however, is sensitive to oxygen, so unprotected pentacene transistors are vulnerable to even normal environments. It is therefore interesting to use parylene (readily a proven biocompatible material) as a pentacene-protecting polymer. Parylene C, a widely used MEMS (micro-electro-mechanical system) material, shows great flexibility (Young's modulus ~ 4 GPA), chemical inertness and biocompatibility (4). Parylene C has been recognized as a USP Class VI material and its intraocular biocompatibility has been

273

studied (2). In fact, parylene has been and is being studied for both encapsulation layer and even as a new gate insulator (5) (6). However, this work reports the first flexible parylene-pentacene electronics where parylene is exclusively used as the substrate, gate insulator and encapsulation layer.

Device Design

Our TFTs assume bottom-contact configurations. The schematic structure is shown in Figure 1. Pentacene is intolerant to exposure to solvents and other liquids (7). It has been demonstrated that the bottom contact configuration gives inferior performance to the top contact configuration for a range of deposition conditions and material thickness (8). However, the top contact configuration requires shadow masks, which introduces difficulties when integrating OTFT processes with standard photolithographic CMOS fabrication technology. We chose the bottom contact because of easy process integration. However, this brings up another problem, the contact resistance of source and drain. To reduce Au contact resistance, very thin Au (30nm ~50nm) without any adhesion layers is used (9). To further reduce contact resistance, we use large contact area. They are 4mm*1mm, 4mm*0.5mm, 1mm*1mm, 1mm*4mm, and 0.5mm*1mm (width/length). Mobility results of these transistors with different geometries of source/drain contacts are shown in the Table 1.

(a)

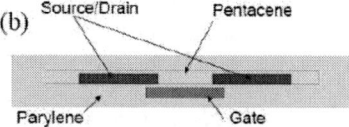

(b)

Figure 1. Isometric (a) and cross-section (b) views of bottom-contact configuration of parylene-packaged flexible pentacene TFTs

Device Fabrication

The basic technology involved in parylene-packaged pentacene TFT is the parylene/metal skin technology (10) (11). The simplified fabrication process is illustrated in Figure 2. The fabrication started with photoresist-coated wafers. A 10-μm parylene-C was first deposited as the substrate. A 1,500-angstron Au with 100-angstron Cr was thermally deposited and patterned to be the gate. A 0.1-μm parylene-C was then deposited as the gate dielectrics. Next, a 500-angstron Au was deposited and patterned as the sources and drain. Then, a 200-nm pentacene (as purchased from Sigma-Aldrich) was thermally-evaporated under high vacuum. A 1-µm parylene-C was deposited as top-protecting layer. Finally, the whole parylene-pentacene TFTs was liftoff from the photoresist in a flexible MEMS form, shown in Figure 3.

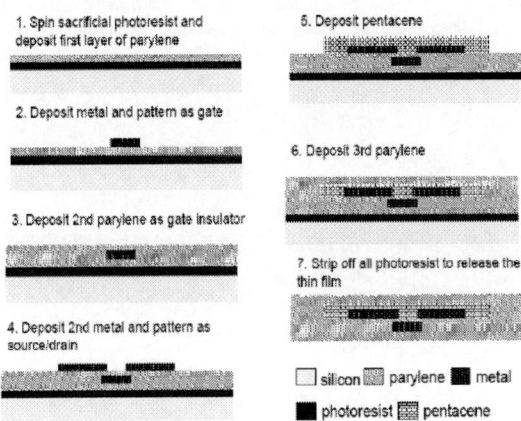

Figure 2. Simplified process flow.

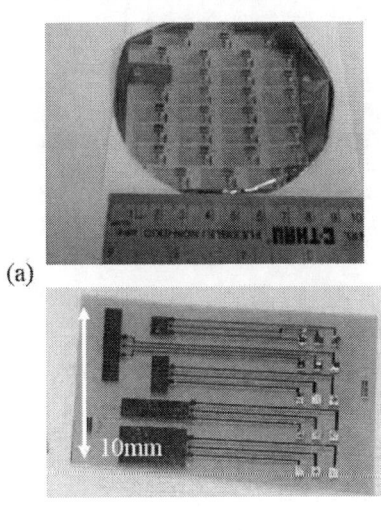

Figure 3. Fabricated Parylene film containing pentacene thin-film transistors. (a) Released film (b) Closer view

Device Characterization

We obtained the drain and gate characteristics of the thin-film transistor with a probe station and the HP4145B semiconductor parameter analyzer. The transistor was measured at room temperature. The mobility of charge carrier (μ) in the saturation regime can be calculated from the drain current given by the equation.

$$I_D = \frac{1}{2}\mu C_i \frac{W}{L}(V_{GS} - V_T)^2 \qquad [1]$$

Take square root of both sides.

$$\sqrt{I_D} = \sqrt{\frac{C_i W}{2L}}\mu(V_{GS} - V_T) \qquad [2]$$

Solving this equation and use a definition of "k" as

$$k = \sqrt{\frac{C_i W}{2L}}\mu \qquad [3]$$

$$\mu = \frac{2L}{WC_i}k^2 \qquad [4]$$

Equating the slope of the plot $\sqrt{I_D}$ versus V_G to "k" determines the μ in the saturation regime.

As mentioned before, we have fabricated OTFTs with source and drain electrodes of different geometries. The mobility results are show in Table 1. From these values, the mobility does not vary a lot for different sizes of source/drain electrodes. However, transistors with too small source/drain electrodes show poor performance, according to our previous observations.

The best transistor we fabricated shows a mobility of 0.084809 cm^2/V-s with an on-off ratio of 10^4. Its drain and gate characteristics are shown in Figure 4.

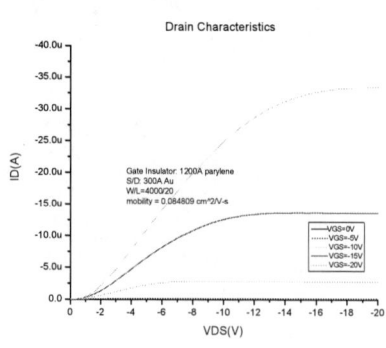

(a)

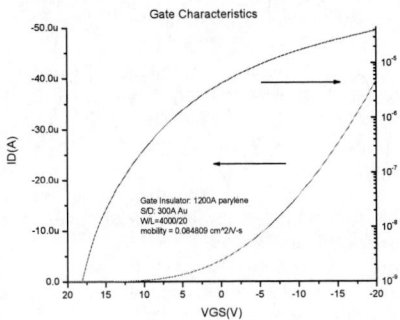

(b)

Figure 4. Drain (a) and gate (b) characteristics of the fabricated transistors. W/L=4000um/2000um

TABLE 1. Mobilities of transistors with different source/drain electrode area (cm²/V-s).

sets\area	4mm*2mm	4mm*0.5mm	1mm*1mm	1mm*4mm	1mm*0.5mm
Set01	0.058707	0.05208	0.046192	0.049484	0.050826
Set02	0.05067	N. A	0.047853	0.050902	0.062866
Set03	0.060771	0.065774	0.062914	0.067255	0.065339
Set04	0.06093	0.061961	0.057696	N. A.	0.051044
Set05	0.047353	0.045827	0.047425	0.050727	N. A.
Set06	0.050896	N. A.	0.050609	0.049328	0.055986
Set07	0.044969	0.042931	0.042062	0.049058	0.047858

Conclusion

We fabricated parylene-packaged pentacene thin-film transistors with fully MEMS-compatible parylene thin-film technology. The pentacene thin-film transistors use parylene as the substrate, the gate dielectric and the encapsulation layer, and take the bottom contact configurations. The fabricated pentacene thin-film transistor has a mobility of 0.084809 cm²/V-s with an on-off ratio of 10^4. Further bio-implantation tests are underway.

Acknowledgements

The authors would like to thank Mr. Trevor Roper for his assistance with equipment and fabrication. We would also thank Tanya Owen, Christine Matsuki and other members of the Caltech Micromachining Laboratory for their assistance.

References

1. Humayun, M.S., "Intraocular retinal prosthesis", Tr: Am Ophth Soc, **99**, 2001.
2. D. C. Rodger, J.D.W., M.S. Humayun, and Y.C. Tai, "Scalable Flexible Chip-level Parylene Package for High Lead Count Retinal Prosthesis", Transducer. 2005.

3. S.C. Lim, S.H.K., J.H. Lee, H.Y. Yu, Y. Park, D. Kim and T. Zyung, "Organic thin-film transistors on plastic substrates", Elsevier Materials Science and Engineering B, **121**, pp211-215, 2005.
4. Wolgemuth, L., "Assessing the performance and suitability of parylene coating", Medical Device & Diagnositic Industry, **22**: p. 42-49. 2000
5. I. Kymissis, C.D.D., and S. Purushothanman, "Patterning pentacene organic thin film transistors", Journal of Vacuum Science & Technology, **B 20-3**: p. 956-959, 2002
6. D. Feili, M.S., T. Doerge, S. Kammer, and T. Stieglitz, "Encapsulation of organic field effect transistors for flexible biomedical microimplants", Elsevier Sensors and Actuators, A, **120**: p. 101-109. 2005
7. D. J. Gundlach, T. N. Jackson, D. G. Schlom, and S. F. Nelson, "Solvent induced phase transistion strain relaxation in thermally evaporated pentacene films", Appl. Phys. Lett, **74**, p3302, 1999
8. F. Garnier, F. Kouli, R. Hajlaoui, and G. Horowitz, "Tunneling at organic/metal interfaces in oligomer-based thin-film transistors", MRS Bull., pp52-56, June, 1997
9. N. Yoneya, M. Noda, N. Hirai, K. Nomoto, M. Wada, and J. Kasahara, "Reduction of contact resistance in pentacene thin-film transistors by direct carrier injection into a-few-molecular-layer channel", Appl. Phys. Lett, **85**, #20, 2004
10. M. Liger, N.P., Y.C. Tai, S. Ho and C.M. Ho. Large-area electrostatically-valved skins for adaptive flow control on ornithopter wings. in Technical Digest, Solid State Sensor and Actuator Workshop. 2002. Hilton Head Island, South Carolina, USA.
11. Y.C. Tai, F.J., Y. Xu, M. Liger, S. Ho and C.M. Ho. Flexible MEMS skins: technologies and applications. in Proceedings, Pacific Rim MEMS Workshop. 2002. Xiamen, China

ECS Transactions, 3 (8) 279-285 (2006)
10.1149/1.2356364, copyright The Electrochemical Society

**The Improvement of Electrical Characteristic of Solution Processed
Triisopropylsilyl Pentacene Organic Thin-Film Transistors Employing
Hexamethyldisilazane Treatment**

Yong-Hoon Kim[a,b], Jae-Hoon Lee[b], Min-Koo Han[b] and Jeong-In Han[a]

[a] Information Display Research Center, Korea Electronics Technology Institute,
Seongnam, Kyunggi 463-816, Korea
[b] School of Electric Engineering and Computer Science, Seoul National University,
Seoul 151-742, Korea

The field-effect mobility of triisopropylsilyl (TIPS) pentacene
organic thin-film transistors (OTFTs) is increased considerably by
employing hexamethyldisilazane (HMDS) treatment on poly-4-
vinylphenol (PVP) gate insulator. A simple spin coating of HMDS
on PVP gate insulator of OTFTs suppressed the void formation
significantly and improved structural morphology of pentacene
film. The field-effect mobility of OTFT where the TIPS pentacene
was spin coated was increased from 0.01 $cm^2V^{-1}s^{-1}$ to 0.1 $cm^2V^{-1}s^{-1}$
after HMDS treatment. The threshold voltage was decreased from -
2.8 V to -2.0 V. We have also fabricated ink-jet printed OTFTs and
investigated the effect of HMDS treatment on ink-jet printed
OTFTs. The field-effect mobility of OTFT was not increased
significantly by HMDS treatment while the threshold voltage of
the device was reduced considerably. The threshold voltage of
untreated OTFT was +11.1V while that of HMDS treated OTFT
was -3.0 V. Our experiment results suggest that the HMDS
treatment improved the characteristics of OTFTs.

Introduction

Organic thin-film transistors have attracted a considerable attention for various
display applications. The pentacene is the most widely used organic material for OTFTs
and exhibits fairly good electrical characteristics [1]. However, the rather poor electrical
reliability may be a critical problem for practical applications and pentacene based
OTFTs may not be a suitable material for large area applications due to rather costly
vacuum deposition process. Recently, various solution based deposition methods such as
ink-jet printing and stamp based printings have attracted a considerable attention [2]. The
advantage of solution based deposition over conventional deposition process is the usage
of low-cost and non-vacuum deposition process, so that loss of costly organic materials
can be significantly reduced.

In solution based OTFTs employing a polymer as a channel layer [3,4], the field-
effect mobility is rather low (< 0.1 $cm^2V^{-1}s^{-1}$) and also the device stability is poor due to
moisture and oxygen [5,6]. High performance and stable soluble organic semiconductor
materials are thus desired for OTFTs. Recently, a new pentacene precursor which can be
patterned by a standard UV photolithography and OTFT with a field-effect mobility of
0.015 $cm^2V^{-1}s^{-1}$ were reported [7]. Also, an organic transistor with a new pentacene
precursor entitled as triisopropylsilyl (TIPS) pentacene was reported of which the field-

279

effect mobility was 0.4 $cm^2V^{-1}s^{-1}$ [8]. Also, an OTFT employing solution processed triethylsilylethynyl anthradithiophene as a channel layer with field-effect mobility around 1 $cm^2V^{-1}s^{-1}$ was reported [9].

It is well known that the gate insulator in OTFT is very important to secure the electrical performance. It has been reported that a surface modification of gate insulator results in higher field-effect mobility of OTFTs [10]. The surface treatment of the gate insulator decreases the order of the pentacene molecular stacks, which indicates that an increased density of flat-lying molecules and improvement of the mobility at the pentacene/octadecyltrichlorosilane interface [11]. The self-assembled monolayer (SAM) treatment is also reported to improve the threshold voltage of OTFTs. The SAMs have a built-in dipole field which depends on the SAM molecule's functional group and the dipole field would alter the mobile charge carrier density in the semiconductor [12]. By employing SAMs with different dipole moments, the threshold voltages of OTFTs have been controlled from the negative to the positive value [12].

The purpose of the paper is to report the effect of HMDS treatment of gate insulator on the electrical and morphological properties of solution processed TIPS pentacene OTFTs. By spin coating very thin (< 10 nm) HMDS on PVP gate insulator, the field-effect mobility of OTFT increased from 0.01 $cm^2V^{-1}s^{-1}$ to 0.1 $cm^2V^{-1}s^{-1}$ in spin coated OTFTs. Also in ink-jet printed OTFTs, the threshold voltage of OTFT was considerably reduced by the HMDS treatment.

Experimental

OTFTs were fabricated with the widely used bottom-gate and bottom-contact geometry. 100-nm-thick silicon oxide barrier layer was deposited on a glass substrate by e-beam evaporation. Then, 100-nm-thick Al-Si gate electrode was deposited by dc sputtering and patterned. On the top of the gate electrode, an organic gate insulator, poly-4-vinylphenol (PVP) was spin-coated and thermally cured (T_{cure} =175 ~ 200°C). The PVP solution was mixed with poly (melamine-co-formaldehyde) methylated in propylene glycol monomethyl ether acetate. For source/drain electrodes, e-beam-deposited Cr (5 nm-thick) with a thermally evaporated Au (50 nm-thick) layer was used. The organic active layers were coated with a thickness of 30 ~ 70 nm either by spin coating or by an ink-jet printing method (30 μm piezo nozzle, Microfab). The concentration of TIPS pentacene solution was fixed at 1wt% to minimize the effect of solution concentration on the electrical properties of the transistors. The solvents used in this experiment were chlorobenzene and anisole, which have boiling points of 132°C and 155°C, respectively.

In order to investigate the effect of HMDS treatment of PVP gate insulator on the electrical properties such as mobility and threshold voltage, the 30 ~ 70 nm thick TIPS pentacene was coated on various gate insulator surfaces such as bare PVP and HMDS treated PVP. The current-voltage (I-V) characteristics of the transistors were measured in dark and air-ambient environment. Also, the morphologies of TIPS pentacene films were studied by atomic force microscopy (AFM).

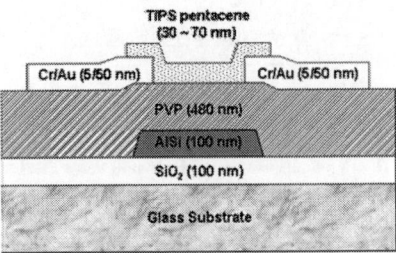

Figure 1. A schematic cross-section of the fabricated OTFT device. The device has bottom-gate and bottom-contact-type geometry and the channel width and length were 400 to 2500 μm and 15 to 25 μm, respectively.

Results and Discussion

Electrical Properties of OTFTs

The I_{DS}-V_{GS} and $\sqrt{I_{DS}}$-V_{GS} curves of OTFTs with TIPS pentacene film spin coated on a bare PVP and HMDS treated PVP are shown in Fig. 2. For HMDS treatment, HMDS solution was added on a thermally cured PVP film. The spin coating was performed for 2 minutes with a rate of 3000 rpm/min. After HMDS spin coating, 30 ~ 70 nm thick TIPS pentacene was also spin coated for 2 minutes with a rate of 2000 rpm/min. As shown in the transfer curves, the field-effect mobility of OTFT with HMDS treatment was 0.1 $cm^2V^{-1}s^{-1}$, while that of OTFT without the treatment was 0.01 $cm^2V^{-1}s^{-1}$. However, it should be noted that the on/off ratio was not altered by the HMDS treatment. The field-effect mobility, on/off ratio and threshold voltage of the devices are summarized in Table I.

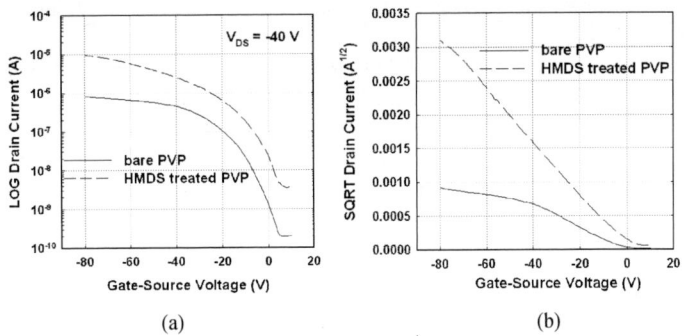

(a)   (b)

Figure 2. (a) The I_{DS}-V_{GS} and, (b) $\sqrt{I_{DS}}$-V_{GS} curves of OTFTs with TIPS pentacene film spin coated on a bare PVP and HMDS treated PVP gate insulator. Here, the V_{DS} was -40 V and the curing temperature of PVP was 200°C.

TABLE I. Electrical properties of OTFTs with spin coated TIPS pentacene.

Condition	Mobility (cm^{2}V^{-1}s^{-1})	on/off ratio	Threshold voltage (V)
Untreated (bare)	0.01	4 x 10^3	-2.8
HMDS treated	0.1	3 x 10^3	-2.0

The I_{DS}-V_{GS} and $\sqrt{I_{DS}}$-V_{GS} curves of OTFTs which were fabricated by employing ink-jet printing method with TIPS pentacene solution are shown in Fig. 3. It may be noted that widely used well-structure was not employed to define the active layer pattern. Instead, 1 to 3 ink drops per device were jetted on the source-drain electrode regions to form the active layer. Unlike spin coated devices, the field-effect mobilities of both untreated and treated devices were almost identical (~ 0.05 cm^{2}V^{-1}s^{-1}). However, a small increase in the on/off ratio and large shift of threshold voltage were observed. For example, the threshold voltage of untreated device was +11.1 V, whereas the HMDS treated device showed a threshold voltage of -3.0 V. The field-effect mobility, on/off ratio and threshold voltage of ink-jet printed devices are summarized in Table II.

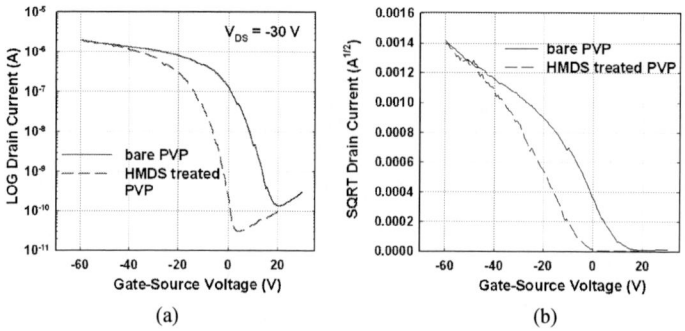

Figure 3. (a) The I_{DS}-V_{GS} and, (b) $\sqrt{I_{DS}}$-V_{GS} curves of OTFTs with TIPS pentacene film ink-jet printed on a bare PVP and HMDS treated PVP gate insulator. The V_{DS} was -30 V and the curing temperature of PVP was 175°C.

TABLE II. Electrical properties of OTFTs with ink-jet printed TIPS pentacene.

Condition	Mobility (cm^{2}V^{-1}s^{-1})	on/off ratio	Threshold voltage (V)
Untreated (bare)	0.045	10^5	+11.1
HMDS treated	0.050	10^5 ~ 10^6	-3.0

The large shift in the threshold voltage (+11.1 V to -3.0 V) between untreated and HMDS treated devices can be explained by the surface potential difference [12]. A shift in the threshold voltage is attributed to the difference in surface potentials of the layer next to the channel layer. Accordingly, each SAMs has a different dipole moment resulting in a variation of surface potentials. The direction of the dipole moment can be positive or negative depending on the molecular structure and the atoms in the head groups [13]. From Fig. 3(a), the drain current at zero gate bias is nearly 3 orders higher in the untreated device compared to the HMDS treated device, showing that there are more mobile charge carriers present in the channel layer. The mobile charge carrier can be calculated by using the equation, $N=C_{ox} \cdot V_{to}/e$, where N is the induced mobile charge carrier density, C_{ox} the capacitance of gate insulator per unit area, V_{to} the turn-on voltage (which is defined as the gate voltage when the drain current begins to increase exponentially) and e the elementary charge. The turn-on voltages obtained from the graph were +20.5 V and +4.0 V for untreated and HMDS treated devices, respectively. With C_{ox} of 7.97 nF/cm^2, the mobile charge carrier densities of untreated and HMDS treated devices are 1.02 x 10^{12}/cm^2 and 1.99 x 10^{11}/cm^2, respectively. The mobile carrier density of untreated device is 5 times higher than the HMDS treated device, reflecting the higher drain current flowing at zero gate bias.

It was reported that when PVP was used as a gate insulator, the conductivity of poly (3-hexylthiophene) (P3HT) can be increased [14]. The increase of P3HT conductivity is attributed to the protonation of P3HT due to the hydroxyl group in PVP and/or due to an interface dipole caused by the phenyl group. In our case the latter one is more likely to occur. The PVP with a dipole moment of 1.45 D [14] can induce mobile charge carriers in the channel layer and thus a high drain current can be flow even at zero gate bias. Previously it was reported that the HMDS treatment has an effect on lowering the dielectric constant of porous silica films [15]. The Si-OH bond in the porous silica films has a large dipole moment resulting in a high dielectric constant. To lower the dielectric constant or the dipole moment, HDMS treatment is applied to replace the Si-OH bond to Si-(CH$_3$)$_3$ bond. Hence in our case, the surface potential of bare PVP is expected to be lowered by the HMDS treatment and in consequences, the number of mobile charge carriers are reduced.

HMDS Treatment Effect on Pentacene Film Morphology

The AFM images of TIPS pentacene films obtained from spin coated on a bare PVP and HMDS treated PVP are shown in Fig. 4. As displayed in the figure, the TIPS pentacene film coated on a HMDS treated PVP has a different film morphology compared to that on a bare PVP. The TIPS pentacene on a bare PVP had an amorphous-like (or small grains) morphology with many voids at the interface contact (Fig. 4(a)). On the other hand, the TIPS pentacene on a HMDS treated PVP had large-sized grains (0.5 ~ 1.5 µm) and also shows molecular steps. It is commonly known that there is a close relationship between the channel layer morphology, molecular ordering and the electrical properties of the transistor [11,16]. Especially, it is important to note that the physical contact between the pentacene grains governs the overall electrical conductivity of the channel during charge accumulation [11].

As indicated in Fig. 4(a), a significant number of voids are present in the TIPS pentacene coated on a bare PVP. These voids lower the overall electrical conductivity of the TIPS pentacene film and, in turn, degrade the field-effect mobility of the device. By a HMDS treatment, the OH-terminated PVP surface is modified to (CH$_3$)$_3$-termination, improving the adhesion between the TIPS pentacene molecules and the gate insulator and results in a TIPS pentacene film with more ordered molecular structure (Fig. 4(b)).

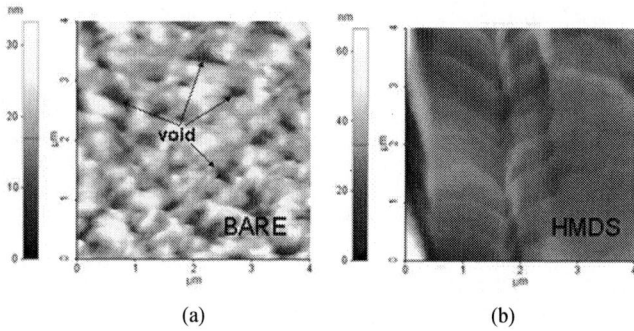

| (a) | (b) |

Figure 4. AFM images of TIPS pentacene films spin coated; (a) on a bare PVP gate insulator, and (b) on a HMDS treated PVP gate insulator. The images were acquired from contact-mode AFM measurement with a size of 4 μm x 4 μm.

Conclusions

We have successfully fabricated TIPS pentacene OTFTs by simple and inexpensive solution processes. Our experimental results show that the HMDS treatment of PVP gate insulator suppressed the void formation significantly and improved structural morphology of pentacene film. When the TIPS pentacene was spin coated, the field-effect mobility of OTFT increased from 0.01 $cm^2V^{-1}s^{-1}$ to 0.1 $cm^2V^{-1}s^{-1}$ by HMDS treatment. Also, the threshold voltage was changed from -2.8 V to -2.0 V. We have also fabricated ink-jet printed OTFTs and investigated the effect of HMDS treatment on ink-jet printed OTFTs. The field-effect mobility of OTFT was not increased significantly by HMDS treatment while the threshold voltage of the device was reduced considerably. The decrease of threshold voltage may be attributed to the surface potential change by HMDS treatment. Our experimental results suggest that the HMDS treatment improved the characteristics of OTFTs.

Acknowledgments

This research was supported by a grant (F0004063) from the Information Display R&D Center, one of the 21st Century Frontier R&D Program funded by the Ministry of Commerce, Industry and Energy of the Korean Government.

References

1. S. F. Nelson, *Appl. Phys. Lett.*, **72**, 1854 (1998).
2. S. K. Volkman, *Mat. Res. Soc. Symp. Proc.*, **769**, H11.7.1/L12.7.1 (2003).
3. Y. S. Yang, S. H. Kim, S. C. Lim, J. I. Lee, J. H. Lee, L. M. Do and T. Zyung, *Appl. Phys. Lett.*, **83**, 3939 (2003).
4. Z. Bao, A. Dodabalapur and A. J. Lovinger, *Appl. Phys. Lett.*, **69**, 4108 (1996).
5. S. Hoshino, M. Yoshida, S. Uemura, T. Kodzasa, N. Takada, T. Kamata and K. Yase, *Appl. Phys. Lett.*, **95**, 5088 (2004).

6. S. K. Park, Y. H. Kim, J. I. Han, D. G. Moon and W. K. Kim, *IEEE Transactions on Electron Devices*, 49, 2008 (2002).
7. A. Afzali, C. D. Dimitrakopoulos, T. O. Graham, *Adv. Mater.*, **15**, 2066 (2003).
8. C. D. Sheraw, T. N. Jackson, D. L. Eaton and J. E. Anthony, *Adv. Mater.*, **15**, 2009 (2003).
9. C. -C. Kuo, M. Payne, J. E. Anthony and T. N. Jackson, IEEE International Electron Devices Meeting, 373 (2004).
10. K. Xiao, Y. Liu, Y. Guo, G. Yu, L. Wan and D. Zhu, *Appl. Phys. A*, **80**, 1541 (2005).
11. M. Shtein, J. Mapel, J. B. Benziger and S. R. Forrest, *Appl. Phys. Lett.*, **81**, 268 (2002).
12. K. P. Pernstich, S. Haas, D. Oberhoff, C. Goldmann, D. J. Gundlach, and B. Batlogg, A. N. Rashid, G. Schitter, *J. Appl. Phys.*, **96**, 6431 (2004).
13. H. Sugimura, K. Hayashi, N. Saito, N. Nakagiri, O. Takai, *Appl. Surf. Sci.*, **188**, 403 (2002).
14. T. G. Bäcklund, R. Österbacka, and H. Stubb, J. Bobacka and A. Ivaska, *J. Appl. Phys.*, **98**, 074504 (2005).
15. S. –I. Kuroki and T. Kikkawa, *Proceeding of 3rd Hiroshima International Workshop on Nanoelectronics for Terra-Bit Information Processing*, P-24, 70 (2004).
16. I. Yagi, K. Tsukagoshi and Y. Aoyagi, *Appl. Phys. Lett.*, **86**, 103502 (2005).

ECS Transactions, 3 (8) 287-292 (2006)
10.1149/1.2356365, copyright The Electrochemical Society

Suppression of OLED Current Error Caused by the Hysteresis of a-Si:H TFT
For AMOLED display

Jae-Hoon Lee[a], Sang-Geun Park[a], Jae-Hong Jeon[b], Joon-Chul Goh[c], JoonHoo Choi[c],
Kyuha Chung[c], and Min-Koo Han[a]

[a]School of Electrical Engineering, Seoul National University, Seoul, Korea
[b]School of Electronics, Hankuk Aviation University, Gyeonggi-do, Korea
[c]LCD R&D center, Samsung Electronics, Gyeonggi-do, Korea

The new a-Si:H TFT pixel circuit, which successfully suppresses
OLED current error caused by the hysteresis of a-Si:H TFT, is
proposed and fabricated. We have also investigated the hysteresis
mechanism of the a-Si:H TFT for AMOLED. When an identical
V_{GS} and V_{DS} is applied to a-Si:H TFT device, the drain current of
a-Si:H TFT is altered due to the hysteresis phenomenon, which
changes the interface trapped charge between an a-Si active layer
and SiN$_x$ gate insulator. When the data voltage of 8V is applied to
the conventional a-Si:H TFT pixel, OLED current is changed from
117nA to 167nA due to the hysteresis phenomenon of a-Si:H TFT.
In our proposed pixel circuit, OLED current error caused by the
hysteresis is successfully suppressed by applying a reset voltage to
the gate node of current-driving a-Si:H TFT before a data voltage
is written, because a reset voltage can enable the starting gate
voltage for a desired one not to be varied.

Introduction

Active matrix organic light emitting diode (AMOLED) employing thin film transistor
(TFT) pixels, such as amorphous or polycrystalline silicon TFT, have attracted
considerable attentions due to high brightness, compactness and wide viewing angle [1].
Although excimer laser annealed poly-Si TFT has been intensively considered a pixel
element of AMOLED display due to its high current driving capability, the non-
uniformity of OLED current in the conventional poly-Si TFT pixel, which is caused by
the fluctuation of excimer laser energy, may be critical problem for AMOLED [1-3].
Recently, hydrogenated amorphous silicon (a-Si:H) TFTs are considered as the pixel
element of AMOLED due to the well proved uniformity of a-Si:H TFT [4-6]. However, it
is well known that the device stability of a-Si:H TFT, such as the degradation of V_{th}, is a
critical problem [7-9]. Hysteresis phenomenon is an inherent characteristics of the a-Si:H
TFTs, which is caused by the considerable amount of interface trap in the SiN$_x$ gate
insulator. Residual image, where the previous display image remains apparent in the
subsequent image due to the hysteresis phenomenon [10], may be an issue observed in
conventional a-Si:H TFT pixel composed of 2-TFT.

The purpose of our work is to propose the new pixel to suppress I$_{OLED}$ variation
caused by the hysteresis of a-Si:H TFT. TFT pixel driving scheme to suppress a
hysteresis effect on I$_{OLED}$ have been scarcely reported. Our experimental results show that

under the proposed pixel driving scheme, I_{OLED} error caused by the hysteresis can successfully be suppressed due to the fixed V_{GS} sweep direction of a-Si:H TFT.

Hysteresis Mechanism in the a-Si:H TFT

The a-Si:H TFTs with an inverted staggered bottom gate type were fabricated by a standard commercial process. The gate metal layer was deposited on Corning 1737 glass by DC sputtering. Triple layer of SiN_x, a-Si:H, n^+ a-Si:H was deposited on the gate by plasma-enhanced chemical vapor deposition (PECVD). After active island patterning, the metal layer was deposited for the source and drain electrode by sputtering. After patterning the source and drain electrode by a wet etching, the n^+ a-Si layer between the source and drain electrode was removed by a dry etching to make an etch back type channel structure. A SiNx was deposited for a passivation. After contact holes are formed, an indium tin oxide (ITO) electrode was deposited and patterned. The characterization of the device is performed by measuring the transfer characteristics in the voltage sweeping mode, while that of the pixel circuit by measuring output current in the sampling mode. All electrical measurements were obtained at room temperature (26°C).

Fig. 1 shows that the a-Si:H TFT structure used for experiment and photograph fabricated on a glass substrate.

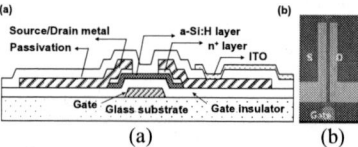

(a)   (b)

Figure 1. Fabricated a-Si:H TFT sample. (a) cross-sectional view, (b) top view (photograph).

Fig. 2 shows the drain current difference observed at the same drain bias condition (10V) according to a variable gate voltage sweep direction. The width and length of TFT are 200μm and 4μm, respectively. Recently, OLED efficiency has improved so that the maximum OLED current is about 2~3μA. The driving current range is from 0A (dark) to 2~3μA (white), which requires a maximum V_{GS} of 5~6V. Thus, we set that V_{GS} range from 0V to 6V to measure hysteresis phenomenon in a practical AMOLED driving range. As shown in Fig. 2, the current difference was observed in the practical V_{GS} range to drive AMOLED display.

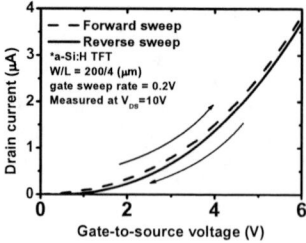

Figure 2. Hysteresis phenomenon of a-Si:H TFT with gate voltage sweep direction.

In order to investigate the cause of the hysteresis phenomenon, we measured the transfer characteristics in the reverse-voltage sweeps with different starting gate voltages, as shown in Fig. 3(a).

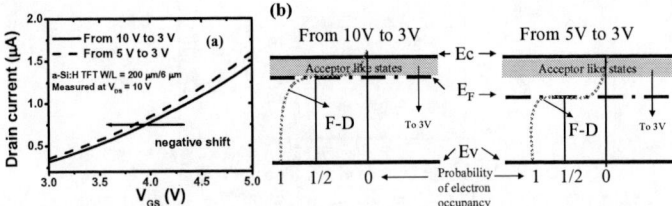

Figure 3. (a) The transfer characteristics for reverse-voltage sweeps with different starting positive gate voltages, (b) The band diagram in accordance with the different positive starting gate voltages. The Fermi-Dirac distribution (F-D) for electron in the interface is also shown. The conduction and valence band energy are denoted as Ec and Ev, respectively. Fermi-level energy is also denoted as E_F.

By employing different starting gate voltages [11], different quantities of charge were trapped so that the transfer curve is shifted with a different starting voltage. The shift in transfer curves caused by hysteresis phenomenon would be attributed to differences in an initial Fermi-level at the starting gate voltage, as shown in Fig. 3(b). As the starting gate voltage became smaller, the parallel shift toward the negative direction in the I_D-V_G characteristic was observed. The hysteresis and shift in transfer curves are attributed to differences in the initial Fermi level at the starting gate voltage. The hysteresis varies the interface trap occupancy, resulting in changes of effective insulator charges. Under the same drain current, a decreased gate charge ($\Delta Q_G < 0$) is the result of an increase in effective interface charges ($\Delta Q_{it} > 0$) between a-Si active layer and gate-insulator layer because the charge in silicon (Q_S) is constant [12]. For a small starting gate voltage in the reverse voltage sweeps (from $V_{GS} = 5$ V in Fig. 3 (a)), fewer acceptor traps are filled, resulting in a $\Delta Q_I > 0$ because the Fermi level is further away from the conduction band. Thus, for smaller gate voltages, the transfer characteristics exhibit a parallel shift toward the negative direction in the reverse gate voltage sweep direction.

In order to investigate the relation between an observed hysteresis and OLED current error, we have measured a current with different starting gate-voltage in both device and well known 2-TFT pixel circuit. Fig. 4 is the output current variation in the device with a different starting gate-voltage, from 0V to 8V. The gate node of current driving TFT in the well known 2-TFT pixel circuit experiences a very fast data voltage transition during one row time (~several ten µsec). Thus, we measured only two gate-source voltages (for example from V_{GS}=0V to 4V or V_{GS}=8V to 4V) at the same drain voltage, as shown in Fig. 4. Fig. 4 shows that the same gate-source (V_{GS}=4V) and drain-source (V_{DS}=10V) voltages exhibits the different output current with the previous gate-source voltage due to hysteresis phenomenon with a different starting gate-voltage. For example, the drain current of 1.67µA was measured at a previous gate-source voltage of V_{GS}=0V, while that of 1.46µA was measured at a previous gate-source voltage of V_{GS}=8V.

ECS Transactions, 3 (8) 287-292 (2006)

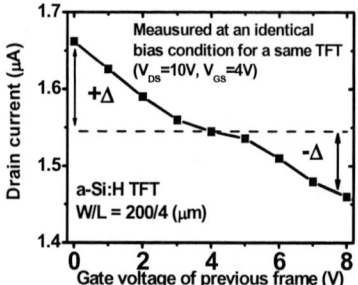

Figure 4. Experimental results of the drain current at the identical V_{GS}, V_{DS}, with previous gate starting voltage at the identical a-Si:H TFT device. It should be noted that the gate voltage during the previous frame can vary the drain current in the present frame.

Suppression of OLED Current Error Caused By Hysteresis of a-Si:H TFT for AMOLED Backplane

In order to eliminate I_{OLED} variation induced by hysteresis of a-Si:H TFT, we propose a reset voltage should be applied before a data for a present frame is written, which enables the V_{GS} sweep direction of current-driving TFT to be fixed.

Fig. 5 shows the proposed a-Si:H TFT pixel driving scheme, which can apply a reset voltage before a data voltage is applied to a pixel, to eliminate I_{OLED} variation due to the hysteresis of a-Si:H TFT. When the proposed driving scheme is employed in the panel, only one TFT (T3) and one clock signal line are required for one row line so that an aperture ratio may not be sacrificed in the pixel array. The core operation of the proposed ac driven pixel is as follows. When a clock signal is a low state (reset voltage), T3 is turned on so that a gate node voltage of T2 would be discharged to a reset voltage. When a clock signal is higher than a data voltage, T3 is turned off due to $V_{GS\_T3} = 0$ so that T2 can supply I_{OLED} in accordance with a data voltage. Therefore, V_{GS} sweep direction of T2 is always forward direction and the gate starting voltages of T2 are independent of the data voltages.

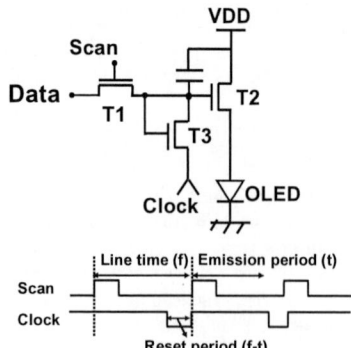

Figure 5. The proposed reset voltage-driven pixel and a timing diagram.

290

We have fabricated and measured the proposed 3-TFT pixel circuit to evaluate the performance to suppress I_{OLED} variation caused by the hysteresis phenomenon of a-Si:H TFT.

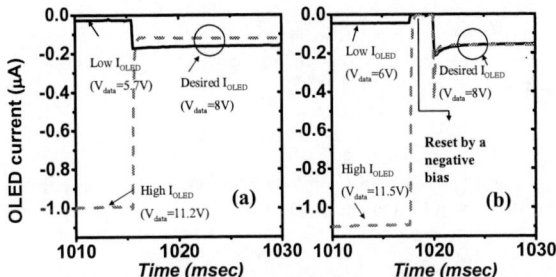

Fig. 6. The effect of the hysteresis in a-Si:H TFT to drive AMOLED pixel circuit.
(a) I_{OLED} in the conventional 2-TFT pixel is varied from 117nA to 167nA due to the hysteresis of a-Si:H TFT. We can measure the I_{OLED} variation at an identical data voltage (8V), from a high data voltage (11.5V) and a low data voltage (6V). (b) I_{OLED} variation in the proposed driving scheme is almost eliminated due to a fixed V_{GS} sweep direction of a-Si:H TFT.

When we apply the data voltage of 8V (middle-current) from both 11.2V (high current, ~1.1 μA) and 5.7V (low current) to the conventional 2-TFT pixel, in order to investigate I_{OLED} variation caused by the hysteresis phenomenon of a-Si:H TFT, I_{OLED} is changed from 117nA to 167nA due to the difference of previous data voltages, as shown in Fig. 6 (a). I_{OLED} variation caused by hysteresis of a-Si:H TFT can be successfully suppressed in the proposed pixel, as shown in Fig. 6 (b). The desired current for 8V data voltage (~ 0.2 μA) in the proposed pixel, from both 11.5V current (~ 1.2 μA) and low current (~ 0.02 μA), is not varied according to the previous data voltages. The desired I_{OLED} of the proposed pixel is independent of the I_{OLED} of previous frame due to a fixed V_{GS} sweep direction, while that of the conventional pixel is dependent on the previous I_{OLED} due to the hysteresis of a-Si:H TFT causing a drain current variation according to V_{GS} sweep direction. Our proposed pixel would exhibit an immunity against the hysteresis of a-Si:H TFT because a reset voltage can be applied to the gate electrode of a-Si:H TFT before a data voltage is applied. Our proposed TFT pixel driving scheme to suppress a residual image caused by the hysteresis would be suitable for TFT backplane for AMOLED.

Conclusion

We have designed and fabricated the new a-Si:H TFT pixel, which successfully suppresses OLED current error caused by the hysteresis of a-Si:H TFT. Our experimental results show that the proposed pixel can suppress OLED current error while the conventional one exhibits OLED current error, which is caused by the hysteresis of a-Si:H TFT. When an identical V_{GS} and V_{DS} is applied to a-Si:H TFT device, the drain current of a-Si:H TFT is altered due to the hysteresis phenomenon, which changes the interface trapped charge between an a-Si active layer and SiN_x gate insulator. When the data voltage of 8V is applied to the conventional a-Si:H TFT pixel, OLED current is changed from 117nA to 167nA due to the hysteresis phenomenon of a-Si:H TFT. In our proposed pixel circuit, OLED current error caused by the hysteresis is successfully suppressed by applying a reset voltage to the gate node of current-driving a-Si:H TFT before a data voltage is written, because a reset voltage can enable the starting gate voltage for a desired one not to be varied.

The proposed pixel would be suitable to suppress the OLED current error caused by the hysteresis phenomenon of a-Si:H TFT.

Acknowledgments

This research was supported by a grant, one of the new growth engine R&D programs funded by the Ministry of Commerce, Industry and Energy of Korean government.

References

1. R.Dawson, etc.at al, *IEEE IEDM*, 1998, 875.
2. S.H.Jung, W.J.Nam and M.K.Han, *IEEE EDL*, **25**,690 (2004).
3. J.H.Lee,W.J.Nam,H.S.Shin, Y.M.Ha, H.S.Choi,C.H.Lee, S.K.Hong and M.K.Han, *IEEE IEDM*, 2005, 953.
4. G.R.Chaji, D.Striakhilev, and A.Nathan, *IEEE EDL*, **26**, 737 (2005).
5. J.H.Lee, J.H.Kim, and M.K.Han, *IEEE EDL*, **26**, 897 (2005).
6. K.S.Shin, J.H.Lee, S.M.Han, I.H.Song, and M.K.Han, Journal of N.Crystalline Solids, **352**, 1708 (2006).
7. M.J.Powell, C.van Berkel, A.R.Franklin, S.C.Deane and W.I.Milne, *Physical Review B*, **45**, 4160 (1992).
8. T.Tsujimura, *Jpn.J.Appl.Phys.*, **43**, 5122 (2004).
9. F.R.Libsh and J.Kanicki, *Appl.Phys.Lett.* **62**, 1286 (1993).
10. B. K. Kim, O. Kim, H. J. Chung, J. W. Chang and Y. M. Ha, *Jpn.J.Appl.Phys.*, **43**, 482 (2004).
11. K.Chatty, B.Banerjee, T.P.Chow and R.J.Gutmann *IEEE EDL*, **23**, 330 (2002).
12. D.A.Neamen, Semiconductor physics and devices, McGraw-Hill, New-York (2003).

ECS Transactions, 3 (8) 293-300 (2006)
10.1149/1.2356366, copyright The Electrochemical Society

Integrated circuits based on amorphous indium-gallium-zinc-oxide-channel thin-film transistors

M. Ofuji[a], K. Abe[a], N. Kaji[a], R. Hayashi[a], M. Sano[a], H. Kumomi[a],
K. Nomura[b], T. Kamiya[b], and H. Hosono[b, c]

[a] Canon Research Center, Canon Inc., 3-30-2 Shimomaruko, Ohta-ku, Tokyo 146-8501, Japan
[b] ERATO-SORST, Japan Science and Technology Agency, and Materials and Structures Laboratory, Tokyo Institute of Technology, 4259 Nagatsuta, Midori-ku, Yokohama 226-8503, Japan
[c] Frontier Collaborative Research Center, Tokyo Institute of Technology, 4259 Nagatsuta, Midori-ku, Yokohama 226-8503, Japan

Five-stage ring oscillators (ROs) composed of amorphous In-Ga-Zn-O (*a*-IGZO) channel TFTs were fabricated at room temperature with no post-deposition annealing. We observed oscillation of ROs with a variety of channel lengths and channel widths. A RO with channel lengths of 10 μm operated at 21.5 kHz (propagation delay of 4.7 μs / stage), when the external voltage of +18 V was supplied. A circuit simulation reproduced the measured output characteristics of ROs qualitatively, and also the simulated propagation delays agreed with the measured ones approximately.

Introduction

There have been diverse researches toward the development of light-weighted flexible active-matrix flat-panel displays at acceptable manufacturing cost. Exploration for new semiconductor materials as channel materials for the thin-film transistor (TFT) backplanes has been one of the research focuses. Major efforts are concentrated on amorphous silicon (*a*-Si) (1), low-temperature polysilicon (2), and organic semiconductors (3). A candidate material is desired (i) to have high carrier mobility to enable high-speed switching and large current drive, (ii) to be amorphous to ensure panel uniformity, and (iii) to be deposited at a moderate temperature to fabricate TFTs directly on flexible (e.g. polymer) substrates.

Oxide semiconductors whose conduction bands are composed of hybridized *s*-orbitals of post-transition metals can exhibit relatively large carrier mobility even in amorphous phases. There are increasing reports on oxide semiconductor thin films, including polycrystalline and amorphous ones (4-14). Some of the present authors have shown that amorphous In-Ga-Zn-O (*a*-IGZO)-channel TFTs have greater carrier mobility (field-effect mobility, μ_{FE}, of 6 – 12 cm^2 V^{-1} s^{-1} (5, 9, 15)) than that of a typical amorphous silicon (*a*-Si) TFT ($\mu_{FE} \lesssim 1$ cm^2 V^{-1} s^{-1}).

A great advantage of *a*-IGZO is that its thin films can be grown by sputtering on unheated substrates, which is applicable to large-area fabrication of TFT backplanes and circuits. Besides, as-deposited films of *a*-IGZO exhibit good characteristics as a semiconductor. The carrier concentration in *a*-IGZO can be controlled by partial pressure of oxygen during sputtering of *a*-IGZO layer and no post-deposition annealing (e.g. 300 – 600 °C) (7) was mandatory. We have also demonstrated *a*-IGZO TFTs fabricated on plastic films (16).

293

In order for TFTs to be applied to the backplanes, dynamic characteristics of the TFTs are as important as their DC characteristics. Ring oscillators (ROs) are widely used to test whether TFTs properly work to charge / discharge capacitive loads (the gate capacitance of other TFTs) repeatedly. Although there are some demonstrations of ROs with organic TFTs (17-19), there is few reports on ROs with oxide TFTs except the In-Ga-O-TFT RO (6). With the aid of a circuit simulation, the simplest test of dynamic characteristics of a TFT is to see whether device parameters extracted from DC characteristics of the TFT can also reproduce measured transient characteristics of a circuit built from the TFTs. This can be done at different levels of accuracy; even the most primitive MOSFET model can be used to an 'order-of-magnitude' evaluation.

In this paper, we demonstrate fabrication of the ROs made with the a-IGZO TFTs. We also attempt a preliminary check on validity of the obtained output characteristics by means of SPICE simulations.

Experimental

We designed enhancement/enhancement-inverter-based five-stage ROs using a-IGZO bottom-gate top-contact TFTs having different combinations of channel length (L) and channel width (W). Figure 1 is a circuit schematic of a RO. In each inverter and RO, we set the L common to the load and drive TFTs ($L = L_{load} = L_{drive}$), which was either 10 or 30 (μm). Beta ratios, β_R ($= (W_{drive} / L) / (W_{load} / L)$), ranged from 5 to 20. We fixed (W_{load} / L_{load}) = 4 and changed only W_{drive}. The drive TFTs had interdigital source-drain electrodes. The source- (and drain-) gate overlap length was 5 μm. Hereafter, we specify inverters and ROs by L and β_R.

The ROs were fabricated as follows: Glass plates (Corning #1737) were used as substrates. Gate electrodes (Ti: 5 nm + Au: 40 nm + Ti: 5 nm) were electron-beam deposited and patterned by a lift-off technique. SiO$_2$ blanket layer (100 nm) was deposited by rf magnetron sputtering (sputtering gas: Ar, sputtering pressure: 0.1 Pa, input power: 400 W, final thickness: 100 nm). Wet etching with buffered HF was employed to make contact holes. Amorphous IGZO layer was deposited by rf magnetron sputtering (sputtering gas: O$_2$ (3.3 vol%) / Ar, sputtering pressure: 0.53 Pa, input power: 300 W, final thickness: 50 nm), and patterned by etching. During these sputtering processes, no intentional control of the substrate temperature was employed. Finally, Ti (5 nm) and Au (200 nm) were electron-beam deposited to form source and drain electrodes and interlayer connections. Figure 2 shows a cross-sectional schematic and a photograph of a fabricated RO.

In order to evaluate DC characteristics of the constituent TFTs and inverters, we also fabricated discrete TFTs and inverters next to each RO, which have the same channel dimensions. The DC characteristics of the discrete devices were measured with a semiconductor parameter analyzer (Keithley 4200SCS).

The output characteristics of the ROs were measured with a high impedance FET probe (Picoprobe 28, GGB industries, input capacitance: 0.04 pF) contacted on the output stage of one of the inverters. We supplied the same voltage to V_{gg} (to the gate electrodes of the load TFTs) as V_{dd} (to the drain electrodes of the load TFTs).

We performed SPICE simulations (20) with a level 1 NMOS model to check whether or not the observed output characteristics were reasonable. Source- and drain-gate overlap capacitances were calculated from mask dimensions.

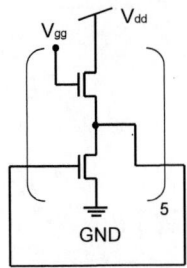

Figure 1. A circuit schematic of an E/E 5-stage ring oscillator.

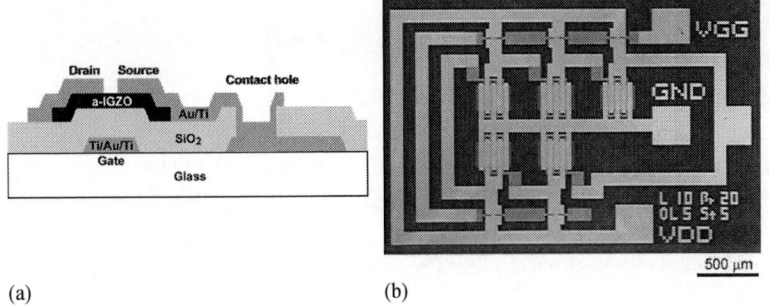

(a)   (b)

Figure 2. (a) A cross-sectional schematic and (b) a photograph of a fabricated RO.

Results & Discussion

Measured characteristics of TFTs, inverters & ROs

Here we present the results from a RO with $L = 10$ (μm), $\beta_R = 5$. First, we show I_{ds}-V_{gs} DC characteristics of the discrete TFTs in Fig. 3. On / off ratios, defined as I_{ds} (at $V_{gs} = +20$ (V)) / I_{ds} (at $V_{gs} = 0$ (V)), were 6.5 x 10^5 and 2.5 x 10^6, for load and drive TFTs, respectively. The TFTs were operated in saturation region near $V_{gs} = +15$ (V), where μ_{FE} and threshold voltage V_{th} were estimated as follows: (load) $\mu_{FE} = 3.5$ (cm^2 V^{-1} s^{-1}), $V_{th} = +7.2$ (V); (drive) $\mu_{FE} = 1.9$ (cm^2 V^{-1} s^{-1}), $V_{th} = +7.0$ (V).

ECS Transactions, 3 (8) 293-300 (2006)

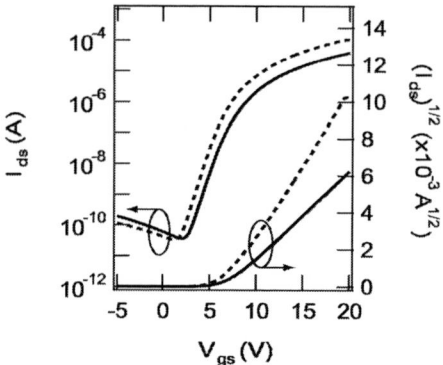

Figure 3. Measured I_{ds}-V_{gs} DC characteristics of discrete TFTs. Solid lines: $L = 10$ (μm), $W = 40$ (μm); dashed lines: $L = 10$ (μm), $W = 200$ (μm). The TFTs are identical to the load and drive TFTs in a RO with $L = 10$ (μm), $\beta_R = 5$.

Figure 4 is the DC characteristics of an inverter with $L = 10$ (μm) and $\beta_R = 5$, measured under voltage supply of $V_{dd} = V_{gg} = +11$ (V). The gain was 1.2, defined by the slope of a tangential line to the measured characteristics at the inversion voltage V_{inv} (where input voltage V_{in} equals to output voltage V_{out}). The gain greater than unity suggests that bistable operation is possible with this inverter.

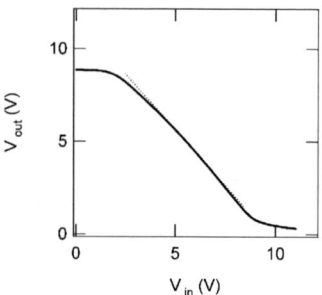

Figure 4. Measured DC transfer characteristics of an inverter ($L = 10$ (μm), $\beta_R = 5$) supplied with $V_{dd} = +11$ (V). The dotted line is a tangent at the inversion voltage, V_{inv}.

The measured output characteristics of a 5-stage RO with $L = 10$ (μm) and $\beta_R = 5$ at the supply of $V_{dd} = +15$ (V) is shown in Fig. 5(a). In Fig. 5(b) are shown the propagation delay time and the output voltage swing measured as a function of V_{dd}. The RO started to oscillate at $V_{dd} \sim +13$ (V). On increasing V_{dd}, the oscillation frequency and the voltage

296

swing increased. At the supply of $V_{dd} = + 18$ (V), the RO operated at 21.5 kHz, corresponding to delay of 4.7 µs / stage. The output voltage swing was 4.9 V_{p-p}.

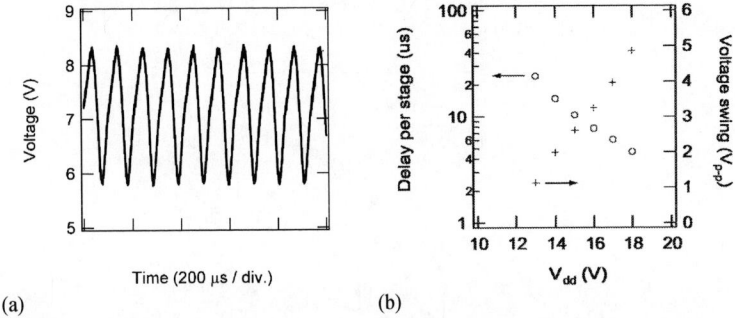

(a)   (b)

Figure 5. (a) Measured output characteristics of a RO ($L = 10$ (µm), $\beta_R = 5$) with $V_{dd} = +$ 15 (V), showing frequency of 9.6 kHz and output voltage swing of 2.6 V_{p-p}. (b): (Circles) measured propagation delay time per stage, and (crosses) measured output voltage swing, of the same RO.

Validation of RO output characteristics by SPICE simulation

The μ_{FE}'s for the discrete TFTs determined in the last section differ from what we have reported (6 – 12 cm^2 V^{-1} s^{-1} (5, 9)). We also fabricated control bottom-gate top-contact TFTs using thermally oxidized doped Si wafer, with a-IGZO layers deposited on top in the same batch as those for the ROs. The doped Si and thermal SiO$_2$ served as a gate electrode and a gate insulator, respectively. We obtained μ_{FE} ~9 cm^2 V^{-1} s^{-1} in saturation region of the control TFTs. Although the reason for the reduced μ_{FE}'s in TFTs with sputtered SiO$_2$ insulators is unclear, the possible explanations are: (i) enhanced carrier scattering at the interface between the sputtered insulator and the active layer, and/or (ii) increased interfacial states due to chemical reactivity of as-sputtered SiO$_2$ surface. However, we consider this gap in μ_{FE}'s could be reduced, since the sputtering conditions for SiO$_2$ insulator have not been optimized yet.

Now we consider to simulate operation of the RO with $L = 10$ (µm) and $\beta_R = 5$ supplied with $V_{dd} = + 15$ (V). Since the median of the output voltage swing is ~ +7 V as shown in Fig. 5, time-averaged gate-source voltage on drive and load TFTs in the RO can be regarded as +7 V and +8 V, respectively. Therefore, we need models that reproduce I_{ds}-V_{gs} characteristics of the actual TFTs in the vicinity of these voltages. In a level 1 NMOS model, the major fitting parameters are the transconductance parameter, KP, and the threshold voltage, VTO. However, there is no single set of (KP, VTO) that can reproduce the whole I_{ds}-V_{gs} characteristics for the measured range of V_{gs}. Figure 6 is the comparison between the measured and the simulated I_{ds}-V_{gs} characteristics for a discrete TFT with the same channel dimensions as drive TFTs ($L = 10$ (µm) and $W = 200$ (µm)). The solid line is a simulated characteristics in which KP and VTO was chosen so that it fits best at 'on' region (near V_{gs} ~ +15 (V)) of the measured characteristics. Clearly, the agreement is poor for the region V_{gs} < 10 (V). The dashed line is another simulation with a different set of KP and VTO; they were found by a linear fit to the measured $(I_{ds})^{1/2}$-V_{gs} plot at V_{gs} = +7 (V). The model agrees well with the measured I_{ds}-V_{gs} characteristics near V_{gs} = +7 (V) as intended. In a similar way, KP and VTO for the load TFTs were found by a fit at V_{gs} ~ +8 (V) to the measured characteristics (not shown).

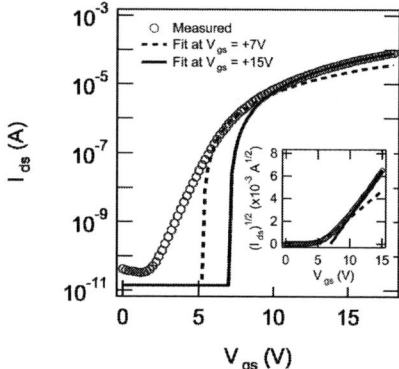

Figure 6. (Circles) measured and (lines) simulated I_{ds}-V_{gs} characteristics for a discrete TFT with L = 10 (μm) and W = 200 (μm). The parameters for simulations are: (solid line) KP = 6.6 x 10^{-8} (A V^{-2}), VTO = +7.0 (V); (dashed line) KP = 2.3 x 10^{-8} (A V^{-2}), VTO = +5.2 (V). The corresponding $(I_{ds})^{1/2}$-V_{gs} characteristics are shown in the inset.

We performed transient simulation of the RO output characteristics using these parameters. Compared to the measured results, we obtained similar voltage oscillation swing (simulated 2.8 V$_{p-p}$ vs measured 2.6 V$_{p-p}$) and no more than double operation frequency (simulated 19 kHz vs measured 9.6 kHz). Similar simulation was made using the same KP and VTO for the other ROs with different β_R's supplied with V_{dd} = + 15 (V). For all of the five ROs examined, the ratios of the measured values of propagation delay per stage to the simulated values were less than 2.5. In Fig. 7 are shown the measured and simulated output characteristics of a RO with L = 10 (μm) and β_R = 20.

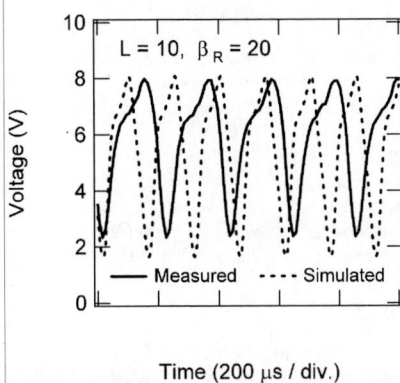

Time (200 μs / div.)

Figure 7. Measured (solid line) and simulated (dashed line) output characteristics of the RO with $L = 10$ (μm) and $\beta_R = 20$ supplied with $V_{dd} = + 15$ (V). The frequencies and the voltage swings are: (measured) 4.7 kHz, 5.6 V_{p-p}; (simulated) 6.6 kHz, 6.2 V_{p-p}, respectively.

The argument is limited within a level of 'order-of-magnitude' evaluation, since we used common KP and VTO values for different ROs, whose measured output characteristics have different median voltages. At this level, we consider the simulated results agree with the experimental ones.

Conclusion

We fabricated ring oscillators (ROs) composed of a-IGZO-channel bottom-gate top-contact TFTs. Since we employed no post-deposition annealing process to control carrier concentration in a-IGZO, our method presented here is promising for fabrication of integrated circuits on flexible polymer substrates.

We observed oscillation in ROs with a variety of channel lengths and channel widths, with minimum channel length down to 10 μm. A RO with channel lengths of 10 μm at the supply of external voltage of +18 V operated at 21.5 kHz, corresponding to propagation delay of 4.7 μs / stage.

We also attempted an 'order-of-magnitude' validity evaluation of the measured output characteristics of the ROs. We performed a SPICE simulation based on level 1 NMOS model that were tuned so they reproduce the measured I_{ds}-V_{gs} profiles near the operating points of the respective TFTs. The simulated output characteristics of the ROs by a simulation qualitatively matched the measured ones, and also the simulated propagation delays were approximately consistent with the experiment.

Acknowledgment

We thank Messrs. Tadahiko Hirai, Jun Sumioka and Kaoru Okamoto for valuable discussion on SPICE simulation.

References

1. C. -S. Yang, L. L. Smith, C. B. Arthur, and G. N. Parsons, *J. Vac. Sci. Technol. B*, **18**, 683 (2000).
2. P. G. Carey, P. M. Smith, S. D. Theiss, and P. Wickboldt, *J. Vac. Sci. Technol. A*, **17**, 1946 (2000).
3. G. H. Gelnick et al., *Nature Mater.*, **3**, 106 (2004).
4. H. Hosono, *J. Non-Cryst. Sol.*, **352**, 851 (2006).
5. H. Yabuta, M. Sano, K. Abe, T. Aiba, T. Den, H. Kumomi, K. Nomura, T. Kamiya, and H. Hosono, *Appl. Phys. Lett.*, in press.
6. R. E. Presley, D. Hong, H. Q. Chiang, C. M. Hung, R. L. Hoffman, and J. F. Wager, *Sol. Stat. Elec.*, **50**, 500 (2006).
7. H. Q. Chiang, J. F. Wager, and R. L. Hoffman, *Appl. Phys. Lett.*, **86**, 013503 (2005).
8. N. L. Dehuff, E. S. Kettenring, D. Hong, H. Q. Chiang, J. F. Wager, R. L. Hoffman, C. -H. Park, and D. A. Keszler, *J. Appl. Phys.*, **97**, 064505 (2005).
9. K. Nomura, H. Ohta, A. Takagi, T. Kamiya, M. Hirano, and H. Hosono, *Nature*, **432**, 488 (2004).
10. E. Fortunato, A. Pimentel, L. Pereira, A. Goncalves, G. Lavareda, H. A'guas, I. Ferreira, C. N. Carvalho, and R. Martins, *J. Non-Cryst. Sol.*, **338-340**, 806 (2004).
11. E. M. C. Fortunato, P. M. C. Barquinha, A. C. M. B. G. Pimentel, A. M. F. Goncalves, A. J. S. Marques, R. F. P. Martins, and L. M. N. Pereira, *Appl. Phys. Lett.*, **85**, 2541 (2004).
12. H. Ohta, K. Nomura, H. Hiramatsu, K. Ueda, T. Kamiya, M. Hirano, and H. Hosono, *Sol. Stat. Elec.*, **47**, 2261 (2003).
13. R. L. Hoffman, B. J. Norris, and J. F. Wager, *Appl. Phys. Lett.*, **82**, 733 (2003).
14. J. Nishii, F. M. Hossain, S. Takagi, T. Aita, K. Saikusa, Y. Ohmaki, I. Ohkubo, S. Kishimoto, A. Ohtomo, T. Fukumura, F. Matsukura, Y. Ohno, H. Koinuma, H. Ohno, and M. Kawasaki, *Jpn. J. Appl. Phys.*, **42**, L347 (2003).
15. H. Kumomi, N. Kaji, K. Abe, H. Yabuta, T. Iwasaki, M. Sano, and T. Den, K. Nomura, T. Kamiya, and H. Hosono, International TFT Conference '06 (ITC '06, Jan. 2006, Kitakyushu, Japan), 6.3.
16. C. Chang, M. Sano, M. Majima, H. Kumomi, K. Nomura, T. Kamiya, and H. Hosono, The 53rd Spring Meeting of The Japan Society of Applied Physics and Related Societies (Mar. 2006, Tokyo, Japan), 22a-P1-44.
17. F. Eder, H. Klauk, M. Halik, U. Zschieschang, G. Schmid, and C. Dehm, *Appl. Phys. Lett.*, **84**, 2673 (2004).
18. W. Fix, A. Ullmann, J. Ficker, and W. Clemens, *Appl. Phys. Lett.*, **81**, 1735 (2002).
19. B. K. Crone, A. Dodabalapur, R. Sarpeshkar, R. W. Filas, Y.-Y. Lin, Z. Bao, J. H. O'Neill, W. Li, and H. E. Katz, *J. Appl. Phys.*, **89**, 5125 (2001).
20. B^2 Spice Version 5, Beige Bag Software.

The Abnormal Degradation Behavior of ZnO TFT under Bias Stress

Chi-Sun Hwang, Sang Hee Ko Park, Sung Mook Chung, Jeong-Ik Lee, Yong Suk Yang, Lee-Mi Do and Hye Yong Chu

IT Convergence & Components Lab., ETRI
161 Gajeong-Dong, Yuseong-Gu, Daejeon, 305-350, KOREA

ZnO TFT was fabricated using ALD deposited materials. The temperature dependence of transfer characteristics and the degradation behavior under gate bias stress was observed. The shift of Vth under gate bias stress was explained through electron trapping and H movement.

INTRODUCTION

Transparent oxide TFT attracts much interest nowadays, because of its possible application to flat panel display devices[1]. The degradation behavior under gate bias stress is critical parameter for the applications such as AMOLED panel, because in such application the driving TFT would be stayed turned-on throughout each frame. It is generally known that a-Si:H TFT is degraded under gate bias stress so that the threshold voltage of the stressed TFT is shifted to the direction of gate bias. An additional compensation circuits are required to use a-Si:H TFT array as the backplane of AMOLED.
We made a transparent ZnO TFT on glass substrate using ALD grown ZnO layer. The change of drain currents under high gate bias were observed at various temperature. The possible origin of abnormal degradation behavior was discussed.

EXPERIMENTS

The ZnO TFT was fabricated in bottom gate and bottom contact structure with ALD grown ZnO as active layer[2-3] . Figure 1 shows the cross-section of ZnO TFT. Commercial ITO glass was used as substrate. The ITO layer was patterned as gate electrode.

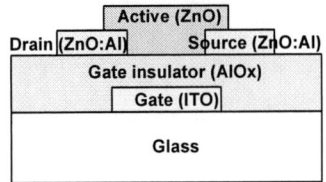

Figure 1. The Cross-section of ZnO TFT.

Al_2O_3 was grown using ALD at 250°C and used as gate insulator with thickness of 170nm. ALD grown Al_2O_3 has some merits as gate insulator, such as very low leakage current at low deposition temperature, excellent smoothness of the surface and rather high dielectric constant. Contact pattern on gate electrode was etched using wet etching method.

ZnO:Al layer deposited using ALD at 180°C is used as source/drain electrode. The resistivity of ZnO:Al layer was high compared to the common transparent conducting oxide such ITO but as an application for ZnO active layer ZnO:Al was suitable as a low barrier junction. The source/drain pattern was fabricated using wet etching method. Thick ZnO:Al layer (300nm) was used to reduce the sheet resistance.

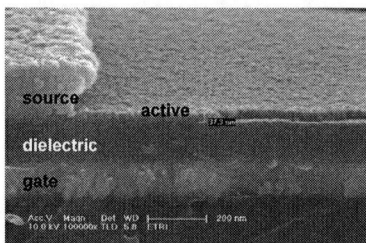

Figure 2. SEM view image of the cross-section of ZnO TFT near souce/drain edge.

Undoped ZnO layer was deposited with ALD at 100°C. The thickness of ZnO layer gave influence on the leakage current of the TFTs. We reduce the ZnO layer up to 70nm to decrease the leakage current. Excellent step coverage of ALD deposited ZnO made it possible to overcome the very steep edge of source/drain layer (fig. 2). The ZnO layer was patterned using wet etch method.

The fabricated ZnO TFT was transferred to vacuum chamber with base pressure of $5x10^{-4}$Torr and constant bias stress was applied and measured the characteristics at various temperature (from RT to 170°C) with Agilent 4155C semiconductor parameter analyzer.

RESULTS

The transfer characteristics of the ZnO TFT were measured at various temperature. The measured drain current showed thermally activated behavior up to 170°C (fig. 3). The maximum value of activation energy of the drain current was ~1.4eV at -8V, the activation energy was decreased as increasing the gate bias, which showed the movement of Fermi level according to the gate bias (fig. 4).

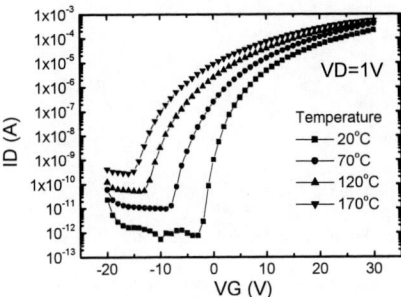

Figure 3. The transfer characteristics of ZnO TFT at various temperature.

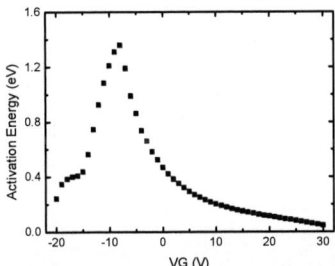

Figure 4. The change of activation energy of drain current with gate bias in ZnO TFT.

Figure 5 showed the change of drain current under constant gate bias 20V with 40V drain bias at room temperature. At first the drain current decreased rapidly and after ~10^3 seconds increased slowly. After ~2hrs the drain current was increased more than initial value. As shown in fig. 6, at elevated temperatures the behavior was similar to at room temperature case.

The gate bias stress caused degradation and improvement of drain current. At low temperature the drain current increased more than initial value, but at high temperature the drain current could not recovered to initial value.

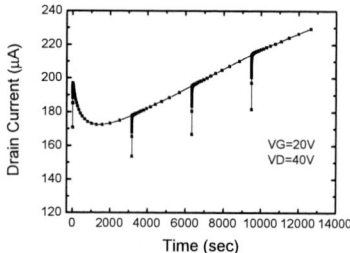

Figure 5. The change of drain current at RT during gate bias stress.

The change of drain current came from the shift of Vth in ZnO TFT. In general, the shift of Vth in TFT was caused by charge trapping into gate insulator or change at density of states in the energy gap of active layer.

The increase of Vth at early stage of bias stress could be explained as electron trapping into gate insulator. The electron trapping would be done through energy barrier, therefore the amount of decrease in drain current increased with elevated temperature. (The fitted activation energy of electron trapping was 0.056eV).

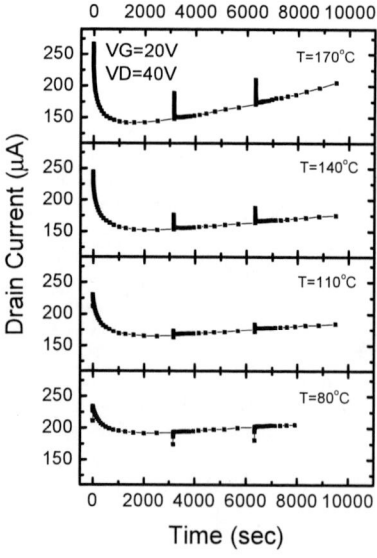

Figure 6. The change of drain current at various temperature during gate bias stress.

It is believed that the H in ZnO acts as dopants[4]. The movement of H in ZnO was studied extensively. In some theoretical studies, ZnO was diffused with potential barrier of 0.5eV[5]. The distribution of H could be affected by gate bias stress, because, due to the charged status of H, the chemical potential of charged H is different according to the position of Fermi level.

The decrease of Vth could be explained through the movement of H. The gate bias stress caused the change of the chemical potential of H and electron trapping into gate dielectric layer, which made redistribution of H throughout the active layer and gate dielectric layer. The interlayer diffusion of H from gate dielectric layer into active layer could be done. The increase of H concentration in active layer gave increase of the doping density in active layer, which would decrease Vth of the ZnO TFT.

In general diffusion process is slower process than charge trapping process. Therefore possible explanation of abnormal degradation behavior of ZnO TFT was following : During gate bias stress, at first the Vth increased by electron trapping process, and then the Vth decrease by diffusion process of H.

The characteristics of ZnO TFT largely depend on the deposition methods of ZnO layer. The study about the dependence of degradation behavior on the deposition methods is under progress. The results will be presented at conference.

CONCLUSIONS

The degradation behavior of ZnO TFT was investigated. During the gate bias stress the drain current was decreased rapidly at first, but increased after 10^3 seconds. The temperature affects amount of annealing behavior. The decrease and increase of drain current came from the shift of Vth, which caused from the electron trapping and H movement.

Acknowledgments

The Korea ministry of Information and Communications financially supported the accomplishment of present work.

References

1. J. F. Wagner, *Science* **300**, 1245 (2003).
2. Sang-Hee Ko Park, Chi-Sun Hwang, Ho-Sang Kwack, Seung-Youl Kang, Jin-Hong Lee, and Hye Yong Chu, IMID'05, 1564 (2005).
3. Chi-Sun Hwang, Sang-Hee Ko Park, and Hye Yong Chu, IDW'05, 1149 (2005).
4. C. G. Van de Walle, *Phys. Rev. Lett.* **85**, 1012 (2000)
5. M. G. Wardle, J. P. Goss, and P. R. Briddon, *Phys. Rev. Lett.* **96**, 205504 (2006)

ECS Transactions, 3 (8) 307-312 (2006)
10.1149/1.2356368, copyright The Electrochemical Society

All-Solution-Processed Organic Thin Film Transistors Fabricated by Non-Piezoelectric Inkjet Printing

I. Takasu, K. Sugi, Y. Nomura, H. Nakao, K. Mori, I. Amemiya, and S. Uchikoga

Advanced Electron Devices Laboratory, Corporate Research & Development Center,
Toshiba Corporation
1, Komukai Toshiba-cho, Saiwai-ku, Kawasaki, Kanagawa, 212-8582, Japan

Non-piezoelectric inkjet printing methods, namely, ultrasonic inkjet printing and electrostatic inkjet printing have been used for fabricating an all-solution-processed organic thin-film transistor. Silver nanoparticle ink is patterned to form source/drain electrodes using the electrostatic inkjet method and high-viscosity poly(3,4-ethylenedioxythiophene) doped poly(styrene sulfonate) (PEDOT:PSS) was patterned as a gate electrode using ultrasonic inkjet printing. Polymer insulator and polymer semiconductor are spin-coated as a gate dielectric layer and a semiconductor layer, respectively. This is the first fabrication of all-solution-processed organic transistor using non-piezoelectric inkjet printing.

Introduction

In recent years, inkjet printing has attracted significant interest in view of its potential for application as a large-area, low-cost, and high-throughput fabrication technology[1,2]. Organic thin-film transistor (OTFT) is a typical application for the inkjet fabrication since all the components of OTFT are solution-processable materials, such as soluble polymer semiconductors. The solution-processable OTFTs have been typically fabricated by piezoelectric inkjet used to form source, drain and gate electrodes. Though, compared with thermal inkjet, piezoelectric inkjet has an advantage in that it employs a non-heating process, which causes no thermal degradation of ink materials such as organic materials and nanoparticle inks, the conventional piezoelectric inkjet printing has a problem, namely, difficulty in expelling high-viscosity inks with high stability.

We have examined non-piezoelectric inkjet printing methods: electrostatic inkjet printing [3] and ultrasonic inkjet printing [4,5]. In the electrostatic inkjet printing, ink droplets, which generally contain charged colloids, are ejected by electrostatic forces between the inkjethead and the substrates. The electrostatic inkjet printing is characterized by its potential for fine patterning since ink-droplet dispersion by air resistance is suppressed by electrostatic forces. In the ultrasonic inkjet printing, ink droplets are ejected by focused ultrasonic beams at the ink surface. The ultrasonic inkjet printing has two principal advantages: nozzle-less structure, which leads to less clogging, and small restriction for ink properties, such as high-viscosity.

The advantages of printing high-viscosity inks for device fabrication are as follows: less ink-spread, no need to stack dots, less infiltration into polymer layers, etc. This work reports on all-solution-processed TFT fabrication using non-piezoelectric inkjet printing. In particular, non-diluted PEDOT: PSS (poly-ethylenedioxythiophene: poly(styrenesulfonic acid) aqueous dispersion (viscosity: 100-350 mPa·s) was used as

307

gate electrode material. In general, commercially available PEDOT:PSS has high viscosity, whereas, using piezoelectric inkjet, it is difficult to achieve reliable printing without dilution. We succeeded in printing non-diluted PEDOT:PSS using an ultrasonic inkjet printing system.

Materials and methods

Silver nanoparticle ink (Ag1TeH from ULVAC Materials, Inc.) and PEDOT:PSS aqueous suspension (Baytron P HCV4 from Starck Co.) were used as materials for the source/drain electrodes and the gate electrodes of the OFET, respectively. As a semiconductor material poly(2,5-bis(3-decylthiophen-2-yl)thieno[2,3-b]thiophene (pBTCT from Merck Chemicals Ltd.) and as a gate insulator material a polymer insulator solution (lisicon D001 from Merck Chemicals Ltd.) were spin coated from solution. Printing of the silver nanoparticle ink and the PEDOT:PSS aqueous dispersion was performed on an electrostatic inkjet system and an ultrasonic inkjet printing system, respectively. Figure 1 (a) shows fundamental setup of the electrostatic system, where a direct current voltage is applied between the nozzle and counter electrode (the target sample). Figure 1(b) shows a fundamental principle of the ultrasonic inkjet printhead, where ultrasonic beam generated by transducers is focused on the free-liquid surface by the acoustic lens and a droplet is expelled. Droplet ejection was observed using high-speed camera (NAC image Technology, Inc., Ultranac FS501). FET characterization was carried out using a Hewlett Packard HP4145B semiconductor parameter analyzer.

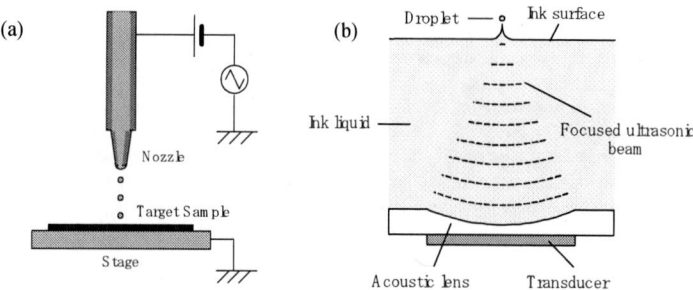

Figure 1. Schematics of (a) electrostatic inkjet printing and (b) ultrasonic inkjet printhead.

Fabrication

The top-gate TFT structure illustrated in Fig. 2. was fabricated using the two types of non-piezoelectric inkjet printing methods. The source/drain electrodes are printed on the glass substrate using the electrostatic inkjet printing with silver nanoparticle ink and cured at 230 °C for 20 min under ambient atmosphere. After the silver source-drain electrodes deposition, the substrate was treated with hexamethyldisilazane (HMDS) and a polymeric semiconductor pBTCT film was deposited over the substrate via spin-coating of xylene solution and cured at 120 °C for 15 min under nitrogen atmosphere. A gate insulator, liscon D001, was then spin-coated and cured at 125 °C for 30 min under nitrogen atmosphere. As a gate electrode, non-diluted PEDOT: PSS (poly-ethylenedioxythiophene: poly(styrenesulfonic acid) aqueous dispersion (viscosity: 100-350 mPa·s) was printed over the channel area using the ultrasonic inkjet printing system.

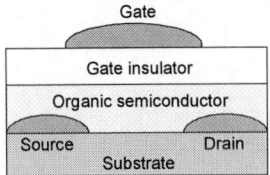

Figure 2.   Schematic cross-section of the top-gate thin film transistor.

Results and discussion

Silver wire and electrode fabrication using electrostatic inkjet method

Printing of silver nanoparticle ink was performed on the electrostatic inkjet system. Electrostatic inkjet printing has an advantage over piezoelectric inkjet and thermal inkjet printing, namely, its ability to eject very fine droplets.   Though we used capillary inkjet head with internal diameter of 40 μm in this experiment, by fabricating a fine nozzle structure, much finer droplets of several microns in diameter can be ejected.   Figure 3 shows AFM images of the printed silver wire. The line width is about 75 μm and the height is about 233 nm.   The sectional view of the wire shows almost a flat shape with no bulge at the edges, which means the "coffee-stain effect" [6] is not observed.   This flat shape may be partially due to the high solid concentration of the silver ink.

(a)   (b)

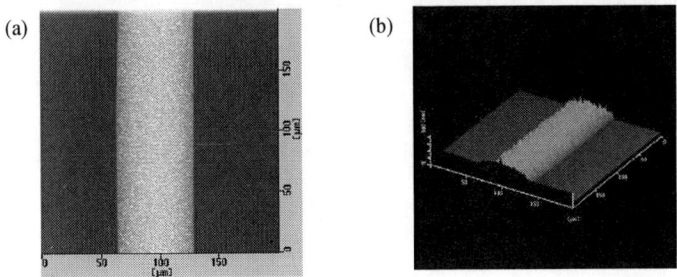

Figure 3. (a) (b)   AFM images of the printed silver wire using the electrostatic inkjet printing system.

Printability of high-viscosity PEDOT:PSS using ultrasonic inkjet method

PEDOT: PSS dispersion is a widely used organic conductor solution. PEDOT: PSS dispersion is suitable for gate electrode material of top-gate OTFT owing to its low cure temperature, which avoids degradation of organic material components. Printing of PEDOT:PSS water dispersion was performed on the ultrasonic inkjet printing system. Ultrasonic inkjet printing has an advantage in terms of printability of high-viscosity droplets. For conventional inkjet printing methods (piezoelectric-type and thermal-type), the viscosity of less than several mPa·s is desirable for stable printing. The time evolution of non-diluted PEDOT:PSS (viscosity: 100-350 mPa·s) droplet formation is shown in Fig. 4 for an RF frequency of 25 MHz. A droplet size is approximately 70 μm. In general, an inkjetted main droplet is followed by a tail, which develops into smaller droplets following the main droplet at lower speed. When the tail is very long, it changes to many small droplets, which uncontrollably diffuse and cause defects. In Fig. 4, the droplet is a sphere with no tail, which is characteristic of inkjet printing using focused ultrasonic waves.

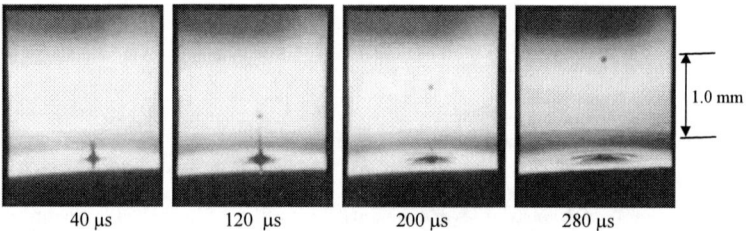

| 40 μs | 120 μs | 200 μs | 280 μs |

Figure 4. Stroboscopic images of PEDOT:PSS droplets formation at an RF frequency of 25 MHz at the time (shown at the bottom of each image) after applying voltage to the transducer.

Fabrication and electrical properties of all-solution-processed OTFT

The polymer semiconductor material selected was a semiconducting copolymer, pBTCT (Fig. 5) [7]. One of the problems of conventional polymer semiconductor is stability against oxidative doping. pBTCT has superior stability due to the introduction of a central cross-conjugated double-bond structure into the polythiophene structure. In order to suppress the penetration of PEDOT: PSS into the gate insulating layer, a cross-linking polymer insulator (lisicon D001) is used as a gate insulator material.

Figure 5. Chemical structure of poly(2,5-bis(3-alkylthiophen-2-yl) thieno[2,3-b]thiophene) (pBTCT). R = $C_{12}H_{25}$

Following the process described above, a top-gate OTFT was fabricated and TFT characteristic (p-type) was obtained. Fig. 6 shows the I_{ds}-V_{ds} plots and I_{ds}-V_g plots obtained by the printed OTFT in air. The transfer curves show typical characteristics for FET, namely, gate voltage dependence and saturation region at high drain voltages. The small offset between the curves in the low drain voltage region may be ascribed to gate leakage [8]. This offset shows a tendency to become more prominent as the semiconductor layer becomes thicker. The W/L for the device was ~10 (W~1.0 mm, L ~100 μm), the extracted field-effect mobility from the saturated regime was 2.5×10^{-3} cm^2/Vs, on/off drain current ratio was 1.8×10^2, and threshold voltage was 3V. In the bottom-contact TFT structure composed of thermally oxidized Si substrates and evaporated gold source/drain electrodes, pBTCT shows the mobility of higher than 0.01 cm^2/Vs and on-off drain current ratio of around 10^5 [7].

The difference in the performance of the TFTs may be due to such as differences of film morphologies and crystallinity of the semiconductor layers, and contact resistances between the semiconductors and the electrodes. It has been pointed out that as gate dielectrics, low-k polymer dielectrics are favorable since randomly disordered polar functional groups in polymer dielectrics induce structural disorder in the organic semiconductor, resulting in lower carrier motilities [9]. The OTFT shows susceptibility to the environment; the on-off drain current ratio decrease at reduced pressure. This phenomenon is also observed in top-gate polymer TFTs having other polymer gate dielectrics [10]. The use of low-k polymer dielectrics may also have an advantage of less susceptibility to the atmospheric conditions such as humidity.

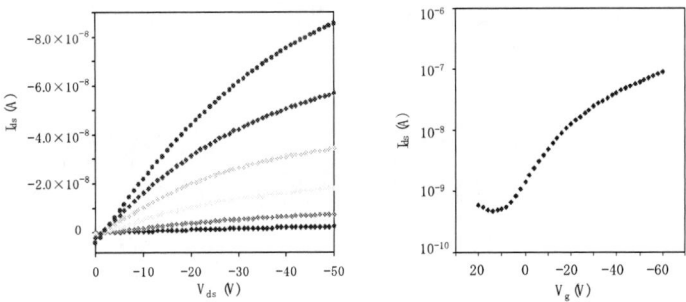

Figure 6. (a) Output characteristics of printed OTFT (gate voltage is varied from 0 V to -50 V in -10 V steps). (b) Transfer characteristics of printed OFET (V_{ds}=-60V).

Conclusion

In summary, by using non-piezoelectric printing methods, electrostatic inkjet printing and ultrasonic inkjet printing, an all-solution-processed OTFT was fabricated. High-viscosity PEDOT: PSS water dispersion was printed as gate electrodes of the OTFT by an ultrasonic inkjet printing system. To the best of our knowledge, this is the first fabrication of all-solution-processed OTFT using non-piezoelectric inkjet printing.

References

1. R. Parashkov, E. Becker, T. Riedl, H.-H. Johannes, W. Kowalsky, Proceeding. IEEE **93**(7), 1321 (2005).
2. T. Kawase, T. Shimoda, C. Newsome, H. Sirringhaus, and R. H. Friend., *Thin Solid Films,* **438-439,** 279 (2003).
3. T. Murakami, S. Hirahara, H. Nagato, and Y. Nomura, Proceeding. NIP 14: International Conference on Digital Printing Technologies, 35, (1998).
4. S. A. Elrod, B. Hadimioglu, B. T. Khuri-Yakub, E. G. Rawson, E. Richley, C. F. Quate, N. N. Mansour, and T. S. Lundgren, *J. Appl. Phys.,* 65, 3441 (1989).
5. I. Amemiya, H. Yagi, K. Mori, N. Yamamoto, S. Saitoh, C. Tanuma, and S. Hirahara, Proc. NIP 13: International Conference on Digital Printing Technologies, 698, (1997).
6. R. D. Deegan, O. Bakajin, T. F. Dupont, G. Huber, S. R. Nagel, and T. A. Witten, *Nature,* **389,** 827 (1997).
7. M. Heeney, C. Bailay, K. Genevicius, M. Shkunov, D. Sparrowe, S. Tierney, and I. McCulloch, *J. Am. Chem. Soc.,* **127,** 1078 (2005).
8. H. Jia, G. K. Pant, E. K. Gross, R. M. Wallace, and B. E. Gnade, *Org. Electron.,* **7,** 16 (2006).
9. J. Veres, S. D. Ogier, S. W. Leeming, D. C. Cupertino, and S. M. Khaffaf, *Adv. Funct. Mater.,* **13,** 199 (2003).
10. T. G. Backlund, R. Osterbacka, H. Stubb, J. Bobacka, and A. Ivaska, *J. Appl. Phys.,* **98,** 074504 (2005).

ECS Transactions, 3 (8) 313-321 (2006)
10.1149/1.2356369, copyright The Electrochemical Society

Thin-Film Transistors in Disordered Semiconductors for High Performance Macroelectronic Circuits

F. Balon and J. M. Shannon

Advanced Technology Institute, SEPS, University of Surrey, Guildford, GU2 7XH, UK

In general the performance of circuits in macroelectronic systems depends on the quality of semiconductor material used to make the devices. In this paper it is suggested that by using a different device structure based on the source-gated concept we can overcome intrinsic material problems and achieve much better device speed and stability in all disordered and poor quality semiconductors. Experimental and modeling results, on the most commonly used disordered material, amorphous silicon, show that high performance thin-film devices and circuits based on the SGT concept are a real possibility.

Introduction

There is a lot of interest in electronic circuits on flexible, inexpensive large area substrates such as polymers, paper, thin glass and even textiles (1). Applications include cheap tags, intelligent clothes, electronic paper, wall coverings, flexible displays and imaging systems. To preserve the integrity and mechanical properties of these substrates, low processing temperatures are essential and inevitably there is a compromise in the quality of the semiconductor in terms of its structure and transport properties.

A range of different technologies for low-cost low-temperature deposition of semiconductors have been explored including solution processing, spin coating, dip coating, screen printing, ink printing, evaporation and low temperature plasma enhanced chemical vapour deposition (2). Semiconductors range from inorganic such as hydrogenated amorphous silicon, amorphous oxides (3), metal chalcogenides (4), organic polymers and plastics and organic-inorganic hybrid semiconductiong materials (5). The low temperatures used in these technologies leads to amorphous or microcrystalline semicoductors containing a high concentration of defects.

The challenge is to engineer a transistor such that it provides high performance even though the quality of the semiconductor is poor in terms of carrier mobility and electrical stability. Both these properties severely limit the range of applications since low mobility restricts the speed of the transistor while instability prevents circuits from being designed for analog systems.

Thin-Film Transistors and Device Requirements

The elementary device used in all electronic systems is a transistor and to date the only thin-film transistor used in macroelectronics is the field-effect transistor (FET) (6). Due to the versatility of thin-film technology the transistor design can vary, but the basic principle of operation remains the same. The voltage on the gate electrode modulates the conductance of the channel between two ohmic contacts (source and drain) and current

313

saturation occurs when the semiconductor at the drain end is depleted of charge carriers. This device has characteristics that are sufficiently good for less demanding applications but in poor quality materials strong channel accumulation of excess carriers and low mobilities lead to instability and slow speed of response.

The role of excess carriers falls from the Stutzmann's work (7) which concludes that defect formation under charge injection is a general feature of disordered semiconductors and postulates that states are created in the disordered network to minimize excess charge carriers in strained or weak bonds. Low carrier mobilities are caused by the carrier transport properties (hopping or tunneling via localized states) in disordered semiconductors. However, carrier velocities can be increased by operating with high internal fields across the short distances between source and drain.

Based on these considerations we conclude that to make stable transistors with a worthwhile frequency response and good device characteristics one has to (i) reduce the excess carrier concentration to reduce defect formation and (ii) operate the device with high internal fields across the active part of a transistor with small dimensions.

Recently a different type of thin-film transistor was introduced, named a source-gated transistor (SGT) (8), in which the principle of operation differs from that of the FET. A major difference between them is that a source comprises a reverse biased potential barrier (Fig.1). The barrier controls the on-state current and saturation is governed by the electrostatics at the source barrier. Interestingly here the "pinch-off" is at the source edge rather than the drain end of a channel. The depletion region is clearly seen in Fig.2 where Atlas simulations showing 2D distributions of the carrier concentration in the SGT for different drain voltages are shown. Furthermore the geometry of the SGT leads to a much smaller susceptibility to short channel effects (9) and higher output impedance because the source barrier is screened from the drain field by the gate (10).

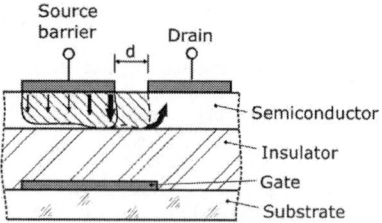

Figure 1. Schematic diagram of a SGT with a depletion region and current flow. The solid line represents the depletion envelope at saturation whilst the dashed line is for a large drain bias. d is the distance between the edge of the source barrier and the drain contact.

The source barrier used in all our studies was a metal-semiconductor Schottky barrier (11). Current transport across the reverse polarized Schottky barrier is based on thermionic and thermionic-field emission (12). Therefore as one might expect this carrier transport is thermally activated which is a disadvantage of the SGT device having a Schottky barrier compared with a FET. However this temperature dependence is low for small barrier heights and higher currents. In Fig.3 the measurement of activation energy is shown. The transistor characteristics were measured from room temperature up to 80°C with the gate voltage ranging from 4V to 24V with a step of 2V. The SGT device had 100nm of a-Si:H and 150nm of SiN, (d = 1.5 micron, W = 2 microns) and a phosphorus barrier modification implant dose of 3×10^{14} cm^{-2}.

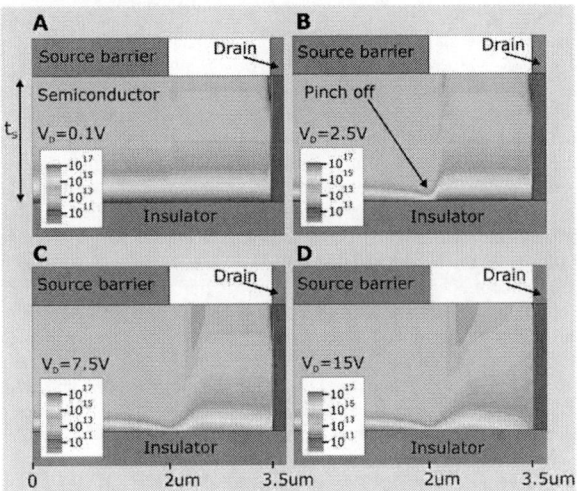

Figure 2. Atlas simulations showing 2D distribution of carrier concentration in the SGT for different drain voltages (a) 0.1V, (b) 2.5V, (c) 7.5V, (d) 15V and V_G=10V. The thickness of the semiconductor t_S (a-Si:H) and the insulator (SiN) is 100nm and 300nm respectively.

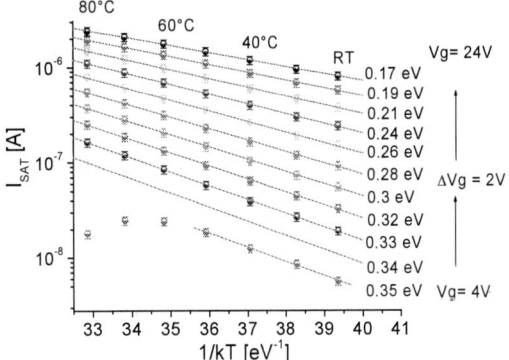

Figure 3. Activation energy measurements on a SGT device showing the effect of the gate voltage on the effective barrier height of the Schottky source barrier.

The activation energy was estimated from the linear fit of $log(I_{SAT})$ plotted as a function of $1/kT$. It is seen that the activation energy decreases with increasing gate voltage as the barrier is pulled down. Also the fact that the source barrier can be pulled down (in the on-state) to 0.17eV with sufficiently high fields ($V_G = 24$V) is interesting because it is within the factor of 2 of the activation energy measured on FET's (0.08eV) in strong accumulation. Alternatively the temperature dependence could be minimised using a different type of barrier (e.g. barriers where the current transport is governed by field-emission processes).

The SGT and FET devices were prepared using thin-film technology and a 4-mask process. Ion implantation was used to modify the effective barrier height of the chromium Schottky contacts (13). For this paper a-Si:H was chosen as a good example of disordered and unstable semiconductor material. Secondly this material is well characterized and all the parameters required for modeling of device behavior are well known (14). The simulations were performed using an improved TCAD SILVACO Atlas (v.8.10.0.R). In comparison with previous results where an effective barrier height was used with low internal fields across the barrier and device (15), here a different model (UST - Universal Schottky Tunneling model (16)) in conjunction with thin highly doped surface layers was used. Therefore, in this paper, the reference barrier height is higher (0.75eV) and internal fields resulting from the doped layer and applied voltage are higher as well. The concentration of carriers in the 10nm doped layer was $1 \times 10^{19} \text{cm}^{-3}$ for all simulations shown here. A modeling parameter D.TUNNEL (specifies the maximum tunneling distance) of 10nm was found to give the best fit between the Schottky barrier field dependence and the experimental results reported in Ref. 17.

Device Stability

Stressing experiments were undertaken at voltages that would be used in typical circuit applications (<15V). The FET and SGT devices were made using the same deposition conditions for the insulator (SiN, 300nm) and a-Si:H (100nm). All experiments were carried out at elevated temperature of 30°C and under continuous DC stressing in order to accelerate the degradation of devices. The majority of experiments involved measurements of the device transfer characteristics over a period of 24 hours of continuous stressing.

One may argue that SGT transistors are more stable than FET's due to a smaller on-current through the device caused by the source barrier. Therefore all results are plotted as a function of the stressing current measured initially in the stressing experiment. Then by varying the stressing conditions for different SGT and FET devices we were able to properly compare these devices. This comparison is plotted in Fig.4 for SGT devices biased $V_{SAT} < V_D < (V_G - V_{TH})$ and $V_D > V_G - V_{TH}$. The FET devices were also biased with $V_D > V_G - V_{TH}$.

It is seen that SGT devices biased with $V_D > V_{SAT}$ are much more stable than the FET for the same value of the stressing current. Also it can be said that for the same device stability the SGT is able to operate at a much higher current than the FET. When the device is biased with $V_D > V_G - V_{TH}$ then the parasitic FET (9) in the SGT is partly depleted, it pinches off at the drain as well as the source and the stability is even better with the change of the current less than 2%.

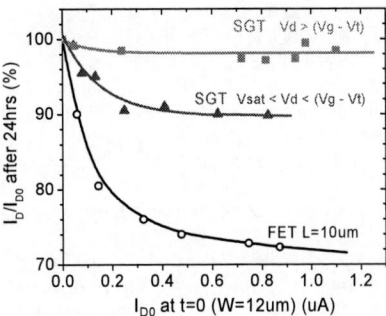

Figure 4. The current through a SGT and a FET in a-Si:H after DC stressing for 24 hours at 30°C. The gate and drain voltage was keep constant during stressing at the value needed to give the current I_{D0} at time (t) = 0. Source-drain separation (d) for curve (▲) SGT ranged between 2 and 4 microns. The SGT devices in curve (■) SGT had $d \sim 1.5$um and were stressed with high drain voltages.

Next we stressed the same SGT device ($W = 2.5$um and $d = 2.2$um) for a much longer period of time (i.e. 5 days) V_D was changed each day (apart from phase 2) while V_G was kept at a constant value (15V). In this case a higher V_G than V_D was chosen to enhance the degradation process (maximum value of $V_D = 12$V). The result is seen in Fig.5.

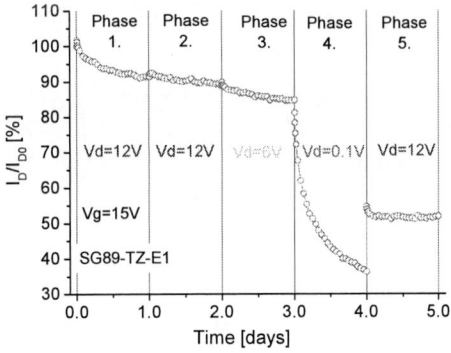

Figure 5. Degradation of the on-current over 5 days for the same SGT with $d = 2.2$um and different V_D.

In the 1st phase ($V_D = 12$V) the on-current has dropped by 7.6% to 92.4% of its initial value (I_{D0}). The same stressing was used also for the 2nd phase and the on-current dropped another 2.7% to 89.7% of I_{D0}. The V_D was changed in the 3rd phase to 6V and the 4.9% change in current was recorded giving subtotal drop to 84.8% of I_{D0}. For the 4th phase V_D was decreased further to 0.1V resulting in rapid decay of current down to 36.3% of I_{D0}. Between the 4th and 5th phase there was a one-day break in the measurement. The device was slowly cooled down to room temperature and stored in an airtight container.

Then the 5th phase was executed with $V_D = 12$V. The relative on-current dropped during the 5th phase was only 2.7% which is exactly the same as during the 2nd phase but what is more interesting is a recovery of the current between the 4th and 5th phase. The recovery was from a value of 36.3% to 54.6% of I_{D0} presumably due to some annealing of dangling bond defects during the one day interval. Fig.5 shows that the rate of degradation depends on the value of drain voltage with higher voltages giving less degradation. This is related to change in the excess carrier concentration as shown below.

Sub-micrometer Device Characteristics

Based on the fact that SGT devices are much stable and they are more immune to short channel effects we can progress further with more advanced structures. The simulated output characteristic of a SGT with thinner layer depositions and sub-micron source-drain separation is shown in Fig.6. This device has a source length of 1um and $d = 250$nm. It is seen that the saturation voltage remains low. The output impedance is very much better than an equivalent FET (19).

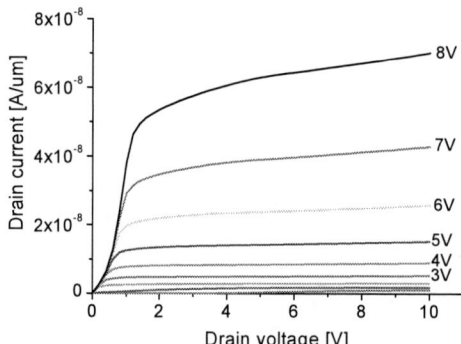

Figure 6. Simulated SGT device output characteristics $d = 250$nm. The thickness of the semiconductor (a-Si:H) and the insulator (SiN) is 40nm and 120nm respectively.

The corresponding carrier distribution and electric field along the semiconductor-insulator interface for the SGT device shown in Fig.6 is plotted in Fig.7. Looking at Fig.7a one can better understand the relation between the excess carrier distribution and improved SGT stability which coincides with the first requirement mentioned above. The lower the excess carrier concentration the better the stability of the device because a rise in the carrier concentration gives rise in defect formation and carrier trapping. It is seen that high drain voltages give low carrier concentrations across the region d. In contrast, the source region of a FET is heavily accumulated thus rapid degradation occurs.

Furthermore, concurrent with a low carrier concentration there is a high electric field (Fig.7b). Starting with the low drain voltage (2.5V) the field peaks underneath the source edge. However for higher voltages (>7.5V) a high field occurs across the whole of d which sweeps carriers towards the drain electrode. This results in higher carrier velocity and the possibility of a faster operating frequency for the SGT whilst retaining good output characteristics.

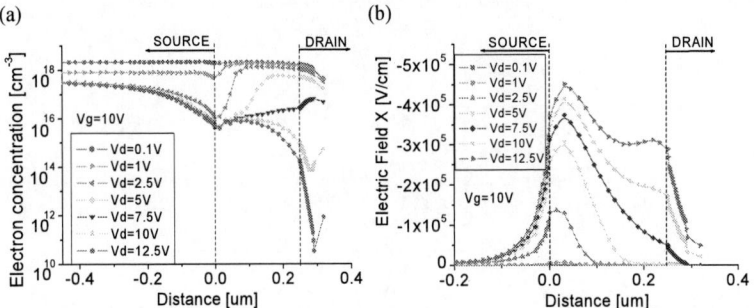

Figure 7. (a) The carrier concentration and (b) internal electric field in x-direction along the semiconductor/insulator interface (dimensions as in Fig.6).

Optimization of Speed and Stability

Fig.7 clearly shows that when the separation d between source and drain is small then high electric fields and low carrier concentration can be obtained throughout the parasitic FET for low drain voltages. Fig.8 shows that as expected the electric field increases and the carrier concentration decreases as d gets smaller. We therefore expect smaller devices to be more stable at a given V_D. For $d = 0.25$um the minimum electric field is $>3\times10^5$ V/cm. Therefore even for a semiconductor with a mobility of ~0.01cm^2/V.sec the transit time limited cut-off frequency f_T is >20MHz.

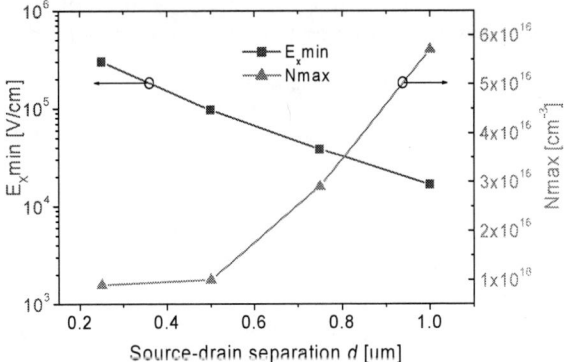

Figure 8. The lowest electric field ($E_X$min) across the region d and corresponding maximum carrier concentration (N_{MAX}) along the semiconductor/insulator interface plotted as a function of d at the same current ($V_D=12.5$V and $V_G=10$V).

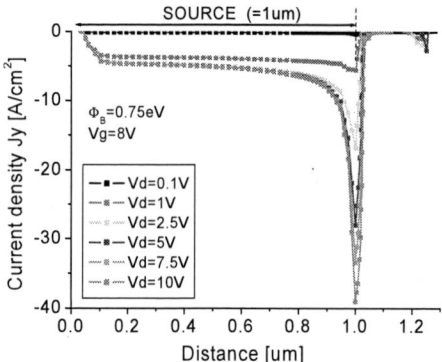

Figure 9. Current density distribution normal to the source barrier showing a peak in the current density at the edge facing the drain (dimensions as in Fig.6).

More precisely of course, we have to include the transit time of carriers across the source region. Fortunately it turns out that a large proportion of the current crosses the source barrier near the edge of the barrier facing the drain (Fig.9). These carriers contribute very small additional transit times because they are swept towards the drain by the high internal field at the edge of the source.

In practice, the frequency response is limited not by transit time but by the RC charging time of the source (18). For the device shown in Fig.6 with a source length of 1 micron the f_T predicted by ATLAS is ~30MHz (μ ~0.4cm^2/V.sec).

Conclusions

The performance of macroelectronic circuits and systems is dependent on the quality of the semiconductor. Disordered semiconductors tend to be unstable and have low carrier mobility and therefore poor frequency response. The stability is related to the excess carrier concentration during operation while frequency response can be improved by operating with high internal fields. Both these parameters can be improved using devices based on the SGT concept rather than conventional thin-film FET's. It is shown that as the SGT gets smaller then internal fields increase and carrier concentrations reduce leading to the possibility a good frequency response whilst maintaining good stability and output characteristics. All these features make the SGT well suited to the realization of high performance macroelectronic circuits using disordered and poor quality semiconductors.

Acknowledgments

This work was supported by the Engineering and Physical Sciences Research Council (EPSRC), Swindon, U.K. The transistors were made at Philips Research Laboratories, Redhill, Surrey, U.K.

References

1. J. B. Lee and V. Subramanian, *IEEE Trans. Electron. Dev.*, **52**, 269 (2005).
2. S. R. Forrest, *Nature*, **428**, 911 (2004).
3 K. Nomura, H. Ohta, A. Takagi, T. Kamiya, M. Hirano and H. Hosono, *Nature*, **432**, 488 (2004).
4. D. B. Mitzi, L. L. Kosbar, C. E. Murray, M. Copel and A. Afzali, *Nature,* **428**, 299 (2004).
5. C. R. Kagan, D. B. Mitzi and C. D. Dimitrakopoulos, *Science*, **286**, 945 (1999).
6. C. R. Kagan, P. Andry, Eds., *Thin-Film Transistors,* Marcel Dekker, New York (2003).
7. M. Stutzmann, in *Amorphous and Microcrystalline Semiconductor Devices*, J. Kanicki, Editor, p.129-187, Artech House, Boston (1992).
8. J. M. Shannon and E. G. Gerstner, *IEEE Electron Dev. Lett.*, **24**, 405 (2003).
9. F. Balon, J. M. Shannon and B. J. Sealy, *App. Phys. Lett.*, **86**, 073503 (2005).
10. J. M. Shannon and F. Balon, *Digest of Technical Papers AM-LCD'04,* The Japan Society of Applied Physics Cat. No.AP041239, 321 (2004).
11. F. Balon and J. M. Shannon, *Solid-State Electronics*, **50**, 378 (2006).
12. E. H. Rhoderick and R. H. Williams, in *Metal-Semiconductor Contacts.* P. Hammond, R. L. Grimsdale, Eds., Oxford University Press, Oxford (1988).
13. J. M. Shannon and E. G. Gerstner, *Solid-State Electron*ics, **48**, 1155 (2004).
14. R. A. Street, Ed., *Hydrogenated Amorphous Silicon,* Cambridge Univ. Press, Cambridge, UK (1991).
15. F. Balon and J. M. Shannon, *J. Electrochem. Soc.,* **152**, G674 (2005).
16. SILVACO ATLAS user's manual, Silvaco International Inc. (2005).
17. J. M. Shannon and K. J. B. M. Nieuwesteeg, *App. Phys. Lett,* **62**, 1815 (1993).
18. F. Balon, *PhD thesis*, University of Surrey, (October, 2005).

CHAPTER 8
NON LCD APPLICATIONS

324

ECS Transactions, 3 (8) 325-331 (2006)
10.1149/1.2356370, copyright The Electrochemical Society

Artificial Retina using Thin-Film Photodiode and Thin-Film Transistor

M. Kimura, T. Shima, and T. Yamashita

Department of Electronics and Informatics, Ryukoku University, Seta, Otsu 520-2194,
Japan

An artificial retina using thin-film photodiodes (TFPDs) and thin-film transistors (TFTs) is proposed, which is an improvement of system-on-panel (SOP) concepts where in-pixel and pixel-to-pixel operations are executed using the TFTs. Device characteristics of the TFPD and TFTs are measured, and they are modeled into circuit simulation. Circuit configurations of a retina pixel and retina array are invented with some improvements, and they are designed. It is confirmed that the artificial retina can operate and achieve edge enhancement. Behavior tolerance against characteristic deviations of the TFTs and photo-sensitivity control of the edge enhancement are evaluated.

Introduction

Advantages of thin-film transistors (TFTs) are large-areal fabrication, low cost, substrate flexibility, etc. In particular, since polycrystalline-silicon (poly-Si) TFTs have high performance comparable with bulk-Si transistors, any kind of electronic circuits can be composed (1). However, conventional integrated drivers utilize only peripheral area (2). If in-pixel and inter-pixel operations are executed in pixel area, system-on-panel (SOP) concepts will become more effective.

In this presentation, an artificial retina using thin-film photodiodes (TFPDs) and poly-Si TFTs is proposed, which is an improvement of the SOP where the in-pixel and pixel-to-pixel operations are executed using the poly-Si TFTs (3-6). Device characteristics of the TFPD and poly-Si TFTs are measured, and they are modeled into circuit simulation. Circuit configurations of a retina pixel and retina array are invented with some improvements, and they are designed. It is confirmed that the artificial retina can operate and achieve edge enhancement, which is one of functions of living retinas. Behavior tolerance against characteristic deviations of the poly-Si TFTs and photo-sensitivity control of the edge enhancement are evaluated. Moreover, if this artificial retina is regarded as an initiative development of artificial organs, since it can be fabricated on flexible, harmless and organic substrates, it is expected to be suitable for living bodies.

Thin-Film Photodiode

The TFPD is fabricated using the same fabrication processes as poly-Si TFTs. There are 2 types of the TFPD: p/i/n TFPD and p/n-/n TFPD. The p/i/n TFPD has an intrinsic layer between a p+ region and n+ region. The p/n-/n TFPD has an n- layer between a p+ region and n+ region.

An actual device structure and actual device characteristics of the p/i/n TFPD are shown in Fig. 1. Dependences of a photo-induced current (Iphoto) on a photo-illuminance (Ephoto) with a variation in a reverse voltage (Vreverse) are shown. Dependences of the Iphoto on the Vreverse with a variation in the Ephoto are also shown. It is found that a

325

dark current exists even under no illuminance, which is due to thermal generation of electron-hole pairs via trap states at an oxide interface and grain boundaries. A transformation efficiency from photons to the electron-hole pairs is as low as 10 %, which is due to recombination also via trap states. In any case, the dependences of the Iphoto on the Ephoto are linear.

The actual device structure and actual device characteristics of the p/n-/n TFPD are shown in Fig. 2. It is found that dark current is much reduced. In any case, the dependences of the Iphoto on the Ephoto are also linear.

Difference in the device characteristics between the p/i/n TFPD and p/n-/n TFPD can be explained as follows although there is difference in sizes of the TFPDs. The p/i/n TFPD has potential gradient in whole intrinsic layer. Therefore, an area for the generation of the electron-hole pairs is large. As a result, values of the Iphoto is large. Moreover, since the potential gradient is dependent on the Vreverse, the Iphoto is strongly dependent on the Vreverse. On the other hand, the p/n-/n TFPD has potential gradient only near a p/n- junction due to a built-in potential. Therefore, an area for the generation of the electron-hole pairs is small. As a result, values of the Iphoto is small. Moreover,

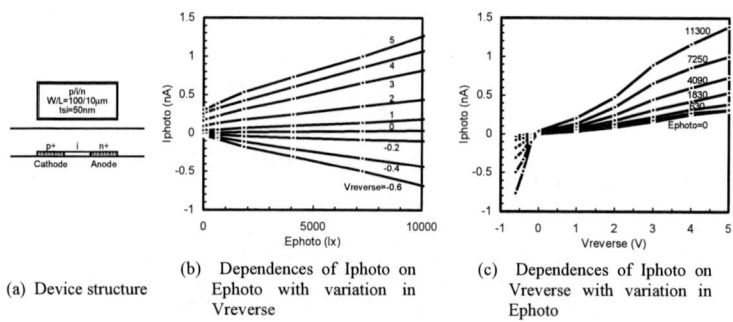

(a) Device structure

(b) Dependences of Iphoto on Ephoto with variation in Vreverse

(c) Dependences of Iphoto on Vreverse with variation in Ephoto

Figure 1. Device structure and device characteristics of p/i/n TFPD.

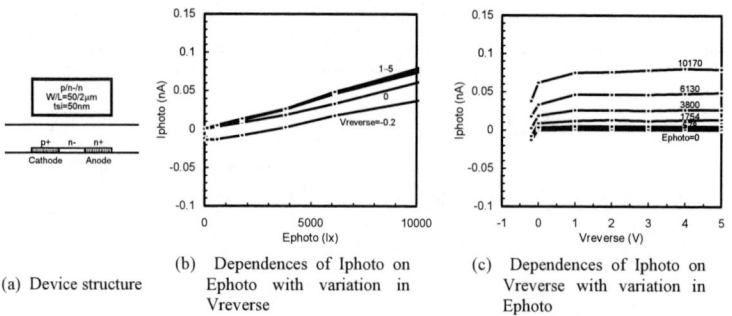

(a) Device structure

(b) Dependences of Iphoto on Ephoto with variation in Vreverse

(c) Dependences of Iphoto on Vreverse with variation in Ephoto

Figure 2. Device structure and device characteristics of p/n-/n TFPD.

since the potential gradient is independent on the Vreverse, the Iphoto is almost independent on the Vreverse. In summary, an advantage of the p/i/n TFPD is the large Iphoto while an advantage of the p/n-/n TFPD is the independent Iphoto. Selection of these TFPDs is a kind of trade-off. The p/i/n TFPD is used as follows. The device characteristics are modeled into circuit simulation using a method developed by the authors, which is a numerical fitting using spline interpolation, and utilized for the following circuit simulation (7-8).

Thin-Film Transistor

An actual device structure and actual device characteristics of the poly-Si TFTs are shown in Fig. 3. It is fabricated as follows. First, an amorphous-Si film is deposited using low-pressure chemical-vapor deposition (LPCVD) of Si_2H_6 and crystallized using XeCl excimer laser to form a poly-Si. Oxygen plasma treatment is performed to improve the poly-Si and its interface. Next, SiO_2 is deposited using electron-cyclotron resonance chemical-vapor deposition (ECR-CVD) for a gate-oxide. A gate metal is deposited and patterned. Phosphorous and boron are doped and thermally activated for source-drain regions. Contact holes are opened, and a source-drain metal is deposited and patterned.

Transfer characteristics and output characteristics are shown. The poly-Si TFTs have high performance, but they are still inferior to bulk-Si transistors. First, their transistor mobility is not very high and their threshold voltage is not very low. Second, their saturation region is not flat in the output characteristics, which might be a problem in analog circuits, such as the artificial retina. Therefore, circuit simulation is necessary to develop them. The device characteristics are also modeled into the circuit simulation using the method developed by the authors (7-8).

Retina Pixel

A circuit configuration and design layout of the retina pixel are shown in Fig. 4. The circuit configuration of the retina pixel is based on an elementary current mirror, but some improvements are added by considering the device characteristics of the TFPDs and poly-Si TFTs and operation of the artificial retina. Although the part for generation of a mirror current (Imirror) consists of two p-type TFTs, the part for a load resistance consists of two n-type TFTs. Sensitivity can be controlled by both bias voltage (Vbias)

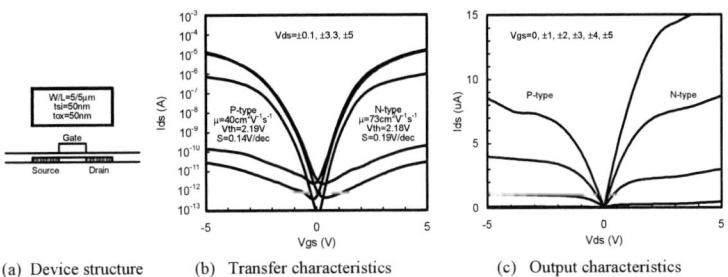

(a) Device structure   (b) Transfer characteristics   (c) Output characteristics

Figure 3. Device structure and device characteristics of poly-Si TFTs.

and adjust voltage (Vadjust). Scales for the TFPD and all TFTs are optimized. The retina pixel including the 1 TFPD and 4 TFTs is designed.

The circuit simulation is performed using the TFPD model and poly-Si TFT model. Simulated circuit characteristics of the retina pixel are shown in Fig. 5. Dependences of an output voltage (Vout) on the Ephoto with a variation in the Vbias are shown. Dependences of the Vout on the Ephoto with a variation in the Vbias are also shown. It is found that the sensitivity can be controlled by the Vbias once a suitable voltage is applied to the Vadjust. When the Vbias is low such as 0 or 0.5 V, since a resistance of a mirror TFT (Tm) is negligible in comparison with a resistance of a adjust TFT (Ta), the Imirror is determined approximately only by the resistance of the Ta, almost constant and low. As the Vbias increases, the Imirror becomes to be determined also be the resistance of the Tm and Ephoto. When the Vbias is high such as 4.5 or 5 V, the Imirror is dependent on the Ephoto in wide range of the Ephoto. The dependences of the Vout also have the similar feature.

Retina Array

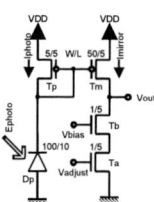

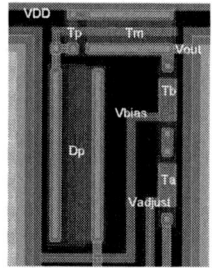

(a) Circuit configuration   (b) Design layout

Figure 4. Circuit configuration and design layout of retina pixel.

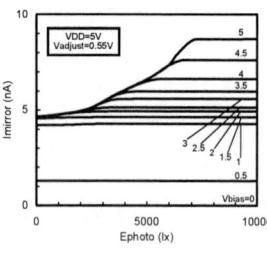

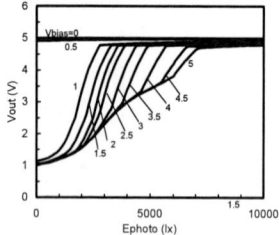

(a) Dependence of Imirror on Ephoto with   (b) Dependence of Vout on Ephoto with
      variation in Vbias   variation in Vbias

Figure 5. Circuit characteristics of retina pixel.

A circuit configuration and design layout of the retina array are shown in Fig. 6. The Vout is connected to the Vbias in adjacent pixels through the pixel-to-pixel capacitor (Cptp). The retina array including the retina pixel and the Cptp is designed. Although large contact pads are located for fundamental evaluation, a principal part is 27300 μm^2, which corresponds to 154 ppi.

Simulated circuit characteristics of the retina pixel are shown in Fig. 7. An Ephoto profile for multiple pixels, Vout profile for multiple pixels and edge enhancement are shown. The Ephoto in a left half of the retina pixels are stairs-like different from the Ephoto in a right half. A suitable voltage is applied to the Vadjust to control the total sensitivity of the retina array. The Vout is not only the output signal but is also applied as the Vbias in adjacent pixels through Cptp. When a pixel is highly illuminated, its Vout is high. When a high voltage is applied as the Vbias in an adjacent pixel, the Vout in the adjacent pixel is decreased, and vice versa. It is found that edge enhancement can be achieved. Difference in the Vout between A and B in Fig. 7(c) is 0.90 V, while difference between C and D is 1.30 V, therefore the edge enhancement is defined as (1.30 V - 0.90 V) / 0.90 V = 44 %.

Simulated behavior tolerance against characteristic deviations of the poly-Si TFTs is shown in Fig. 8. Deviations in drain currents, such as ±5 % and ±10 %, are randomly

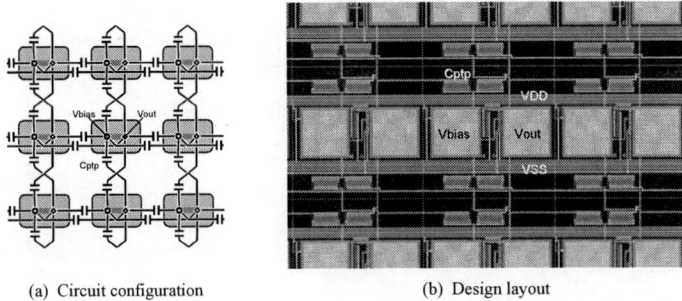

(a) Circuit configuration   (b) Design layout

Figure 6. Circuit configuration and design layout of retina array.

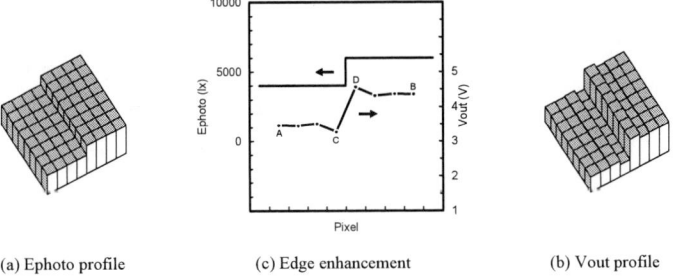

(a) Ephoto profile   (c) Edge enhancement   (b) Vout profile

Figure 7. Circuit characteristics of retina array.

distributed on multiple pixels. The Ephoto profile for multiple pixels is the same as that in Fig. 7(a). The Vout profiles along each pixel column shift and are plotted in the graphs. The edge enhancement is correctly confirmed in the case of the deviation of ±5 %. The edge enhancement is more than 40 % for all neighboring pixels at an illuminance edge. On the other hand, the edge enhancement is roughly confirmed in the case of the deviation of ±10 %. The edge enhancement is more than 40 % for half of the neighboring pixels at the illuminance edge. Therefore, it is necessary to reduce the characteristic deviations of the poly-Si TFTs to within ±5 %.

Simulated photo-sensitivity control of the edge enhancement is shown in Fig. 9. By adjusting the Vadjust, a level of the Ephoto where the edge enhancement is achieved can be controlled. By applying 0.55 V to the Vadjust, the level of the Ephoto of 4000 - 6000 lx is most enhanced, while by applying 0.56 V, the level of the Ephoto of 5000 - 7000 lx is most enhanced.

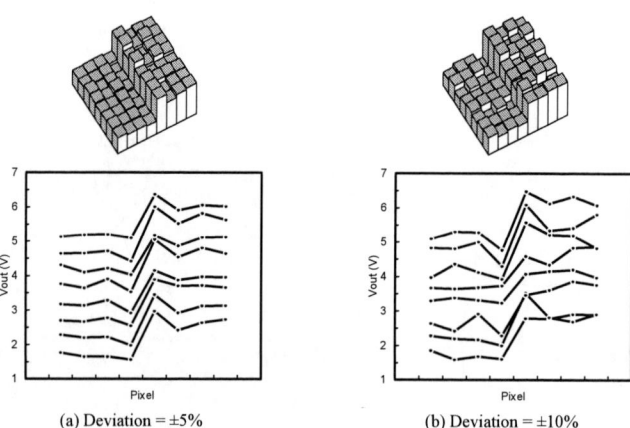

(a) Deviation = ±5%   (b) Deviation = ±10%

Figure 8. Behavior tolerance against characteristic deviations of poly-Si TFTs.

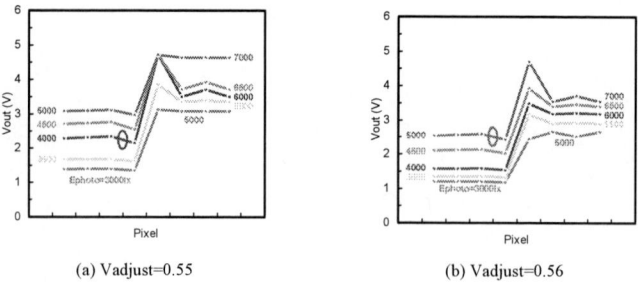

(a) Vadjust=0.55   (b) Vadjust=0.56

Figure 9. Photo-sensitivity control of edge enhancement .

Conclusion

The artificial retina using the TFPDs and poly-Si TFTs was proposed, which is an improvement of the SOP where the in-pixel and pixel-to-pixel operations are executed using the poly-Si TFTs. The device characteristics of the TFPD and poly-Si TFTs were measured, and they were modeled into the circuit simulation. The circuit configurations of the retina pixel and retina array were invented with some improvements, and they were designed. It was confirmed that the artificial retina can operate and achieve the edge enhancement. The behavior tolerance against the characteristic deviations of the poly-Si TFTs and photo-sensitivity control of the edge enhancement were evaluated.

Acknowledgments

The authors wish to thank Prof. Hajimu Kawakami of Ryukoku University, Drs. Hiroyuki Hara, Tomoyuki Okuyama, Satoshi Inoue and Tatsuya Shimoda of Seiko Epson Corporation, Silvaco International, Silvaco Japan, Cadence Design Systems, Cadence Design Systems Japan, Cybernet Systems, and Jedat. This research is partially supported by collaborative research with Seiko Epson Corporation, and a research project of the Joint Research Center for Science and Technology at Ryukoku University.

References

1. N. Karaki, T. Nanmoto, H. Ebihara, S. Inoue and T. Shimoda, SID '05, 1430 (2005).
2. H. Ohshima and S. Morozumi, IEDM '89, 157, (1989).
3. M. Kimura, T. Shima, T. Okuyama, S. Utsunomiya, W. Miyazawa, S. Inoue and T. Shimoda, AM-LCD '05, 323, (2005).
4. M. Kimura, T. Shima, T. Okuyama, S. Utsunomiya, W. Miyazawa, S. Inoue and T. Shimoda, IDW '05, 261, (2005).
5. M. Kimura, T. Shima, T. Okuyama, S. Utsunomiya, W. Miyazawa, S. Inoue and T. Shimoda, Jpn. J. Appl. Phys. **45**, 4419 (2006).
6. M. Kimura and T. Shima, AM-FPD '06, 73, (2006).
7. M. Kimura, Satoshi Inoue and Tatsuya Shimoda, IEEE Trans. CAD. IC & Syst. **21**, 1101 (2002).
8. M. Kimura, S.. Inoue and T. Shimoda, IEICE Trans. Electronics **E86-C**, 63 (2003).

Nonvolatile Amorphous Silicon Thin Film Transistor Memories with the a-Si:H Embedded Gate Dielectric Structure

Helinda Nominanda and Yue Kuo

Thin Film Nano and Microelectronics Research Laboratory, Texas A&M University, College Station, TX 77843-3122, USA

An a-Si:H TFT memory with an a-Si:H layer embedded in the gate dielectric structure has been studied. Fabrication steps for this floating gate structure include a simple one-pump down PECVD and a two-photomask lithography process. Hysteresis of the transfer characteristic demonstrates the memory function. A steady difference between the threshold voltage of the "write" and "erase" state demonstrates the programmability of the memory. A charge retention time of more than 3600 secs has been shown. This new structure could expand the application of conventional a-Si:H TFTs.

Introduction

The non-volatile memory MOSFETs have been prepared through different methods, such as varying the transistor structure or applying a specific gate dielectric material. For instance, the use of polysilicon floating gate and embedded nanocrystalline silicon has exemplified the memory effect through the physical trapping and detrapping of tunneling carriers [1]. Polarization of dielectric material such as ferroelectric has also resulted in such effect [2].

The traditional role of a-Si:H TFTs has been as pixel switching devices in AMLCDs, imagers, sensors, etc. Additional functions can be integrated into these products if non-volatile memory a-Si:H TFTs were available. In fact, functional floating gate a-Si:H TFTs were demonstrated by embedding aluminum into a composite gate dielectric consisting of Si_3N_4 and SiO_2 layers [3]. However, the process requires breaking of the vacuum during the gate dielectric preparation and also necessitates separate evaporation and etching steps. Such a complex fabrication scheme might adversely impact the quality of film interfaces.

In this paper, authors studied a new type of floating gate a-Si:H TFT that included a SiN_x/a-Si:H/SiN_x gate dielectric structure.

Experimental

Self-aligned, inverted staggered, tri-layer TFTs can be fabricated with a simple 2-photomask process on Corning 1737 substrates [4]. Modification of the gate dielectric structure utilizing similar fabrication processes was executed to achieve a floating gate a-Si:H TFT structure. The modification included embedding intrinsic a-Si:H in the middle of the original gate dielectric. The modified gate dielectric was then composed of 150 nm SiN_x/0, 4, 7, or 9 nm thick intrinsic a-Si:H/150 nm SiN_x layers, which were deposited by plasma enhanced chemical vapor deposition (PECVD) method. A 50 nm thick intrinsic a-Si:H layer for the channel, and 250 nm thick SiN_x for the top passivation layers were deposited subsequently. All PECVD deposited layers were carried out in a parallel-plate reactor (Applied Materials AMP Plasma II) at a substrate temperature of 300°C and an rf-frequency of 13.56 MHz without breaking the vacuum. After the "five-layer" deposition, the top channel stop SiN$_x$ layer was defined by a backlight exposure method by using the gate as the self-aligned mask.

A PECVD microcrystalline n$^+$ a-Si:H layer, was then deposited as the source/drain ohmic contact layer. The gate, source and drain electrodes were made of 1000 Å thick DC-sputtered molybdenum (Mo). The n$^+$ layer was reactive ion etched with a 13.56 MHz RF generator using a mixture of CF$_4$ (2 sccm) and Cl$_2$ (8 sccm) at 100 mTorr and 300 W for about 2 minutes. The finished TFT was annealed in air at 250°C for 1 hour to repair the plasma etch induced damage [5]. TFTs were characterized with the Agilent 4155C semiconductor parameter analyzer. Figure 1 shows the cross-section schematic and the top view of the finished TFTs.

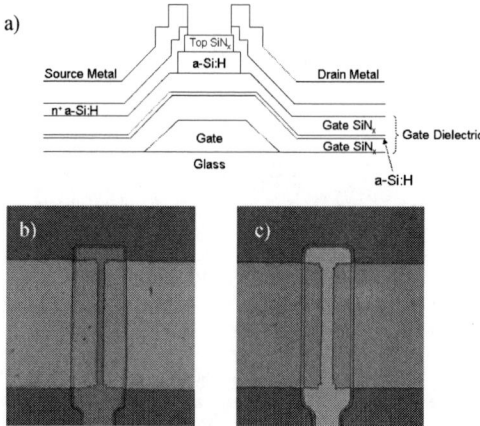

Figure 1. a) Schematic of a-Si:H TFT memory structure b) functional TFT with 9 nm embedded a-Si:H layer c) conventional TFT.

Results and Discussion

Both the control TFT (without an embedded a-Si:H layer) and the TFT with an embedded a-Si:H layer in the gate dielectric show normal transistor behaviors. For instance, high I_{on}'s and I_{on}/I_{off} ratios of more than 10^6 have been observed. Figure 2 shows the $I_d^{0.5}$ vs. V_g curves of TFTs with different embedded layer thickness. Threshold voltage V_t of the TFT was extracted from the intersection of the curve with the x-axes at the saturation region.

From Fig. 2, a shift of the V_t to a more positive value with the increase of a-Si:H layer thickness was observed, e.g., 2.4, 2.7, 3.3, and 6.9 V for the 0, 4, 7, and 9 nm embedded films, respectively. The sub-threshold slope S was also found to follow similar trend with increasing embedded layer thickness. The shift of V_t and the increase of S could be contributed to the defect density state at the interface of a-Si:H and the gate SiN$_x$ layer [6,7]. Since the TFTs with embedded a-Si:H have two additional a-Si:H/SiN$_x$

interfaces, the higher value of V_t and S compared with those of the control TFT could be contributed to the defect density from these extra interfaces.

The effect of the quality of the a-Si:H/SiN$_x$ interface on V_t has also been related to the thickness of the a-Si:H film [8]. For instance, a good quality interface along the a-Si:H side can be achieved with a thicker a-Si:H layer, possibly due to the cleaner environment and the more complete reaction of the precursors as a result of the longer deposition time [8]. On the other hand, a thick a-Si:H layer is prone to the incorporation of particles at the interface due to thepowdery deposition region [8]. Therefore, an optimum a-Si:H thickness is trivial for the formation of better quality a-Si:H/SiN$_x$ interfaces.

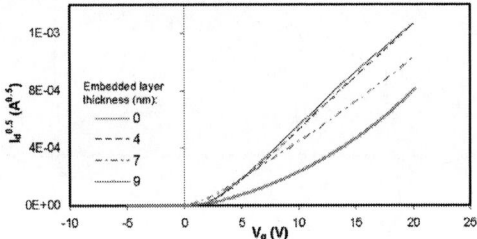

Figure 2. V_t shifts of TFTs with different embedded a-Si:H layer thickness, $V_d = 10$ V, W/L = 2.4.

Figure 3 shows the memory characteristic of the TFTs with different embedded a-Si:H layer thickness. The characteristic was observed from the hysteresis of the I_d - V_g curves. The curves were obtained by forward sweeping of the gate voltage V_g from –20 V to +20 V, which was immediately followed by backward sweeping to –20 V. Both sweepings were performed at a 0.1 V interval with drain voltage $V_d = 10$ V.

The control TFT shows a clockwise hysteresis between the forward and the backward curves with a V_t difference of 0.8 V between the two curves. The TFTs with the embedded dielectric layers show consistent increase of hysteresis gap with the increase of embedded layer thickness. The V_t difference also follows a similar trend, e.g., 1 V and 6.8 V for TFT embedded with 3 and 7 nm a-Si:H layers, respectively.

The behavior of charge accumulated at the interface is directly affected by the polarity of the starting V_g sweep. For instance, for a backward sweeping of an n-channel TFT, the highly positive V_g translates to the filling of the donor-like electron traps since the Fermi level is positioned closer to the conduction band [9]. The increase of donor-like traps resulted in a decrease of positive charges, which translates to the shift of the transfer characteristics to a more positive side, hence the clockwise direction of the hysteresis.

The hysteresis changed direction at the 7 nm a-Si:H embedded TFT. For instance, at the positive V_g region, the backward sweep resulted in a larger drain current than that of the forward sweep. This could be related to the delay in the inversion layer formation. The release of trapped electrons at the embedded a-Si:H site to the channel/gate SiN$_x$ interface might correspond to delay of inversion layer formation. The inversion layer was fully formed during large negative V_g value, i.e., toward the end of the backward sweep and resulted in the larger drain current.

The maximum amount of trapped charges is limited by the Coulomb blockade phenomena. Other than the barrier layer thickness, Coulomb blockade is also related to the conducting layer thickness since it directly affects the film's conductance [10]. The conductance of the floating gate, in turn, is related to the energy level of the floating gate, which determines the possibility for electrons tunneling.

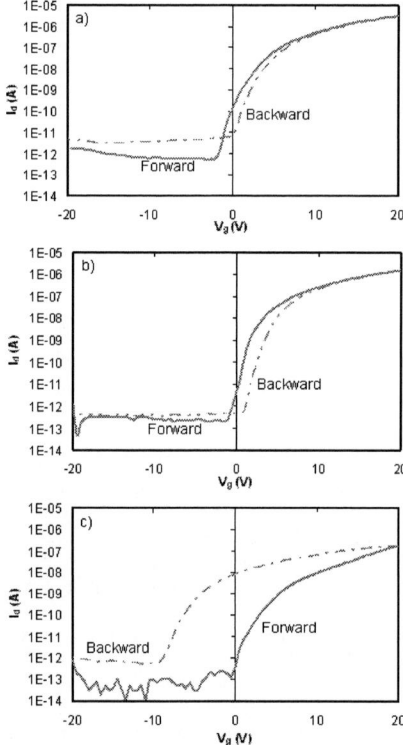

Figure 3. Hysteresis of TFT with a) no embedded layer, b) 3 nm and c) 7 nm embedded a-Si:H, W/L=2.4, $V_d = 10$ V.

The programmable property of the TFT memory can be observed by comparing the transfer characteristics before and after a period of gate bias. Figure 4 shows the transfer characteristics of an a-Si:H TFT with a 9 nm thick a-Si:H layer embedded in the gate SiN_x. The "virgin" state was from the TFT before being stressed, the "write" state was from the same TFT after being biased at $V_g = 10$ V for 1 second and the "erase" state was after being biased at $V_g = -10$ V for another second. Both source and drain electrodes were grounded during gate bias. The transfer curves were measured at $V_d = 10$ V.

Fig. 4 shows that the "write" state shifted the curve toward the right side of the "virgin" state curve and the "erase" state shifted the curve back to the "virgin" state direction. The shift of the "write" state was due to the delay of the formation of accumulation layer to compensate for the trapped electrons resulted from the positive V_g bias. For the "erase" state, the negative V_g bias of the same magnitude as the positive V_g state was enough to detrap the electrons and to recover the transfer characteristic of the "virgin" state. A V_t difference between the "write" and "erase" states was 0.4 V at $V_g = 20$ V.

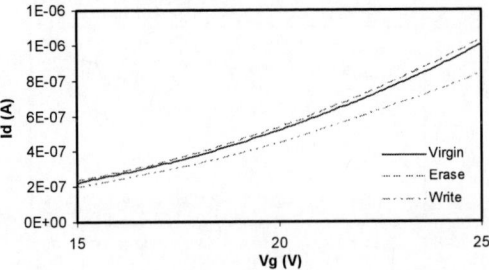

Figure 4. The transfer characteristics of TFTs with a 9 nm embedded a-Si:H layer. W/L = 2.4

Figure 5 shows further insight into the programming characteristic of the TFT in Fig. 4. The V_t shift is defined as the difference between the V_t of the "virgin" state and that of the "write" or "erase" state. For the "write" state, V_t shift increases with increasing writing time. This could be related to the increase of time for writing, which corresponds to the increase of trapped charges. For the "erase" state, the magnitude of V_t shift becomes smaller as the erase time increases, which is due to the detrapping of charges in the long erase time.

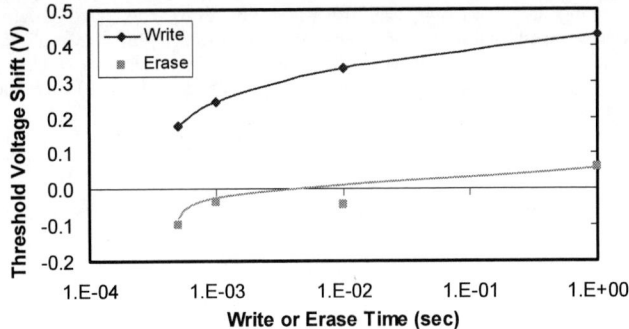

Figure 5. V_t shift as a function of "write" or "erase" time for the TFT of Fig. 4. V_g for write = 10 V, V_g for erase = -10V.

Figure 6 shows the charge retention property of the TFT of Fig. 4. The "write" and "erase" states follow the same condition as the ones used for Fig. 4. The V_t was extracted every 120 seconds until 3600 seconds. The "erase" state and the V_t extractions immediately followed after the completion of the "write" state measurement. A final difference between the V_t of the two states was 1.1 V.

The good retention characteristic for this time scale was possible due to the low leakage current from the gate SiN_x layer. A longer time scale would suffer from eventual charge leaking processes that involve the combination of electron tunneling, hole injection, charge redistribution, etc [11]. This would need further examination.

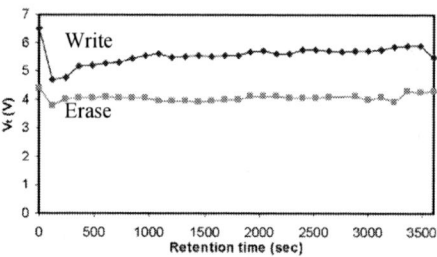

Figure 6. Charge retention characteristic of the TFT in Fig. 4.

Conclusion

The floating gate a-Si:H TFT memory based on embedding an a-Si:H layer into the SiN_x gate dielectric has been demonstrated and studied. The fabrication process includes a one-pump down 5-layer PECVD deposition step and a 2-photo mask lithography process. The TFT's memory characteristic was demonstrated from the hysteresis of the transfer characteristics due to electrons tunneling to the gate dielectric layer. The hysteresis gap increased with the thickness of the embedded a-Si:H layer. The Coulomb blockade phenomenon is related to the embedded a-Si:H film thickness, which affects the charge trapping and detrapping capacity. The TFTs also show good programming characteristics, manifested in the shift of V_t to a more positive value at "write" state, and the shift back to equilibrium value at "erase" state. A large programming window on this kind of TFT has been obtained. A charge retention time more than 3600 secs has been demonstrated.

Acknowledgments

The authors thank Jiong Yan for metal depositions, Guojun Liu for the RIE works, and Tyler Martin for proofread this manuscript.

References

1. L. Guo, E. Leobandung, and S. Y. Chou, *Appl. Phys. Lett.*, **70** (7), 850 (1997).
2. H. I. Hanafi, S. Tiwari, and I. Khan, *IEEE Trans. Elec. Dev.*, **43** (9), 1553 (1996).
3. S. G. Burns, H. R. Shanks, A. P. Constant, C. Grubber, D. Schmidt, A. Landin, C.Thielen, F. Olympie, T. Schumacher, and J. Cobbs, PV 1994-35, p. 370, The Electrochemical Society Proceedings Series, Pennington, NJ (1994).

4. Y. Kuo, *J. Electrochem. Soc.*, **138** (2), 637 (1991).
5. Y. Kuo and S. Lee, *Appl. Phys. Lett.*, **78** (7), 1002 (2001).
6. Y. Kuo in *Thin Film Transistors, Materials and Processes, Volume 1: Amorphous Silicon Thin Film Transistors,* Yue Kuo, Editor, p. 248, Kluwer, New York (2004).
7. H. Iizuka, F. Masuoka, T. Sato, and M. Ishikawa, *IEEE Trans. Elec. Dev.*, **ED-23**, 379 (1976).
8. S. Yamamoto and M. Migitaka, *Jpn. J. Appl. Phys.*, **32**, 462 (1993).
9. A. Nathan, P. Servati, K. S. Karim, D. Striakhilev, and A. Sazonov in *Thin Film Transistors, Materials and Processes, Volume 1: Amorphous Silicon Thin Film Transistors,* Yue Kuo, Editor, p. 83, Kluwer, New York (2004).
10. S. Monfray, A. Souifi, F. Boeuf, C. Ortolland, A. Poncet, L. Militaru, D. Chanemougame, and T. Skotnicki, *IEEE Trans. Nanotech.*, **2** (4), 295 (2003).
11. Y. Hu and M. H. White, *Sol. State Elec.*, **36** (10) 1401, (1993).

340

ECS Transactions, 3 (8) 341-347 (2006)
10.1149/1.2356372, copyright The Electrochemical Society

Sensitivity of Suspended-Gate Polysilicon TFTs to charge variation and application to DNA recognition

F. Bendriaa, M. Harnois, F. Le Bihan, A.C.Salaün,
T. Mohammed-Brahim, O. Bonnaud

G. M., IETR, Université de Rennes I, 35042 Rennes Cedex, France

Suspended-Gate Polysilicon Thin Film Transistors, namely
SGTFT, are fabricated using low temperature surface micro-
machining process. When using sub-micronic gap, high field
effect induces large shift of the transfer characteristics due to
any charge variation inside the gap. The device was used in
measurements of pH of different solutions with high sensitivity
(~250 mV/pH), 4 times the theoretical Nernstian response and
the usual value given by the known transducers. This high
sensitivity, un-reached until now, is explained by the charge
distribution induced by the high applied field. Application of
the device is extended to DNA recognition with high and rapid
answer. Whatever the application, SGTFT is sensitive to charge
variation. Selectivity can be insured by the detection principle
that is based on the couple DNA/complementary DNA in the
case of DNA recognition.

Introduction

Suspended-gate-field-effect-transistor based sensors were extensively investigated
as gas sensors (1,2) and Ion Sensitive FETs (ISFETs) as pH sensors. These devices
use the work function variation as sensitive parameter. However the work function
parameter varies very weakly when the adsorbed charge changes. For example, the
ISFET sensitivity is limited to the Nernstian response. We show here the possibility to
increase highly the sensitivity by introducing a field effect as additional parameter.

Used device is a silicon air-gap Thin Film Transistor that we realized previously.
The device fabrication uses a low temperature surface micro-machining process (3).
Besides the low temperature advantage, the important particularity of the process is
the possibility to fix the gap at any value that can be much lower than 1 μm. In such
low gap, the field effect due to the applied gate voltage is so important that it can
influence the charge distribution in the ambience and then the adsorption phenomenon
and the work function variation.

To be used in liquid, the device is electrically insulated using silicon nitride. The
process was previously optimized to insure a good mechanical maintain of the bridge
during the numerous times where it is dipped.

Experimental

Eight masks were designed for the fabrication of the TFT. The starting n-type
silicon wafer is thermally oxidized. Then source and drain regions are opened to be
doped using a boron diffusion. After an oxide etching, the subtrate is oxidized and
then a LPCVD silicon nitride film is deposited. fabrication process is summarized as
follows:

  1) starting material: n-type Silicon wafer,

341

2) thermal oxidation,
3) first photo masking, oxide etching to determine source and drain regions,
4) doping of silicon using boron or phosphorus,
5) second photo masking and oxide etching,
6) thermal oxidation and LPCVD silicon nitride deposition,
7) LPCVD Germanium deposition,
8) third photo masking : germanium plasma etching to form the regions of the drain and source contacts and the bridge anchors,
9) LPCVD silicon nitride deposition,
10) fourth photo masking: silicon nitride and the previous silicon dioxide / silicon nitride stack of layers are patterned simultaneously to open the drain and source windows,
11) LPCVD polycrystalline silicon deposition and LPCVD silicon nitride deposition
12) fifth photo masking: polycrystalline and silicon nitride etching,
13) sixth photo masking and silicon nitride etching,
14) aluminum deposition,
15) seventh photo masking: aluminum etching to form the source, drain, and gate pads
16) eighth photo masking: the contacts wires are insulated using a photo resist
17) sacrificial germanium layer is easily etched using H_2O_2 only

A scheme of the cross-section of the final SGTFT structure is shown in Figure 1. Figure 2 is a Scanning Electron Microscope up-view of the structure. Figure 3 is an up-view of SGTFT and the final structure. The suspended triple layer bridge is well highlighted in Figure 2.

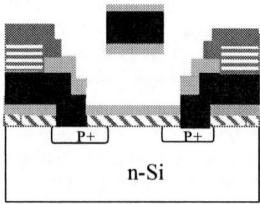

⠀⠀⠀⠀⠀⠀ Silicon dioxide
⠀⠀⠀⠀⠀⠀ Silicon nitride
⠀⠀⠀⠀⠀⠀ Aluminum
⠀⠀⠀⠀⠀⠀ Doped polysilicon
⠀⠀⠀⠀⠀⠀ Resist

Figure 1: Cross-section of the final TFT's structure

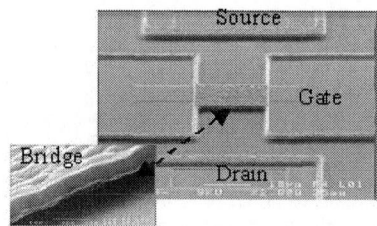

Figure 2 : SEM. up-view of the TFT's structure. The suspended gate is highlighted in the insert.

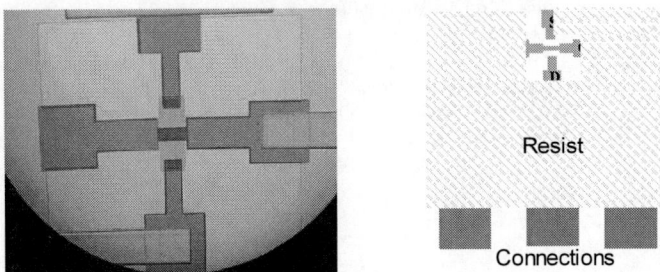

Figure 3 : up view of the SGFET (left) and the final total structure (right)

Results

SGTFT as Gas Sensor

SGTFT was first checked as gas sensor. Effect of NH_3 on the N-type TFT's characteristics is shown here as an example. An increasing negative shift of the threshold voltage with NH_3 content is observed in Figure 4. Its large value shows a very high sensitivity that can be compared to some shifts given in the literature, 71 mV with 50 ppm (4) or 140 mV with 1000 ppm (5).

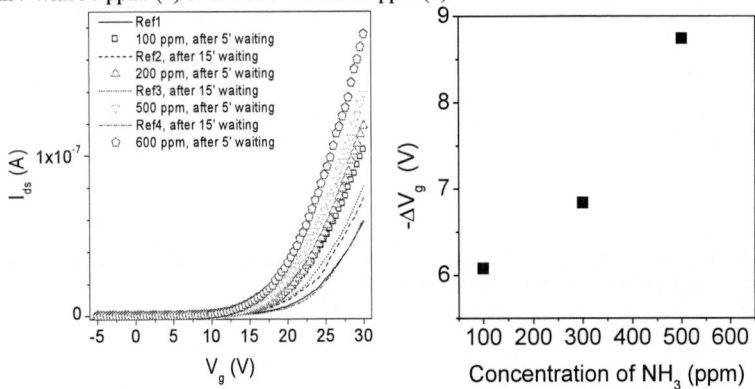

Figure 4: Transfer characteristics of N-type SGTFT (left) and the shift of the gate voltage value at a drain current of $5x10^{-8}A$ (right) when the device is submitted to increasing content of NH_3

Figure 4: Typical transfer characteristics of more than 30 drive-TFTs S_2 (W/L=40/10). Inset shows the simple p-type pixel circuit used here.

SGTFT as pH meter

P-type SGTFT structure was checked as pH meter. Followed procedure consists first on plotting transfer characteristics of transistor that are dipped in highly de-ionised water (DI). Then transistor is dipped in solution with a fixed pH and transfer characteristics is replotted. There after, transistor is rinsed in DI during 5 minutes. Afterwards, transistor is dipped in a solution with another pH and transfer characteristics is plotted. The procedure is then renewed for different pH values.

An increasing positive shift of the threshold voltage with pH value is observed in Figure 5. This is consistent with a decrease of H^+ charges content at high pH. A very high pH sensitivity (255 mV/pH at a drain current value of -0.1 mA) can be noted. It is more than 4 times the theoretical Nernstian value (58.2 mV/pH at 20°C).

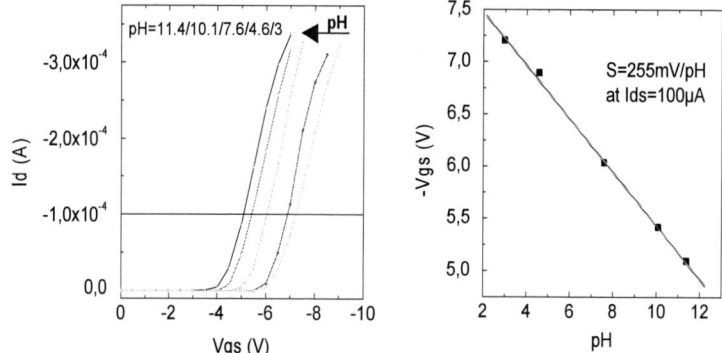

Figure 5: Transfer characteristics of P-type SGTFT (left) and the shift of the gate voltage value at a drain current of -0.1 mA (right) when the device is submitted to increasing pH values from 3 to 11.4

SGTFT in DNA recognition

SGTFT can be used also as biological sensor. Here, specific DNA (Desoxyribo Nucleic Acid) sequence detection is presented as an example.

Procedure starts with an activation of Si_3N_4 inner surface by diving transistors in a solution of glutaraldehyde. An amino-substituted oligonucleotide, ODN, is grafted on this functionalised surface. Transfer characteristics of P-type SGTFT, dived in Phosphate Buffered Saline (PBS) solution, are plotted to detect the grafted ODN. The presence of the grafted ODN is confirmed by the positive shift of the SGTFT transfer characteristic (Figure 6). Positive shift of P-type transistors means presence of negative charge that is ODN charge in this case.

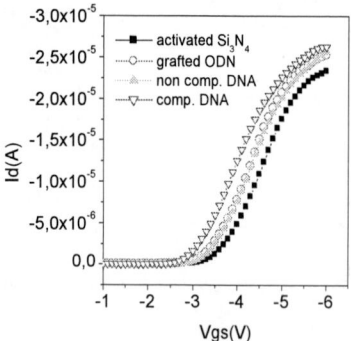

Figure 6: Transfer characteristics when SGTFT is dipped in PBS after activation of Si_3N_4 by glutaraldehyde, after grafting of ODN, after hybridisation trying of non complementary and complementary DNA.

Next, hybridisation with non complementary DNA (5nM concentration) is checked by its effect on transfer characteristics. Transfer characteristic does not shift as shown in Figure 6; that means that no charge variation is induced by the add of non complementary DNA. Hybridisation does not occur with non complementary DNA.

So hybridisation with complementary DNA (5nM concentration) is checked. Figure 6 shows large 0.4V positive shift of the transfer characteristics, induced by the hybridisation of complementary DNA. The present shift is very large in comparison with previous results obtained using ISFETs (6,7). Typically, The shift is in the range of one hundred or less mV when using such ISFETs.

Hence, present SGTFT structure is shown as very sensitive and direct DNA transducer.

Discussion

The precedent results showed a very high sensitivity of the SGTFT structure to any charge variation in the ambience under the bridge.

To understand the origin of the very high sensitivity, the effect of the field between the gate and the semiconductor can be considered. Due to the low gap, 500 nm here, electrical charges in the air-gap are submitted to very high field. Particularly, for pH sensor, positive charges, H^+ for example, accumulate on the bottom of the bridge when negative gate voltage is applied. Desertion of H^+ from the surface of the silicon nitride gate insulator leads to reduced voltage needed to create the channel of the P-type transistor. Then, the effect of the field is to reduce the threshold voltage. In usual ISFET, the field effect is very low. Solution that is initially neutral on any point, stays neutral. Only charges that can be in vicinity of the silicon nitride adsorb that leads to a threshold voltage shift.

So, in the present device, shift of the threshold voltage is due, for one part, to the field effect that induces new charge distribution in the air-gap and for the other part to the charge adsorption at the surface of silicon nitride.

The threshold voltage V_{TH} of SGTFT can be expressed by:

$$V_{TH} = \Phi_{MS} + 2\varphi_F + \frac{Q_{SC}}{C} - \frac{1}{Ce_{ox}} \int_0^{e_{ox}} x\rho(x)dx \quad (1)$$

where Φ_{MS} is the difference between the work function of the gate material and the semiconductor, φ_F is the Fermi level position in the semiconductor versus the mid-gap, Q_{SC} is the space charge in the semiconductor, C is the total capacitance between the gate material and the semiconductor, e_{ox} is the thickness of the insulator that is the stack of air-gap, Si_3N_4 and SiO_2 here, $\rho(x)$ is the charge in the insulator at a distance of x from the gate.

Any variation of the ambience in the air-gap leads to a variation of the total charge in the insulator and a possible variation of its distribution. Moreover some chemical reactions at the inner surface of the gap can occur leading to a variation of Φ_{MS}.

Usually only the last variation is considered in ISFET. However when a high field due to the very low gap is present, the distribution of the charge in the ambience varies leading to a variation of $\rho(x)$. Moreover the high field can influence the adsorption by pushing the charges on the surface.

All these effects lead to a variation of Φ_{MS} but also of the last term in the V_{TH} expression. Then, V_{TH} variation can be very large if the effects of a high field are considered.

Field effect on the sensitivity can be experimentally checked.

Experimental evidence of the field effect on the sensitivity

To illustrate, the effect of the high field on the pH sensitivity, SGTFTs are fabricated by using 0.8 μm air-gap that is larger than the previous 0.5 μm one. Figure 7 shows the linear plot of Vgs versus pH for this SGTFT. The sensitivity, 90 mV/pH, is much lower than the previous 255 mV/pH. Then, the large field effect can occur only with very high fields. Little bit higher air-gap leads to large decrease of the sensitivity. Then, the beneficial effect of the high field on the sensitivity is experimentally evidenced.

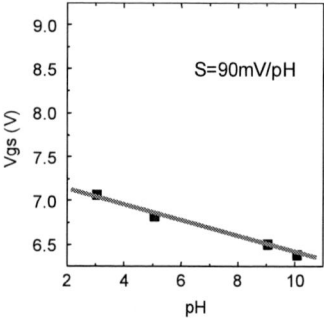

Figure 7: Gate voltage as a function of pH for 0.8 μm air-gap SGTFT. The slope of the linear plot gives a sensitivity of 90mV/pH

Experimental evidence of the effect of the new charge distribution induced by the high field

In the previous qualitative description, the high sensitivity was explained from both the adsorption as in usual ISFET and the effect of the new charge distribution induced by the high field. It is possible to separate experimentally these both effects by using salt solutions that do not change the pH and do not introduce any adsorption at the surface. Salt solutions of KCl and NaCl and basic solution KOH are prepared with exactly the same concentration. pH does not change when using salt solutions as KCl and NaCl. When plotting transfer characteristics of SGTFT that is dipped in these solutions, only effect of the field on the charge distribution will be observed. In the presence of KOH, pH change and then both effects of the new distribution of charges inside the gap and of the adsorption will be observed. Figure 8 shows the transfer characteristics of SGTFT dipped in DI water and in solutions of KCl, NaCl and KOH with the same concentration.

Similar shift of the characteristics is shown in presence of KCl or NaCl with the same concentration. This shift is only due to the distribution of charges induced by the field inside the gap. V_{TH} shift is induced by the variation of the last term of equation (1). Same charge content gives same shift.

With same concentration KOH solution, extra shift is observed. It can be due to the pH of KOH and then to the charges that adsorb at the surface of silicon nitride (first term in equation (1)). Then in presence of KOH the shift is due to both the charge distribution and the adsorbed charge. Both origins contribute to the high pH sensitivity of the present device.

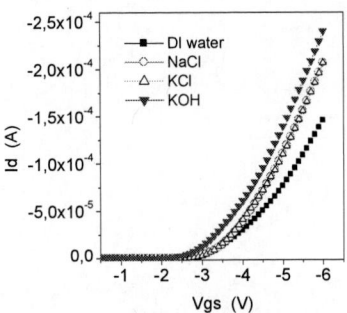

Figure 8: Transfer characteristics when SGTFT is dipped in DI water, and in solutions of NaCl, KCl or KOH. These 3 solutions have the same concentration.

Conclusion

Suspended-Gate TFTs with sub-micron gap showed their ability to detect with very high sensitivity any charge variation inside the gap. This sensitivity is some times the usual value given by the known transducers.

Through different experiments, this high sensitivity is explained form charge distribution inside the gap due to the high field effect. New distribution contributes to the shift of the transistor characteristics. Moreover, the high field increases the eventual adsorption at the surface by pushing or removing charges.

References

1. B. Flietner, T. Doll, J. Lechner, M. Leu, I. Eisele, *Sensors and Actuators B* **18-19**, 632 (1994).
2. M. Burgmair, H.P. Frerichs, M. Zimmer, M. Lehmann, I. Eisele, *Sensors and Actuators B* **95**, 183 (2003)
3. H. Mahfoz-Kotb, A.C. Salaun, T. Mohammed-Brahim, O. Bonnaud, *IEEE Trans. Electr. Dev. Lett.* **24**, 165 (2003)
4. A Karthigeyan, R P Gupta, K Scharnagl, M Burgmair, S K Sharma, I Eisele, *Sensors and Actuators B* **85,** 145 (2002)
5. Z Gergintschew, P Kornetzky and D Schipanski, *Sensors and Actuators B*, **35-36,** 285 (1996)
6. E. Souteyrand, J.P. Cloarec, J.R. Martin, C. Wilson, I. Lawrence, S. Mikkelsen, M.F. Lawrence, *J. Phys. Chem. B* **101**, 2980 (1997)
7. F. Pouthas, C. Gentil, D. Côte, U. Bockelmann, *Appl. Phys. Lett.* **84**, 1594 (2004)

348

ECS Transactions, 3 (8) 349-359 (2006)
10.1149/1.2356373, copyright The Electrochemical Society

**Process Technology for High-Resolution AM-PLED Displays on Flexible Metal Foil
Substrates**

T.-K. Chuang[a], M. Troccoli[a], P.-C. Kuo[a], A. Jamshidi[a], J. A. Spirko[b], M. K. Hatalis[a],
K. Klier[b], I. Biaggio[c], A. T. Voutsas[d], T. Afentakis[d], and J. W. Hartzell[d]

[a] Department of Electrical and Computer Engineering, Lehigh University, Bethlehem,
Pennsylvania 18015, USA
[b] Department of Chemistry, Lehigh University, Bethlehem, Pennsylvania 18015, USA
[c] Department of Physics, Lehigh University, Bethlehem, Pennsylvania 18015, USA
[d] Sharp Laboratories of America, Camas, Washington 98607, USA

> The first successful integration of poly-Si thin-film-transistor
> (poly-Si TFT) backplane with polymer light-emitting diodes
> (PLEDs) onto a flexible stainless-steel foil is described, and a
> high-resolution (230 DPI) monochrome active-matrix polymer
> light-emitting diode (AM-PLED) display is demonstrated. The
> process technology required to implement this high-resolution
> AM-PLED display onto a flexible metal foil substrate is discussed.
> This technology primarily consists of the preparation of flexible
> metal foils, fabrication of the poly-Si TFT backplane, and
> integration with top-emitting PLEDs.

Introduction

Active-matrix organic and polymer-based light-emitting diode (AM-OLED and AM-
PLED) displays are extensively being pursued by many groups worldwide as the next
generation displays because of their light-weight, self-emission, wide viewing-angle, low
power and capability of being readily compatible with flexible substrates. To realize
large-area, high-resolution flexible displays, integration of thin-film electronic devices,
with OLEDs or PLEDs onto a flexible substrate is required. Both polymers and metal
foils are of interest as the flexible substrate material.

In the case of polymer substrates, the fabrication of TFTs is limited to low processing
temperatures. In contrast metal foils have no such restriction, for instance, stainless steel
can withstand process temperature as high as 1000 °C. Thus in the case of metal foils,
low cost silicon crystallization methods, such as furnace annealing or rapid thermal
processing (RTP), can be employed as alternatives to laser crystallization process that is
necessary for polymer substrates. Other high temperature processes, beyond the silicon
crystallization, such as gate dielectric preparation, dopant activation, silicidation, etc.,
also pose no problem on metal foils. The fabrication of both hydrogenated amorphous
silicon TFTs and polysilicon poly-Si TFTs onto flexible stainless steel foils has been
demonstrated to be readily compatible with these substrates (1,2,3).

The fabrication of poly-Si TFTs onto metal foils is of particular interest for AM-
OLED/PLED displays as the high carrier mobility in polysilicon enables the integration
of display driver circuits, the reduced threshold-voltage shift results in stable display
performance, and the excellent barrier of the metal foil to permeation of water and
oxygen results in long display lifetime. The great dimensional stability of the steel foil
substrates enables the implementation of circuit designs with minimum feature size of 1
μm or less. Both n-MOS and p-MOS poly-Si TFT devices can be fabricated so that low

349

power CMOS circuit architectures can be utilized for the integrated display drivers and other integrated electronic system components. We have recently demonstrated the fabrication of very small poly-Si TFTs, onto a flexible steel foil substrate, having a channel with width and length both equal to one micron and having an effective electron mobility of 358 cm^2/Vs (4). High performance, highly integrated circuits are then feasible because of significantly reduced channel length.

Because of the opaque nature of the metal foil substrate, top-emitting PLED or OLED structure is required. To achieve efficient top-emitting diodes, several requirements need to be met such as a transparent cathode with a low work function, a reflective and high work function anode, and a planarized topography. The spin-coating processability of the conjugated light-emitting polymers (5,6) offers a low-cost, two-layer, thin-film advantage with reduced manufacturing complexity. The fabrication of small-molecular organic light-emitting diodes (OLEDs) and the integration with poly-Si TFT backplanes onto metal foils has been previously reported by other authors (7,8,9).

In this paper, we discuss the fabrication of poly-Si TFTs and PLEDs onto metal foils. The integration of these devices and the demonstration of a high-resolution AM-PLED display having the VGA format are also presented in this paper. Topics that are discussed in more detail have included: metal foil substrate preparation, poly-Si TFT backplane fabrication, top-emitting PLED development and integration onto the TFT backplane. All poly-Si TFT devices, driving circuits, and PLEDs reported in this work were fabricated on 150-mm diameter stainless-steel (type 304) substrates.

Experimental

Substrate preparation

One of the challenges in the case of metal foil substrates is to obtain a good surface finish. Metal foils are prepared by a rolling process which results in surface roughness greater than 100 nm. Such roughness can cause significant yield problems and thus it needs to be reduced significantly. The planarization and simultaneous passivation of the metal surface by spin coating dielectric materials has been previously reported (7,8). In this work we have pursued a different approach where the metal foil substrate is polished by chemical-mechanical means to a final surface roughness of ~1 nm. This approach requires no further planarization layers, and thus the surface roughness of the steel foil becomes no longer a yield-limiting factor. Fig. 1 shows an atomic force microscope image of the final foil surface quality.

Before poly-Si TFTs are fabricated on 100-μm-thin steel foils, insulating layers need to be deposited on the steel foils to isolate the devices from the conductive metal substrate. In this work we investigate several insulating layers which were deposited on both sides in order to balance the film stresses and reduce substrate bending. In order to ensure that our coating process has resulted in a complete isolation of the metal surface we have used the following approach. First the insulating layers are deposited on the steel foil and then annealed at 900°C. Aluminum is then evaporated through a shadow mask that defines many 1-cm^2 pads on top of the insulating layer. Shorts between the 1-cm^2 aluminum pads and the steel foil substrate are then identified by measuring the resistance between each of the top square pads and the steel substrate. In total, 56 sites are tested for each steel foil and we verified that no shorts are present.

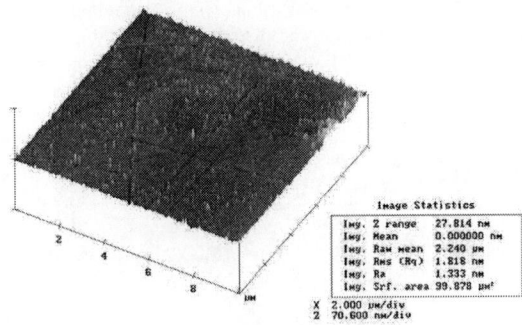

Figure 1. A 3-D picture of the 100-µm steel foil surface measured by an atomic force microscope (AFM).

Poly-Si backplanes

The poly-Si TFTs investigated in this work had the self-aligned poly-Si gate structure, shown in Fig. 2. To realize this structure, first an amorphous silicon thin film, 50-nm thick, is deposited and then crystallized by excimer laser (10). After patterning and etching of the poly-Si semiconductor regions, a silicon dioxide film is deposited by PECVD to a thickness of 100 nm at 300°C to serve as gate dielectric. After depositing and n+ doping a 200-nm thick poly-Si gate layer, patterning and dry etching of the gate electrodes was performed. Then doping of the source and drain regions was performed by ion implantation. Activation of the dopant was accomplished by thermal annealing at 700°C for 3 hours, and this was the highest processing temperature throughout the entire fabrication. We found that the steel foil still maintained good dimensional stability even after exposure to this high temperature.

After the dopant activation anneal, silicidation of the source, drain, and poly-Si lines was performed by first depositing 10 nm of nickel by sputtering followed by a thermal anneal at 400°C to complete the process. Plasma hydrogenation for 10 minutes was then performed, followed by the deposition of a triple layer dielectric structure, consisting of $SiO_2/Ta_2O_5/SiO_2$, to serve as the device passivation layer, and as insulaor between the gate lines and the Metal-I. Afterwards the contact windows were opened on top of the source and the drain, followed by a lift-off process that defined the aluminum/nickel (3000 Å/500 Å) Metal-I stack (11). A combination of PECVD oxide and SOG were then used to planarize the surface topography prior to the preparation of the Metal-II which was to serve as the PLED anode electrode. After the planarization step, contact windows were opened on top of Metal-I. Metal-II was consisting of a dual layer stack both deposited by e-beam evaporation, first 20 nm of titanium were deposited onto the oxide to improve the surface adhesion followed by the deposition of 100 nm gold which acts as the anode in the PLED structure.

The standard two TFTs plus one capacitor circuitry, as shown in Fig. 3, was implemented at each pixel using PMOS transistors. In this pixel design, T2 behaves as a voltage-controlled current source, and T1 is the control switch. Thus there are three lines going into each pixel (V_{data}, V_{add}, and V_{DD}). This circuit works as follows: an analog voltage is programmed in the data line during addressing time (when T1 is turned ON by the address line). This voltage activates T2 allowing a controlled current to flow through it. When the addressing period is over, T1 is turned OFF and C stores the data voltage. This allows T2 to keep supplying the current during the non-addressing time.

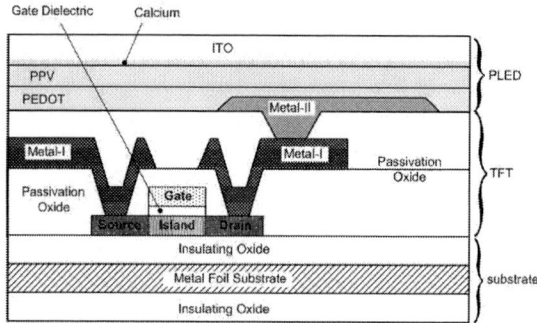

Figure 2. A schematic cross-sectional view of the integrated TFT/PLED stack (not drawn to scale).

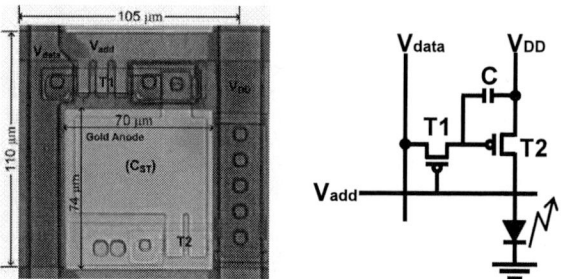

Figure 3. An optical microscope photograph (left) of the AM-PLED pixel on the steel foil is shown. T1 (triple-gate) and T2 (dual-gate) are the switching and the driving transistor, respectively. The equivalent pixel driving circuit is shown on the right.

Top-emitting PLEDs

The use of two organic semiconducting layers to improve the efficiency of small-molecular OLEDs was first demonstrated by Tang and co-workers (12,13). The principle of using more than one semiconducting layer allows the optimization of the energy barriers for electron- and hole-injection at the metal/semiconductor and semiconductor /semiconductor interfaces within the device. Brown et al. then disclosed an efficient bi-layer PLED structure, which incorporated the hole-transporting polymer, (poly(ethylenedioxythiophene)/poly(styrene sulphonic acid), PEDOT:PSS) and the light-emitting polymer (poly(4-4'-diphenylene diphenylvinylene), PDPV) sandwiched by two electrodes (14). In the current study, this bi-layer PLED structure is adopted. Due to the opaque nature of the steel foil, emitted light is reflected from the substrate. Therefore the conventional PLED structure needs to be converted into a top-emitting PLED (15,16), which consists of a low work function transparent cathode, a hole-transporting polymer, a light-emitting polymer, and a high work function reflective anode. After completion of the top-emitting PLEDs the samples were transferred into a dry nitrogen glove box where a novel flexible encapsulation developed by our group (17) was applied.

Transparent cathode. The semitransparent cathode contacts needed for top-emitting PLEDs can be formed by using thin layers of low work function metals or compounds, such as LiF, CsF, etc., (18,19). These thin single-layer metal or compound films, however, cannot provide efficient electron injection and have high sheet resistance, leading to rather low quantum efficiency (<0.01%) and high drive voltage. Therefore, a double-layer cathode, consisting of a thin layer of low work function metal, Ca, to form the ohmic contact with the polymer and a thick metal layer, ITO, to conduct the current, is then employed. A soluble alkoxy-phenyl-PPV has been found to make a useful metal-on-polymer (MOP) contact (20,21) with thermally evaporated Ca (or Ba). In this work we used 5-nm-thick Ca that was evaporated at a base pressure of 1×10^{-7} Torr. Afterwards a 150-nm ITO was deposited by sputtering. Other compound cathodes consisting of a thermally evaporated metal capped with sputtered ITO for transparent OLEDs (TOLEDs) have also been reported in the literature (22).

In the case of top-emitting PLEDs, the conductivity, transparency, and uniformity of ITO are crucial for PLEDs performance. The deposition rate of ITO is found to be proportional to the RF power. In order to reduce the ion bombardment onto the light-emitting polymer while sputtering (3), the RF power was set as low as possible. The optimized ITO film was deposited by an RF power of 50W with an argon pressure of 5×10^{-3} Torr for 30 minutes, which resulted in a deposition rate of 0.086 nm/s and showed a maximum transmittance of 85% between 550 nm and 600 nm (Fig. 4). The peak of our PLEDs electroluminescence (EL) falls within that wavelength region, so that the light output is maximized. ITO was sputtered in the absence of oxygen to avoid oxidizing Ca and darkening the film (23). The size of our sputtering ITO target and its distance to the substrate were 150 mm diameter and 8 cm respectively.

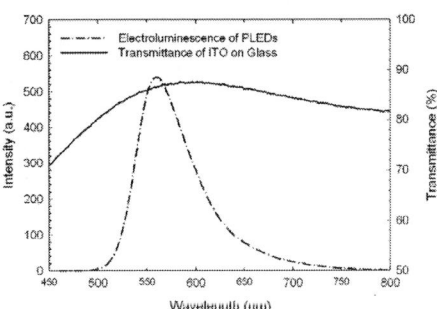

Figure 4.   Solid line represents the transmittance of our ITO film, within the visible wavelength region; dashed line represents the electroluminescence spectrum of our PLED.

Light-Emitting Polymer. Our PLED structure consisted of a hole-transport layer and a light-emitting layer, both applied by spin casting. Prior to the spin casting of the PEDOT:PSS hole-transport layer, the entire surface was treated by oxygen plasma at 0.18 Torr, with a DC power density of 0.137 W/cm^2 for 2 minutes, which improves the adhesion of PEDOT:PSS onto the substrate. PEDOT:PSS was then spin-coated and baked in a vacuum oven at 120°C. The light-emitting layer, phenyl-substituted poly-phenylenevinylene (PPV), was then spin-cast and dried in a vacuum chamber of 1×10^{-3} Torr for 12 hours. The final thicknesses of PEDOT:PSS and PPV were 160 nm and 60 nm respectively.

Anode. At the anode side, a reflective surface with a high work function property is desirable. After a few preliminary experiments in which we compared Au, ITO and Ni as the high work function materials, Au was chosen for the following reasons: (a) Au provides a *smooth surface*, which might make a polymer layer adhere to the anode in a well-defined fashion. Au has a surface roughness of 1.5 nm; (b) Au has a *high work function*, close to the well-treated ITO; (c) Au is chemically inert except to the sulfur functionalities of the anchored hole-transporting polymer. In particular, Au is very resistant to PEDOT:PSS, in contrast to ITO which has been found to be susceptible to attack by PEDOT:PSS, probably due to the acidity of PEDOT:PSS (24); (d) Au surface is reflective. The emissive polymer adopted in this study has the peak emission at 560 nm, which can be reflected by the yellowish Au surface; (e) Au possesses a *high electric conductivity*, ranking the third among the elements.

TABLE I. Comparisons of Anode Performance.

Anode	Light Turn-On Voltage (V)	Visible under the ambient light (V)
Au	5.0	6.4
ITO	6.5	7.4

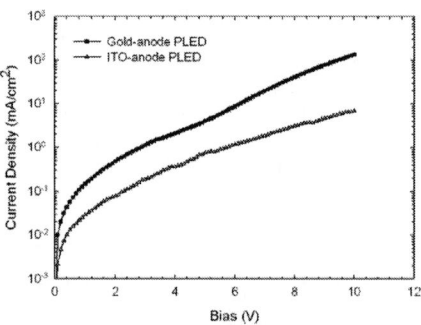

Figure 5. Current Density – Bias (across PLEDs) characteristic of gold- and ITO-anode PLEDs is presented.

As discussed above, we have found that Au has many comparable properties to or even better than ITO, particularly in the processing stability and the device performance. For a top-emitting PLED, the anode does not need to be transparent, but reflective. In addition, ITO requires accurate control of the deposition and the post-deposition treatment (25,26) in order to have the appropriate properties, for instance, high work function, high transmittance, low resistance, etc.. Guan et al. also suggested Au anode, because of its excellent surface resistivity (27).

In a Ca-cathode PLED, while Au is employed as the anode the light turn-on voltage is 5.0 V, which is 1.5 V less than when ITO is used as the anode (Table I). The light output is proportional to the current density if the device quantum efficiency is assumed to be the same for both Au- and ITO-anode PLEDs. Therefore it is concluded that Au is a more efficient hole-injector than ITO, as per Fig. 5. Uniform light emission from pixel to pixel was observed for this Au/PEDOT:PSS/PPV/Ca/ITO PLED (Fig. 6 and Fig. 7).

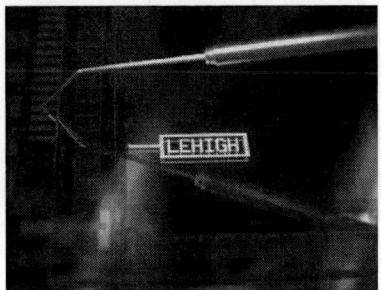

Figure 6. A picture of a gold-anode PLED in operation with "LEHIGH" letters is shown. This PLED was encapsulated by glass with perimeter sealants.

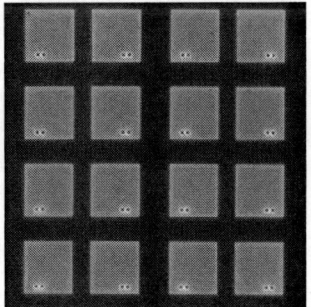

Figure 7. A picture of operating PLED pixels is shown.

Results and Discussions

The transfer characteristics of PMOS transistors on the steel foil measured at a V_{DS} bias of -0.1 V are shown in Fig. 8. A mobility of 42 cm²/Vs could be extracted from the single-gate TFT with W/L ratio of 2. A dual-gate TFT was used to drive the PLED as such devices had improved saturation behavior compared to single gate devices. The Id-Vd characteristics of the dual gate PMOS transistors are shown in Fig. 9. The effect of gate structure on device performance is shown in Table II where single, dual and triple gate structures are compared. As discussed previously, triple gate structure was used in the pixel address TFT because of reduced leakage current. The characteristics of a pixel integrated TFT and PLED devices are shown in Fig. 10.

TABLE II. Device characteristics of PMOS with varied dimensions

W/L	Mobility	V_{th}	S.S.	I_g
12/6	42	-1.64	1.35	7.8×10^{-12}
24/(2×6)	34	-1.52	0.95	1.9×10^{-11}
36/(3×6)	36	-2.67	1.19	2.4×10^{-11}

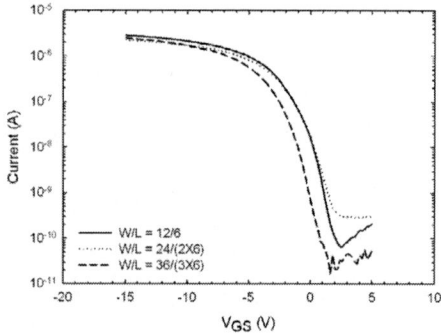

Figure 8. Transfer characteristics of a single-gate, a dual-gate, and a triple-gate PMOS TFTs on the steel foil with different dimensions but with the fixed W/L ratio. Current – Voltage curves were measured at $V_{SD} = 0.1$ V.

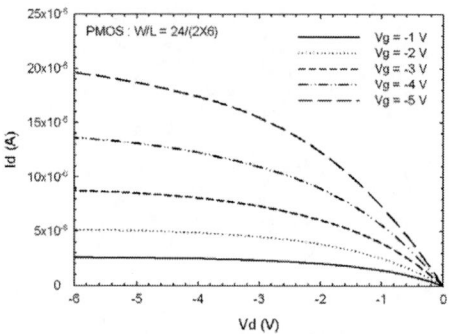

Figure 9. Id-Vd curves of the dual-gate PMOS TFT are shown with varied Vg.

The light turn-on voltage (V_L) of this active-matrix PLEDs observed in the dark room was 4.0 V, while the current density as a function of the applied voltage (V_{PLED}) increases dramatically (triangle dots in Fig. 10). The light emission turned very visible at 5.0 V under ambient lighting conditions. According to our observation, at $V_{PLED} = 7.0$ V the electroluminescence of the PLED was high enough to display clear and bright images. At this given V_{PLED}, the current was only ~1.0 μA for a pixel with an emissive area of 70 μm×74 μm. The full-ON current consumption is 10 μA per pixel which is well within the capabilities of the driving TFTs based on the I-V characteristics of the PMOS TFTs. Corresponding to the pixel size of 105 μm×110 μm, a 45% of the aperture ratio of pixels was obtained in the current design. Fig. 11 demonstrates the completed AM-PLED display in operation.

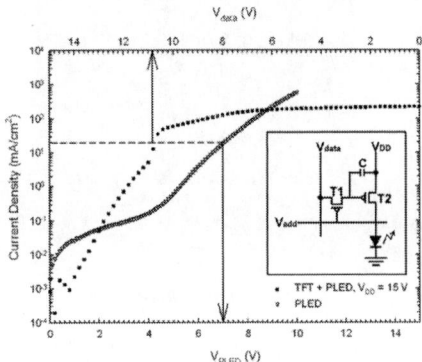

Figure 10. Current – Voltage characteristics of the PLED (versus the PLED voltage, V_{PLED}) and the integrated PLED/PMOS TFTs on the steel foil (versus the data voltage, V_{data}) while the switching TFT (T1) was turned on. 15 V of V_{DD} was applied to T2. The inset schematic represents the pixel driving circuit.

Figure 11. A top-view of operating 3.3" diagonal, 230 ppi (VGA-resolution, 640×480) top-emitting AM-PLED display on a steel foil.

Random pixel defects were observed, but no line defect. However, on the AM-PLED display, non-uniformity from pixel to pixel was observed within a few pixels distance (Fig. 12). The latter observation is attributed to the characteristic variation of the transistors, i.e. the variation of threshold-voltage and/or field-effect mobility, because uniform light emission was observed on the PLED-alone devices (Fig. 7), which were not driven by TFTs.

Figure 12. A picture of operating PLEDs' pixels is presented. The emissive area of the PLED is 70 μm×74 μm.

Conclusions

A complete set of process technologies for the preparation of the steel foil substrates, the fabrication of poly-Si TFTs backplanes, and PLEDs, as well as the flexible encapsulation have been developed and demonstrated. In support of this technology, we have successfully demonstrated a top-emitting 230 ppi AM-PLED displays on a lightweight, rugged, and flexible steel foil substrate. Fine surface finishing of the steel foils is achievable and requires no further planarization layer, so that the surface roughness of the steel foil becomes no longer a yield-limiting factor. This accomplishment not only validates the compatibility of the poly-Si technology for high-resolution flexible AM-PLED displays, but also implies that a variety of large-area microelectronics could be implemented onto the flexible steel foil, thereby benefiting by the advantages of high-temperature processability and dimensional stability. To our best knowledge, this is the world's first AM-PLED display on the flexible steel foil substrate.

Acknowledgments

The authors would also like to thank Prof. R. Pearson and Dr. S. Preis of Dept. of Materials Science and Engineering for the development and consultation concerning the encapsulation techniques, Dr. H. Najafov in Dept. of Physics at Lehigh University for many helpful discussions, and US Army Research Laboratory for funding.

References

1. S.D. Theiss and S. Wagner, *IEEE Electr. Dev. Lett.*, **17**, 578 (1996).
2. T. Afentakis, M. Hatalis, A.T. Voutsas, and J. Hartzell, *IEEE Trans. Electr. Dev.*, **53**, 815 (2006).
3. C.C. Wu, S.D. Theiss, G. Gu, M.H. Lu, J.C. Sturm, S. Wagner, and S.R. Forrest, *IEEE Elect. Dev. Lett.*, **18**, 609 (1997).
4. M. Troccoli, A. J. Roudbari, T.-K. Chuang, and M. Hatalis, *Journal of SSE*, submitted (2006).

5. J.H. Burroughes, D.D.C. Bradley, A.R. Brown, R.N. Marks, K. Mackay, R.H. Friend, and P.L. Holmes, *Nature* (London), **347**, 539 (1988).
6. D. Braun and A. Heeger, *Appl. Phys. Lett.*, **58**, 1982 (1991).
7. D.U. Jin et al., *SID 06 DIGEST*, **37**, 1855 (2006).
8. A. Chwang et al., *SID 06 DIGEST*, **37**, 1858 (2006).
9. J. H. Cheon et al., *IEEE Trans. Electr. Dev.*, **53**, 1273 (2006).
10. A.T. Voutsas, A.M. Marmorstein, and R. Solanki, J. Electrochemical Society, **146**, 3500 (1999).
11. S.K. Saha, R.S. Howell, and M.K. Hatalis, *J. Appl. Phys.*, **86**, 1 (1999).
12. C.W. Tang and S.A. VanSlyke, *Appl. Phys. Lett.*, **51**, 913 (1987).
13. C. W. Tang, S. A. VanSlyke, and C. H. Chen, *J. Appl. Phys.*, **65**, 3610, (1989).
14. T. M. Brown, J. S. Kim, R. H. Friend, and F. Cacialli; R. Daik and W. J. Feast, *Appl. Phys. Lett.*, **75**, 1679 (1999).
15. D.R. Baigent, R.N. Marks, N.C. Greenham, R.H. Friend, S.C. Moratti, and A.B. Holmes, *Appl. Phys. Lett.*, **65**, 21 (1994).
16. M.-H. Lu, M.S. Weaver, T.X. Zhou, M. Othman, R.C. Kwong, M. Hack, and J.J. Brown, *Appl. Phys. Lett.*, **81**, 21 (2002).
17. S. Preis et al., patent pending.
18. T. M. Brown, R. H. Friend, I. S. Millard, D. J. Lacey, J. H. Burroughes, and F. Cacialli, *Appl. Phys. Lett.*, **79**, 174 (2001).
19. L. S. Hung, C. W. Tang, M. G. Mason, P. Raychaudhuri, and J. Madathil, *Appl. Phys. Lett.*, **78**, 544 (2001).
20. X. Gong, D. Moses, A. J. Heeger, S. Liu, and A. K.-Y Jen, *Appl. Phys. Lett.*, **83**, 183 (2003).
21. H. Becker, H. Spreitzer, W. Kreuder, E. Kluge, H. Schenk, I. Parker, and Y. Cao, *Adv. Mater.*, **12**, 42 (2000).
22. P. E. Burrows, G. Gu, S. R. Forrest, E. P. Vicenzi, and T. X. Zhou, *J. Appl. Phys.*, **87**, 3080 (2000).
23. J.C.C. Fan and J.B. Goodenough, *J. Appl. Phys.*, **48**, 3524 (1977).
24. K. W. Wong, H. L. Yip, Y. Luo, K. Y. Wong, and W. M. Lau; K. H. Low and H. F. Chow; Z. Q. Gao, W. L. Yeung, and C. C. Chang, *Appl. Phys. Lett.*, **80**, 2788 (2002).
25. J. S. Kim, M. Granstrom, R. H. Friend, N. Johansson, W. R. Salaneck, R. Daik, W. J. Feast, and F. Cacialli, *J. Appl. Phys.*, **84**, 6859 (1998).
26. M. G. Mason, L. S. Hung, C. W. Tang, S. T. Lee, K. W. Wong, and M. Wang, *J. Appl. Phys. Lett.*, **86**, 1688 (1999).
27. Y. Guan, M. A. Matin, and T. M. Stephen, *Electr. Lett.*, V. 38, No. 15 (2002).

360

Uniform OLED-pixels Using Microcrystalline Silicon TFTs for Active-Matrix Addressing

A. Gaillard [a,b], R. Rogel [a], S. Crand [a], T. Mohammed-Brahim [a], C. Prat [b], P. Leroy [b]

[a] G. M., IETR, Université de Rennes I, 35042 Rennes Cedex, France
[b] Thomson R&D France, CS 17616, 35576 Cesson-Sévigné, France

> Pixel circuits using undoped microcrystalline silicon as active layer have been investigated for the first time by LPCVD technique and implemented in organic light-emitting diodes (OLED) technology based active-matrix (AMOLED). AMOLED technology needs stable and uniform pixel circuits. In this way, key processes introduced here in OLED-pixel fabrication are low grain size to average the number of grain boundaries in the channel region of thin film transistor (TFT) and O_2 plasma treatment to improve the gate SiO_2/Si interface. P-type TFTs exhibit sub threshold voltage improvement of 20% and a hole mobility higher than 5 $cm^2/V.s$. Furthermore, drive current and light intensity of OLED-pixels show uniformity better than 10%.

Introduction

In the light of recent advances in organic flat-display technology, active-matrix OLED (AMOLED) is challenging the currently dominant active-matrix LCD technology. OLEDs are more bright and consume much less power than LCD's. The OLED-pixel brightness is directly related to the current flow through the component during the whole frame period. As a result, drive current have to be precisely controlled. Lack of control leads to non uniform luminance from one pixel to the other over the display area. Competing technologies to insure this control are low-temperature poly-silicon (LTPS) TFT based on laser crystallization and amorphous silicon (a-Si:H) TFT that is the more mature manufacturing technology. Non-uniformity in LTPS and instability in a-Si:H have been reported (1-2) and are current issues. Indeed, LTPS suffers from laser shot-to-shot variation of energy whereas gate bias-stress in amorphous TFT induces shift of the characteristics. Moreover due to the low mobility of carriers, a-Si:H pixels need higher driving voltage. In both cases, a lot of TFTs with many compensating cells are used in pixels circuits to compensate the drift of the characteristics (a-Si:H) or the lack of uniformity (LTPS). Our approach is to provide an alternative solution in active-pixel realization by achieving an efficient process with both stable features and uniformity so that simple pixel circuits can be used. This can be done by using low grain size polysilicon processed at high enough temperature to be free of hydrogen, that is the main origin of the parameters shift. Polycrystalline silicon amorphously deposited and then solid phase post-crystallized at 600°C is well known to fulfill these requirements. However, the long crystallization time, about 8 hours, constitutes a drawback. Alternative can be silicon, directly deposited in the crystallized state at the same temperature. Grain size is lower but we can hope for a sufficiently high mobility to maintain an acceptable _ drive current using a simple pixel circuit design. The outline of this paper is as follows : first a discussion is opened on µc-Si TFT pixel fabrication and their electrical characteristics as a function of processing conditions. A detailed attention is given to the contribution of oxygen plasma treatment on the uniformity. Finally the electrical and luminous uniformity using this technology is evaluated.

Experimental

Silicon layer deposition

Direct deposition of microcrystalline silicon, that is free from hydrogen is performed at 600°C using Low Pressure CVD reactor. The quality of the material is strongly dependent on the deposition pressure (3). Figure 1 shows a Levinson treatment (4) of the transfer characteristics of TFTs using either μc-Si films deposited at 600°C and 0.1 or 10 Pa, or amorphously deposited and then solid phase crystallized polysilicon film. When using such treatment, the slope of the linear part of the curves usually decreases when the material quality increases. Then, Figure 1 shows the higher quality of material deposited at 0.1 Pa. However and as expected, the quality of solid phase crystallized material is higher. The quality of the 0.1 Pa is also highlighted by the preferential {220} orientation of crystals (Figure 2) that is known to produce electrically inactive grain boundaries. Finally, SEM observations show an average grain size of 50nm that is enough low in comparison of the channel size of our TFTs.

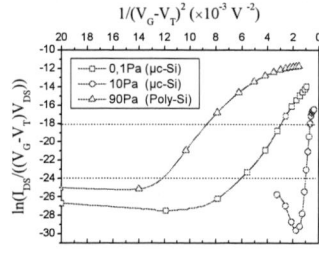

Figure 1: Levinson treatment of transfer characteristics of TFTs made of crystallized deposited films at 0.1 Pa and 10Pa and of solid phase crystallized polysilicon (Poly-Si). Slope of the linear part of the curves decreases when the material quality increases

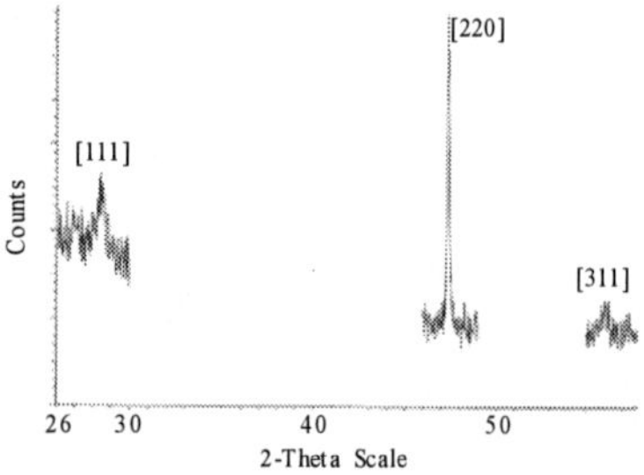

Figure 2 : XRD analysis of microcrystalline silicon film deposited at 600°C and 0.1 Pa. [220] orientation is well highlighted

Microcrystalline silicon back plane

Pixel circuits and matrix are fabricated using an in-house μc-Si single layer TFT process described as follows (Figure 3).

Initially, 300-nm thick silicon oxide film is deposited at 430°C for preventing impurity diffusion from glass substrate. Subsequently, an undoped silicon layer (150nm) followed by a boron doped layer is deposited without interruption by LPCVD. Patterning of the μc-Si active islands is followed by a clean process and resulting oxide is removed with dilute HF. An oxygen plasma exposure is then performed at room temperature and 80-nm thick APCVD SiO_2 gate dielectric is deposited. Contact via are opened and chromium and insulator are deposited and then patterned. In this process, no post treatment such as hydrogenation is employed. A transparent anode (ITO) is then patterned in order to insure a conventional bottom light emission. Organic layers are deposited through a shadow mask followed by a common electrode deposition. This last step does not excess 150°C. Finally, a conventional glass encapsulation method is used to protect OLED from diffusion of moisture and oxygen into the cathode and the light-emitting layer.

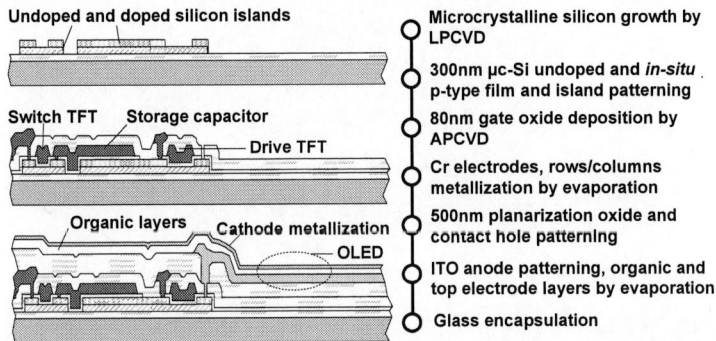

Figure 3: Main process sequence used in this work for µc-Si pixel fabrication

Results

µc-Si drive-TFT characteristics

A simple two-TFT pixel circuit is integrated to drive a bottom emission OLED structure (Figure 4). The switching S1-TFT is similar to the one used in liquid-crystal-TFT while the driving S2-TFT supplies enough current to drive the OLED. A p-type-TFT is implemented to avoid OLED characteristics dependence. Some parameters extracted from the transfer characteristics of more than 30 S2-TFTs (Figure 4) are summarized in Table I.

TABLE I. Summary of typical parameters of LPCVD top-gated p-type drive-TFT obtained from various materials.

Deposition Conditions	Threshold voltage V_T (V)	Mobility µ (cm²/V.s)	Sub threshold Slope S (V/dec)	Interface trap density (cm⁻²)
90 Pa, amorphous, post-crystallized SPC	-8	62	0.7	$2.89\ 10^{13}$
10 Pa, as-deposited µc-Si	-16	0.25	2.6	$1.15\ 10^{14}$
0.1 Pa, as-deposited µc-Si	-11	5.2	1.7	$7.42\ 10^{13}$

Table I confirms the higher quality of 0.1Pa material compared to the 10Pa silicon film. The threshold voltage and the sub threshold slope are lower and the mobility higher.

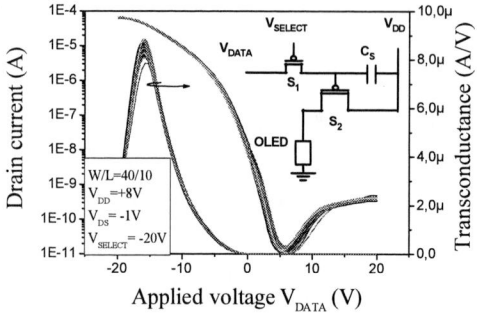

Figure 4: Typical transfer characteristics of more than 30 drive-TFTs S_2 (W/L=40/10). Inset shows the simple p-type pixel circuit used here.

Oxygen plasma treatment effect

In the previous characteristics, it can be important to highlight the effect of the oxygen plasma step. Conventional TFTs without oxygen plasma treatment are also realized for a comparison (Figure 5.). Hole mobility is 60% lower than the previous value (5.2 cm^2/V.s). The higher mobility when using oxygen plasma treatment is probably due to the improvement of silicon / gate insulator interface by a plasma oxidation of silicon surface that leads to a thin SiO_2 layer (2nm-thick oxide was measured on a silicon wafer as reference). The sub threshold slope (S) improvement of 20% with O_2 plasma treatment indicates an effective reduction of the number of interface traps. In comparison to previous works carried out on polysilicon (5-7), here the O_2 plasma effect is particularly significant due to the high concentration of traps in the initial material. Moreover, it is particularly efficient with time exposure.

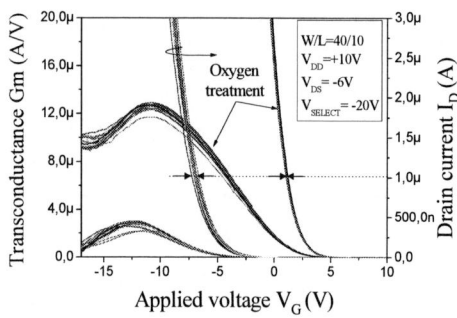

Figure 5: Transfer characteristics and Tranconductance of 15 μc-Si TFTs fabricated without and with oxygen plasma treatment. Treatment was carried out at 2mPa for 40mn.

Uniformity Measurements

Electrical behavior

Experimental results show an excellent electrical uniformity particularly for low level of the output current (<10%) (Figure 6).

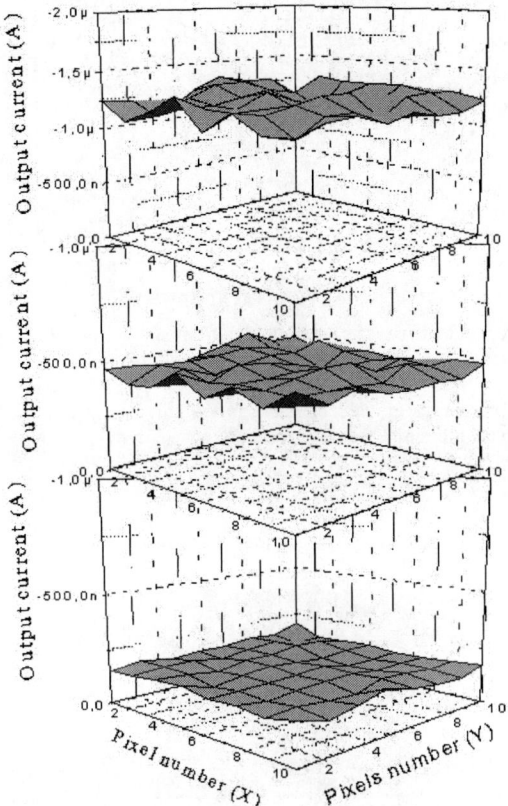

Figure 6: Spatial representation of the evolution of three output current levels (V_{DS}=-5V) measured on 8x8 neighbor pixels. The total area is 180mm². High current uniformity is shown at low current level.

Significant non-uniformity occurs for current of few microamperes. It is related to the short channel effect of the drive-TFT (10μm). As it is important to maintain such low TFT size in order to keep high the pixel aperture in bottom emission, OLEDs with high efficiency and low threshold are required.

Brightness effect

OLEDs with high efficiency and low threshold voltages (2V-3V) are obtained (Figure 7). Moreover very low voltage variations on neighbor organic-diodes are measured in dc driving (~30mV at 500nA). Then, at least locally, OLEDs can be considered as enough uniform.

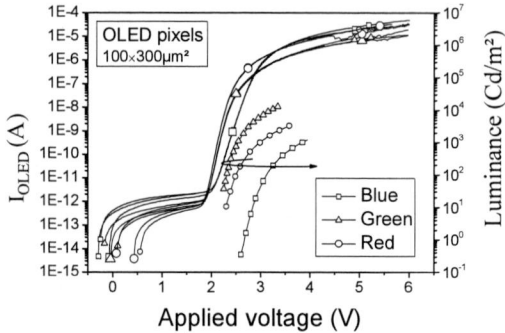

Figure 7: Voltage dependence of measured current and luminance variation of typical OLED-pixels (here three diodes per color separated by 8mm).

A monochromic (green color) demonstration unit (1.6 inch in visible diagonal) is realized to reveal more pixel to pixel contrast.

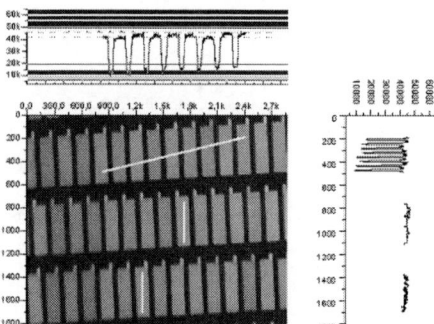

Figure 8: Top-view details of the illuminated 63ppi AMOLED based on μc-Si TFT pixel circuits in gray level conversion. Variation of intensity on scanned pixels highlights the uniformity of neighbor pixels.

In order to estimated the variation of light intensity among neighbor pixels, a gray level conversion has been performed (Figure 8). Scanned pixels confirm the previous electrical behavior and reveals a variation of luminous intensity lower than 10%.

Conclusion

An OLED-pixel process based on an attractive and alternative LPCVD microcrystalline silicon technology have been demonstrated using a simple circuit scheme. An appropriated interface pre-treatment have shown a significant improvement of electrical performances and dispersion suggesting a high defect reduction into such material. Both high mobility of p-channel TFT ($5cm^2/V.s$) and high electrical stability of OLEDs allow p-type μc-Si TFT implementation on active-matrix. This is very promising in the goal of low cost and simple technology flat panel displays.

References

1. S. Higashi, D. Abe, K. Miyashita, T. Kawamura, *Proc. SID 2003*, 1302 (2003)
2. A. Nathan, A. Kumar, K. Sakariya, P. Servati, S. Sambandan, D. Striakhilev, *IEEE J. Sol. State Circuits*, **39**, 1477 (2004)
3. T. Mohammed-Brahim, M. Sarret, D. Briand, K. Kis-Sion, L. Haji, O. Bonnaud, D. Louër, A. Hadjadj, *Phil. Mag.* **B 76**, 193 (1997)
4. J. Levinson, F.R. Shepherd, P.J. Scanion, W.D. Westwood, G. Este and M. Rider, *J. Appl. Phys.* **53,** 1193 (1982)
5. H. N. Chern, C. L. Lee, T. F. Lei, *IEEE* **ED40**, 2301 (1993)
6. K. Mourgues, F. Raoult, L. Pichon, T. Mohammed-Brahim, D. Briand, O. Bonnaud, *Materials Research Society Symp.* **Proc. 471**, 155 (1997)
7. H. Watakabe, Y. Tsunoda, N. Andoh, T. Sameshima, *J. Non Crys. Sol.*, 1321 (2002)

Author Index

Aarnink, T.	185	Hamamura, H.	35
Abe, K.	293	Hamshidi, A.	237
Afentakis, T.	237, 349	Han, J.	279, 279
Ahn, T.	57, 69	Han, M.	137, 279, 287
Amemiya, I.	307	Han, S.	137
Arias, A.	229	Harnois, M.	341
		Hartzell, J.	237, 349
Balon, F.	313	Hatalis, M.	237, 349
Beenakker, K.	167	Hatano, M.	35
Bendriaa, F.	341	Hatayama, T.	173
Biaggio, I.	349	Hayama, H.	3
Bonnaud, O.	341	Hayashi, R.	293
Boogaard, A.	185	He, M.	167
Bramante, N.	23	Holleman, J.	185
Brunets, I.	185	Hosono, H.	293
		Hou, C.	203
Celler, G. E.	81	Hu, C.	203
Chang, K.	255	Hwang, C.	301
Chen, H.	143		
Cheng, H.	143	Ishihara, R.	167
Cheng, I.	249		
Choi, B.	137	Jamshidi Roudbari, A.	349
Choi, J.	287	Jeon, J.	287
Choi, S.	137	Ji, I.	137
Chu, H.	301	Jung, S.	57
Chuang, T.	237, 349		
Chung, I.	57, 63, 69	Kaji, N.	293
Chung, K.	287	Kakiuchi, H.	215
Chung, S.	301	Kamiya, T.	293
Crand, S.	361	Kattamis, A. Z.	249
		Kawachi, G.	45
Daniel, J.	229	Kim, C.	57
Deane, S.	23	Kim, C.	57
Do, L.	301	Kim, C.	149
		Kim, E.	149
Esmaeili Rad, M.	93	Kim, K.	149
		Kim, S.	63
Farmakis, F. V.	75	Kim, Y.	69
Fish, D.	23	Kimura, M.	325
Fuyuki, T.	101, 107, 173, 195	Kirimura, H.	195
		Kishikawa, R.	207
Gaillard, A.	361	Kller, K.	349
George, D.	23	Ko Park, S.	301
Giraldo, A.	23	Kontogiannopoulos, G. P.	75
Gleskova, H.	249	Kouvatsos, D. N.	75, 87
Goh, J.	287	Kovalgin, A. Y.	185

Krusor, B.	229	Nomura, Y.	307
Kumomi, H.	293		
Kuo, P.	237, 349	Oepts, W.	23
Kuo, Y.	333	Oesterlin, P.	185
Kuranaga, T.	207	Ofuji, M.	293
Kwang Sik, H.	69	Ogata, T.	207
Kwok, H.	263	Oh, J.	63
Kwon, H.	149	Ohkura, M.	35
		Ohmi, H.	215
Lagally, M. G.	81	Okuda, M.	195
Le Bihan, F.	341		
Lee, D.	255	Papaioannou, G. J.	75, 87
Lee, H.	57	Park, J.	137
Lee, J.	279, 287	Park, S.	287
Lee, J.	301	Prat, C.	361
Lee, J.	149	Price, S.	255
Lee, K.	137		
Lee, S.	179	Rogel, R.	361
Lee, Y.	143	Ruzyllo, J.	255
Leroy, P.	361	Ryu, M.	149
Lifka, H.	23		
Lin, C.	203	Salaun, A.	341
Lo, H.	273	Sano, M.	293
Lujan, R.	229	Sazonov, A.	93
		Schmitz, J.	185
Ma, Z.	81	Serikawa, T.	107
Matsumoto, M.	119	Setsuo, N.	119
Matsumura, M.	127	Shanmugasundaram, K.	255
Matsumura, M.	35	Shannon, J. M.	313
Meng, C.	179	Shima, T.	325
Metselaar, W.	167	Sokolov, Y. V.	113
Mimura, A.	173	Son, G.	149
Mitani, T.	207	Spirko, J.	237, 349
Miyamoto, T.	107	Steer, A.	23
Miyatake, N.	101	Street, R.	229
Miyazawa, T.	35	Suemitsu, M.	119
Mohammed-Brahim, T.	341, 361	Sugawara, Y.	107, 173
Mori, K.	307	Sugi, K.	307
Mori, Y.	101	Suo, Z.	249
Moschou, D. C.	75	Syun, I.	119
Motooka, T.	207		
Munetoh, S.	207	Tachibana, H.	101
Murata, K.	101	Tai, M.	35
		Tai, Y.	273
Nakao, H.	307	Takami, Y.	127
Nam, D.	57	Takasu, I.	307
Nathan, A.	93	Toyota, Y.	35
Nominanda, H.	333	Troccoli, M.	237, 349
Nomura, K.	293	Tsai, C.	143

Tsai, M.	179
Tuyoshi, U.	119
Uchikoga, S.	307
Ueno, H.	107
Uraoka, Y.	101, 107, 173, 195
Voutsas, A. T.	11, 75, 87, 157, 237, 349
Wagner, S.	249
Warabisako, T.	127
Watanabe, H.	215
Wu, Y. S.	203
Yamamoto, Y.	11
Yamashita, I.	195
Yamashita, T.	325
Yang, J.	63, 69
Yang, M.	63, 69
Yang, P.	179
Yang, Y.	301
Yano, H.	173
Yara, T.	119
Yasutake, K.	215
Yasutake, T.	119
Yoo, J.	137
Young, N.	23
Yuan, H.	81